CYCLOPHANES

Monographs in Supramolecular Chemistry
Series Editor: J. Fraser Stoddart, *University of Birmingham, U.K.*

This series has been designed to reveal the challenges, rewards, fascination, and excitement in this new branch of molecular science to a wide audience and to popularize it among the scientific community at large.

No. 1 Calixarenes
By C. David Gutsche, Washington University, St Louis, U.S.A.

No. 2 Cyclophanes
By François Diederich, University of California, U.S.A.

Forthcoming Titles

Cyclodextrins
By J. Fraser Stoddart, University of Birmingham, U.K. and R. Zarzycki, University of Sheffield, U.K.

Crown Ethers and Cryptands
By G. W. Gokel, University of Miami, U.S.A.

Monographs in
Supramolecular
Chemistry

Series Editor
J. Fraser Stoddart

Cyclophanes

François Diederich
*University of California
Los Angeles, U.S.A.*

British Library Cataloguing in Publication Data
Diederich, François N.
  Cyclophanes.
  1. Aromatic compounds
  I. Title II. Royal Society of Chemistry III. Series
547.6

ISBN 0-85186-966-1

© The Royal Society of Chemistry 1991

All Rights Reserved
No part of this book may be reproduced or transmitted in any form
or by any means – graphic, electronic, including photocopying, recording,
taping or information storage and retrieval systems – without written
permission from The Royal Society of Chemistry

Published by The Royal Society of Chemistry,
Thomas Graham House, Science Park, Cambridge CB4 4WF

Typeset by Servis Filmsetting Ltd., Manchester
and printed and bound in Great Britain by Bookcraft (Bath) Ltd.

Preface

Molecular recognition is the study of polymolecular entities ('supramolecular complexes') and assemblies formed between two or more designed chemical species and held together by intermolecular forces, the non-covalent bond. For their pioneering studies in molecular recognition, *Charles J. Pedersen, Jean-Marie Lehn*, and *Donald J. Cram* were awarded the 1987 Nobel Prize in Chemistry. Molecular recognition stands for a new area in organic-chemical studies. For more than 150 years, organic chemists were predominantly concerned with the nature of the covalent bond in organic molecules. Today, the exploration and utilization of the non-covalent bond has advanced into the center of interest in organic chemistry. This exciting and dynamic research is recognized worldwide as an important intellectual and technological frontier.

Biochemical phenomena have provided inspiration for much of the current work in chemical molecular recognition. The fascinating properties of enzymes, antibodies, membranes and their receptors, carriers, and channels rest on the controlled and efficient use of weak intermolecular interactions. Selective recognition and material transport, high catalytic activity, fast conductance of electrical impulses from the brain to nerve terminals, and replication processes all rely on the reversible formation of complexes and assemblies held together by non-covalent bonding.

Studies of recognition and catalysis in designed supramolecular complexes may provide microscopic level answers to important open questions in the biological sciences. However, a broader motivation for these investigations is the strong desire to generate a full understanding of weak non-covalent interactions in ground and transition state complexes. With such understanding, chemists should not only be able to design synthetic systems with fascinating properties observed in natural systems but also create novel organic chemistry of great interest to both academia and industry. Selected examples for new technological perspectives generated by fundamental research in molecular recognition are: (1) efficient homogeneous solution catalysts, (2) receptors and molecular sensors with unprecedented sensitivity, (3) separation techniques, (4) electronic devices for information and energy storage and transfer, (5) polymers and mesophases with unusual electro-optical properties, and (6) tools for the mapping of the human genome and for investigating the origin of protein folding.

Cyclophanes ('bridged aromatic compounds') represent *the* central class of synthetic receptors (hosts) in molecular recognition. All types of substrates (guests), from inorganic and organic cations and anions to neutral molecules, have been complexed by tailor-made cyclophanes. In these association processes, all known modes of intermolecular binding interactions have been used. A majority of the crown ethers and cryptands and all of the spherands and cavitands for cation complexation are cyclophanes. For example dibenzo-18-crown-6, one of the first crown ethers reported in 1967 by *Pedersen* is a hexaoxa[7.7]orthocyclophane (the numbers in brackets define the length of the oligoether-chains between the two *ortho*-bridged benzene rings). The calixarenes are referred to as metacyclophanes. This extensive list illustrates the considerable problematics of writing a short monograph on molecular recognition by cyclophanes. Within the given frame, it would be unreasonable trying to discuss all types of cyclophanes used in host–guest complexation studies. Fortunately, this monograph appears in a series in which other volumes contain specific chapters on cyclophane chemistry. C. David Gutsche has marvelously covered calixarene chemistry in Volume 1 of this series. Complexation phenomena involving crown ethers, cryptands, spherands, and cavitands are the subject of a forthcoming volume by George W. Gokel. Therefore, these receptors and the principles of alkali and alkaline earth cation as well as primary ammonium ion binding are not discussed in this volume.

This monograph will focus on the principles of supramolecular complexation and catalysis rather than serve as an encyclopedia for all the cyclophanes prepared for molecular recognition studies. It should provide a state-of-the-art view on the intermolecular forces that hold together supramolecular complexes of cyclophanes with neutral and charged organic molecules in the liquid phase. Many studies in this monograph specifically demonstrate the characteristics and advantages of bridged aromatic compounds as receptors. To provide the space for an in-depth discussion of the principles of molecular recognition, cyclophane synthesis has been almost entirely omitted. The reader is referred to the original literature to learn more about the preparation of the rich variety of molecular architecture covered in this monograph. Among the 650 literature references in this volume, more than 360 refer to publications that appeared between 1985 and 1990.

Although the modeling of biological interactions and processes represents one of the important targets in supramolecular chemistry, such work has by no means been limited to cyclophane hosts or other binders. In a great variety of studies, the nature of weak non-covalent interactions in biological systems and mechanistic details of enzymatic reactions have also been modeled by using cyclophanes without any receptor properties. Following a brief historic outline of cyclophane chemistry, *Chapter 1* provides interesting examples for this ongoing biomimetic research. For an excellent extensive treatise of the historical aspects of cyclophane chemistry and cyclophane chemistry in general, the reader is referred to previously produced work. The monograph 'Bridged Aromatic Compounds' by Brandes H. Smith serves as *the* encyclopedia of cyclophane chemistry prior to 1964. Two multi-author volumes on 'Cyclophanes', edited by Philip M. Keehn and Stuart M. Rosenfeld, provide a quantitative overview of cyclophane chemistry prior to 1983. These volumes also describe the role of cyclophanes in early supramolecular chemistry. In addition, two volumes on 'Cyclophanes', edited by Fritz Vögtle, have appeared in the series *Topics in Current Chemistry* (Vols. 113 and 115, 1983).

This monograph will not focus on the content of these previously published books, but rather, it will present the latest developments in cyclophane chemistry. Finally, the reader is referred to an article by Fritz Vögtle and Peter Neumann [*Tetrahedron*, 1970, **26**, 5847] to obtain nomenclature rules that were specifically developed for bridged aromatic compounds. This cyclophane nomenclature, initially proposed by Donald J. Cram, is very attractive and superior to the IUPAC nomenclature due to its simplicity in naming the smaller cyclophanes. However, it is less useful for naming larger cyclophane receptors and therefore will not be extensively utilized in this monograph.

In the following four chapters, molecular recognition by cyclophanes is discussed in great detail. The various studies that are presented are arranged according to the major driving forces that lead to host–guest complexation. The inclusion complexes formed between cyclophanes and apolar molecules in water (*Chapter 2*) and in organic solvents (*Chapter 3*) are predominantly stabilized by van der Waals and solvophobic forces. Coulombic charge–charge attraction and ion–dipole interactions hold together the complexes between cyclophanes and charged organic molecules (*Chapter 4*). Hydrogen-bonding interactions are essential for the stability of the complexes of small neutral substrates in *Chapter 5*.

Chiral recognition processes in supramolecular complexes are described in *Chapter 6*. In the following *Chapter 7*, an analysis is given for the origin of solvent effects in molecular recognition. *Chapter 8* illustrates how the knowledge of host–guest binding has been successfully applied to generate supramolecular function, reactivity, and catalysis. At the heart of this final chapter are catalytic cyclophanes, synthetic homogeneous solution catalysts whose mechanisms follow the enzymatic systems in their mode of action.

To conclude these words of introduction, I wish to thank my enthusiastic and skilled coworkers for their contributions to the development of cyclophane receptors and catalysts; their names are found in references throughout the monograph. Without the generous support from the Max-Planck-Society and the Deutsche Forschungsgemeinschaft during my time at the Max-Planck-Institute for medical research in Heidelberg and from the National Science Foundation, the Office of Naval Research, and the National Institutes of Health now at UCLA, these contributions would not have been possible. Finally I warmly thank David B. Smithrud for reading the final manuscript.

François Diederich
Los Angeles
August 1990

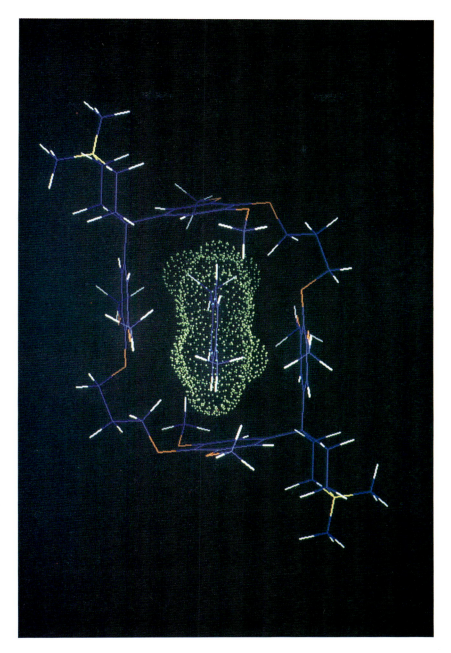

This supramolecular structure represents the low-energy geometry of the *p*-xylene complex of a tetraoxa[5.1.5.1]paracyclophane receptor. The structure was generated in a Monte-Carlo search of conformational space using the AMBER force-field and the BATCHMIN program in the MACROMODEL molecular modeling package by W. Clark Still, Columbia University (I. Chao, F. Diederich, J. Evanseck, K. N. Houk, unpublished results).
Reproduced with permission of the ACS, 1990. [The structure appeared on the cover page of *J. Chem. Educ.*, 1990, **67**, Issue 10 (October).]

Contents

Chapter 1 From Ansa Compounds to [2.2]Paracyclophane to Cyclophane Complexes   1

  1.1 The Origins of Cyclophane Chemistry   1
  1.2 Molecular Shape: An Experimental Calibration of the Space Occupancy of Atoms and Functional Groups   3
  1.3 Probing Weak Non-covalent Interactions between Chromophores in Rigid, Geometrically Defined Cyclophane Frames   6
      1.3.1 Transannular Interactions in Cyclophanes   6
      1.3.2 Models for Intermolecular Excimers and Exciplexes   7
      1.3.3 Donor–Acceptor Cyclophanes as Models for Intermolecular Charge-transfer Complexes   15
  1.4 Modeling Interactions and Reactions in Biological Systems: Stereochemical Control Imposed by Cyclophane Skeletons   18
      1.4.1 Purinophanes   19
      1.4.2 Flavinophanes   20
      1.4.3 Pyridinophanes   24
      1.4.4 Porphyrinophanes   33
  1.5 Cyclophane Complexation: From π–π to Inclusion Complexes   44
      1.5.1 Charge-transfer Complexes of Tetracyanoethylene in Solution   44
      1.5.2 Cyclophane Clathrates   46

Chapter 2 Inclusion Complexes of Neutral Molecules in Aqueous Solution   52

  2.1 Introduction   52
  2.2 The First Cyclophanes for the Inclusion of Apolar Substrates   53
  2.3 Structures of Cyclophane Receptors for Apolar Inclusion Complexation   60

	2.4	Aggregation Behavior of Cyclophanes in Aqueous Solution	66
	2.5	Complexes of Polycyclic Aromatic Hydrocarbons	74
	2.6	Complexes of Naphthalene Derivatives	80
		2.6.1 Determination of Stoichiometry and Stability of Solution Complexes	80
		2.6.2 Complexation of Neutral Naphthalene Derivatives	81
	2.7	Complexes of Benzene Derivatives	86
		2.7.1 Complexation of Flat Benzene Derivatives	86
		2.7.2 Complexes of [$m.n$]Paracyclophanes	95
	2.8	Complexes of Heteroaromatic Substrates	98
	2.9	Complexes of Aliphatic Substrates	100

Chapter 3 Apolar Complexation in Organic Solvents 106

3.1	Introduction	106
3.2	Complexation of Polycyclic Arenes	107
	3.2.1 Structural Factors Determining the Complexation Strength	107
	3.2.2 The Strength of Molecular Complexation of Arenes in Water and in Organic Solvents is Predictable by Linear Free Energy Relationships	108
3.3	Electron Donor–Acceptor Interactions Stabilize Inclusion Complexes of Aromatic Guests	113
3.4	The Cryptophanes: Shape-selective Inclusion Complexation of Methane Derivatives	120
3.5	From Cavitands to Hemicarcerands to Carcerands: Increasing the Barriers for Escape of Encapsulated Organic Molecules	125
	3.5.1 Cavitands for Small Linear Guests	125
	3.5.2 Strong Dimer Formation between Kite-type Molecules in Organic Solvents	127
	3.5.3 Carcerands: Closed-surface Hosts that Imprison Guests behind Covalent Bars	131
	3.5.4 Hemicarcerands: Guest Exchange with High Structural Recognition and Activation Free Energies	135
3.6	Apolar Complexation Strength in Binary Solvent Mixtures	137

Chapter 4 Cyclophane Complexes of Charged Organic Guests — 140

- 4.1 Introduction — 140
- 4.2 High Guest Selectivity in the Inclusion Complexation of Charged Aromatic Compounds — 142
- 4.3 Ion–Dipole Effect as a Force for Molecular Recognition — 146
- 4.4 Attractive Coulombic Interactions at the Origin of Molecular Cation and Molecular Anion Complexation — 150
  - 4.4.1 Complexes of Onium Ions — 150
  - 4.4.2 Anion Receptors — 153
- 4.5 Cyclophane Subunits as Apolar Binding Sites in Polytopic Receptors — 157
- 4.6 From Second-sphere Coordination to Catenane Formation — 164
  - 4.6.1 Second-sphere Coordination — 164
  - 4.6.2 Complexes of the Herbicides Diquat and Paraquat — 165
  - 4.6.3 From Paraquat Complexation and Paraquat-cyclophanes to a [2]Catenane — 170
- 4.7 Azaparacyclophanes at the Interface between Molecular Receptors and Micellar Aggregates — 176

Chapter 5 Hydrogen-bonded Complexes of Cyclophanes and Small Neutral Molecules — 182

- 5.1 Introduction — 182
- 5.2 Small Neutral Molecule Binding to Crowns and Benzo-annelated Derivatives — 183
  - 5.2.1 Complexation by Unfunctionalized Macrocyclic Polyethers — 183
  - 5.2.2 Complexation Assisted by Intraannular Acidic Groups Attached to Aromatic Rings of the Receptor — 185
  - 5.2.3 Urea Binding to Macrocyclic Polyethers Assisted by Co-complexation of Electrophilic Cations — 188
- 5.3 Concave Functionality: Cyclophanes for Phenol Complexation — 189
- 5.4 Carbohydrate Complexation by an Octahydroxy[1.1.1.1]Metacyclophane — 198
- 5.5 Cyclophanes for Nucleotide Base Recognition — 202
- 5.6 Barbiturate Receptors — 214

5.7	Imidazole Recognition: A Remarkable Relation between Solvent Size and Complexation Strength	217

Chapter 6 Cyclophanes for Chiral Molecular Recognition — 222

- 6.1 Introduction — 222
- 6.2 Chiral Recognition in Organic Solvents — 224
  - 6.2.1 Analytical Resolution of Bromochlorofluoromethane by a Cryptophane — 224
  - 6.2.2 Enantioselective Complexation of Chiral Amides by a Macrobicyclic Hydrogen-bonding Receptor — 226
- 6.3 Chiral Recognition in Aqueous Solutions — 228
  - 6.3.1 Diastereomeric Complex Formation by an Optically Active Tetraaza[6.1.6.1]paracyclophane — 228
  - 6.3.2 Enantiospecific Binding of Chiral Onium Guests — 229
  - 6.3.3 On the Origin of Cavity-filling Conformations of Cyclophanes — 231
  - 6.3.4 Chiral Recognition of Naproxen Derivatives — 241

Chapter 7 Solvent Effects in Molecular Recognition — 246

- 7.1 Introduction — 246
- 7.2 Attractive Host–Guest Interactions — 246
- 7.3 The Formation of an Apolar Cyclophane Complex in Solution: A Complex Picture — 247
- 7.4 Enthalpically Controlled Tight Molecular Complexation in Water — 252
  - 7.4.1 The Role of Solvent Cohesive Interactions — 253
  - 7.4.2 The Role of London Dispersion Interactions — 254
  - 7.4.3 Hydrophobic Bonding — 254
- 7.5 Apolar Complexation in Organic Solvents: A Simple General Model for Solvation Effects on Apolar Binding — 257
- 7.6 Functional Group Solvation Determines the Stability of Cyclophane Complexes — 259

Chapter 8 Supramolecular Function, Reactivity, and Catalysis — 264

- 8.1 Introduction — 264
- 8.2 Carrier-mediated Transport of Arenes through Water — 265
- 8.3 Redox-dependent Binding Ability of a Flavinophane — 269
- 8.4 Catalytic Cyclophanes: Supramolecular Reactivity and Catalysis — 273
  - 8.4.1 Cyclophanes with Esterase Activity: From Nonproductive to Productive Binding — 276
  - 8.4.2 Supramolecular Catalysis of the Benzoin Condensation — 281
  - 8.4.3 Supramolecular Catalysis of the Oxidation of Aromatic Aldehydes — 286
  - 8.4.4 A Porphyrin-bridged Cyclophane as a Model for Cytochrome P-450 Enzymes — 290
- 8.5 Concluding Remarks — 296

Indexes — 298

CHAPTER 1
From Ansa Compounds to [2.2]Paracyclophane to Cyclophane Complexes

1.1 The Origins of Cyclophane Chemistry

In 1951, Cram and Steinberg[1] reported on the synthesis of compound **1** in which two benzene rings are held face-to-face by methylene bridges and named it [2.2]paracyclophane. Independently, a preparation of **1** was also described by Brown and Farthing.[2] The attractive naming of **1** as [2.2]paracyclophane led to the development of the cyclophane nomenclature[3,4] for bridged aromatic compounds, *e.g.* compounds **2–4**. According to this systematization, all molecular receptors with at least one aromatic ring bridged by at least one aliphatic *n*-membered bridge ($n \geq 0$) may be called cyclophanes.

Under different names, cyclophanes had been known for decades prior to the preparation of **1**.[3] In the 1930s, Albert Lüttringhaus, one of the pioneers of macrocyclic chemistry, prepared bridged aromatic compounds such as **5** which he later classified as *ansa* compounds.[5] The latin word *ansa* means 'handle' and characterizes the bridges as handles on a ring system. For a complete historical account of the early work on bridged aromatic compounds in the 1930s and 1940s, the reader is referred to the 1964 monograph by Smith.[3] Many of these early developments have paved the way for today's molecular recognition studies with cyclophane receptors. In the 1930s and 1940s, R. Adams, W. H. Carothers, A. Lüttringhaus, V. Prelog, K. Ziegler, and other distinguished scientists explored the stereochemical requirements for the formation of macrocycles and analyzed the stereochemistry of functionalized derivatives.[3] A considerable amount of the synthetic methodology now applied to the preparation of macrocyclic receptors was developed at that time.[3] Some of the molecular structures that were prepared during the

---

[1] D. J. Cram and H. Steinberg, *J. Am. Chem. Soc.*, 1951, **73**, 5691.
[2] C. J. Brown and A. C. Farthing, *Nature (London)*, 1949, **164**, 915.
[3] B. H. Smith, 'Bridged Aromatic Compounds', Academic Press, New York, 1964.
[4] F. Vögtle and P. Neumann, *Tetrahedron*, 1970, **26**, 5847.
[5] A. Lüttringhaus and H. Gralheer, *Liebigs Ann. Chem.*, 1942, **550**, 67.

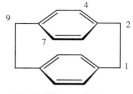

1 [2.2]Paracyclophane

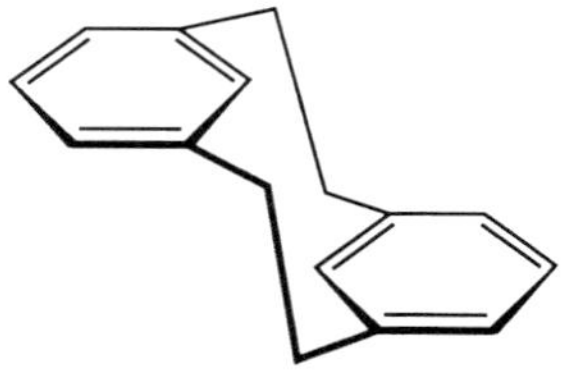

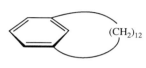

2 [2.2]Metacyclophane

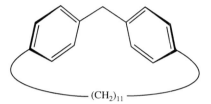

3 [11.1]Paracyclophane

4 [12]Metacyclophane

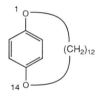

5 1,14-Dioxa[14]paracyclophane

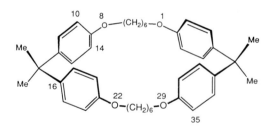

6 15,15,36,36-Tetramethyl-1,8,22,29-tetraoxa[8.1.8.1]paracyclophane

early days of cyclophane chemistry, *e.g.* **6**,[6] closely resemble those that are used today in the complexation of small molecules.

This introductory chapter establishes the link between the early cyclophane research and the present studies of cyclophanes as molecular receptors. Donald J. Cram more than anyone else personifies the close relationship that exists between past and present work: he has been a pioneer in the early chemistry of small cyclophanes and their subsequent development into molecular hosts.[7] This chapter also illustrates the ties that exist between host–

---

[6] A. Lüttringhaus and K. Buchholz, *Ber. Dtsch. Chem. Ges.*, 1940, **73B**, 134.
[7] (*a*) D. J. Cram, in 'Cyclophanes', eds. P. M. Keehn and S. M. Rosenfeld, Academic Press, New York, 1983, Vol. 1, pp. 1; (*b*) D. J. Cram, 'From Design to Discovery', ed. J. I. Seeman, American Chemical Society, Washington, DC, 1990.

guest chemistry and biomimetic cyclophane research that is not related to molecular recognition. Studies in both areas have considerably advanced the microscopic level understanding of weak intermolecular interactions in general and in biochemical processes.

1.2 Molecular Shape: An Experimental Calibration of the Space Occupancy of Atoms and Functional Groups

A correct, experiment-based calibration of space-filling molecular models and of computer-generated models is crucial for the successful design of a molecular receptor.[8] Since the 1930s, investigations into the space occupancy of atoms and functional groups and the generation of strain caused by violation of van der Waals radii have been at the center of cyclophane studies.

Lüttringhaus and coworkers studied the racemization of **7** and **8** as a function of the length of the cyclophane bridge.[5,9,10] The racemization of the optically active conformational isomers ('atropisomers') of these dioxa[n]paracyclophanes (*e.g.* **7a** ⇌ **7b**) occurs by rotation of the aromatic ring through the plane of the bridge. With decreasing bridge size, this rotation becomes increasingly hindered. The compression that occurs between the aromatic rings with their functional groups and the aliphatic bridge in the transition state of the racemization becomes so severe that stable enantiomers can be isolated. Thus, at room temperature, the enantiomers of acid **7** with $n = 10$ are completely stable against racemization, whereas the acid with $n = 12$ cannot be separated into enantiomers. The transition state for the racemization of **8** is less sterically crowded, since the benzene hydrogens can more easily rotate through the plane of the bridge than the bulkier bromo substituent in **7**. At $n = 10$, **8** could not be separated into enantiomers, but the acid having $n = 8$ was completely stable to racemization at ambient temperature. These results provided an experimental calibration of the atomic radii in the Stuart space-filling molecular models used at that time.[10]

The energetic requirements for racemization have also been investigated in the series of [$m.n$]paracyclophanes prepared by Cram and coworkers.[11] Enantiomers of the [4.3]paracyclophanecarboxylic acid **9a** are stable, whereas the optically pure [4.4]cyclophane **9b** readily racemizes at room temperature. In the racemization transition states, the two benzene rings presumably take a perpendicular alignment, and the hydrogen atoms of the unsubstituted benzene ring penetrate into the π-cloud of the second ring.

With the advent of ^{1}H NMR spectroscopy, studies on the dynamic properties of cyclophanes greatly intensified. The application of dynamic nuclear magnetic resonance (DNMR) methods to the conformational analysis

---

[8] D. J. Cram, *Angew. Chem.*, 1986, **98**, 1041; *Angew. Chem. Int. Ed. Engl.*, 1986, **25**, 1039.
[9] A. Lüttringhaus and H. Gralheer, *Liebigs Ann. Chem.*, 1947, **557**, 112.
[10] A. Lüttringhaus and G. Eyring, *Liebigs Ann. Chem.*, 1957, **604**, 111.
[11] D. J. Cram, W. J. Wechter, and R. W. Kierstead, *J. Am. Chem. Soc.*, 1958, **80**, 3126.

of cyclophanes has been reviewed by Mitchell.[12] For example, Vögtle and coworkers investigated the steric hindrance to the macroring inversion in the metacyclophane **10** ($n = 3-12$) which provided an experimental measure of the relative space-filling size of the substituent X.[13] For the conformational process **10a⇌10b**, which was monitored by DNMR analysis of the benzylic AB system, the free energies of activation $\Delta G^{\ddagger}$ at various chain lengths n were measured as a function of the size of X. From the measured $\Delta G^{\ddagger}$ values, the following relative scale for the effective size of substituents in the dynamic process **10a⇌10b** was derived: $SO_2CH_3 > COOCH_3 > SCH_3 > I > OCH_3 > Br > CN > CH_3 > Cl, NO_2 > NH_2 > OH > F > H >$ lone pair (in the analogous pyridinophane).

In host–guest complexation studies, information about the free cavity space in cyclophanes available for potential guest inclusion is highly desirable. Whitlock and coworkers studied by DNMR the dynamics of the passage of functional groups through the molecular cavity of **11** to experimentally evaluate the space available for a potential guest inclusion.[14] The tetraester **11** can exist as *pseudogeminal* and *pseudoortho* isomers; their interconversion occurs by ring inversion which must involve an ester passing through the cavity.[14a] Monitoring the interconversion of the diastereotopic OCH_2 protons in the cyclophane bridges by DNMR, the rates of ring inversion were studied as a function of the size of the ester group that rotates through the cyclophane cavity. The size of methyl, ethyl, and *n*-propyl esters is small compared to the cavity space resulting in low $\Delta G^{\ddagger}_{298}$-values of 13.4, 15.3, and 17.2 kcal mol^{-1}, respectively, for the *pseudoortho–pseudogeminal* interconversion. The neopentyl ester, however, is too bulky and cannot pass through

R = CH_3
C_2H_5
n-C_3H_7
CH_2-$CH(CH_3)_2$
CH_2-$C(CH_3)_3$

pseudoortho **11**   *pseudogeminal* **11**

---

[12] R. H. Mitchell, in 'Cyclophanes', eds. P. M. Keehn and S. M. Rosenfeld, Academic Press, New York, 1983, Vol. 1, pp. 239.
[13] H. Förster and F. Vögtle, *Angew. Chem.*, 1977, **89**, 443; *Angew. Chem. Int. Ed. Engl.*, 1977, **16**, 429.
[14] (*a*) B. J. Whitlock and H. W. Whitlock, Jr., *J. Am. Chem. Soc.*, 1983, **105**, 838; (*b*) A. B. Brown and H. W. Whitlock, Jr., *J. Am. Chem. Soc.*, 1989, **111**, 3640.

the cavity ($\Delta G^{\neq}_{298} > 35$ kcal mol^{-1}). At $\Delta G^{\neq}_{298} = 23.3$ kcal mol^{-1}, the free energy of activation for the passage of the isobutyl ester appears to be at the transition between small and large groups. It was therefore concluded that the cavity of **11** closely approximates the size of an isobutyloxycarbonyl group.

1.3 Probing Weak Non-covalent Interactions Between Chromophores in Rigid, Geometrically Defined Cyclophane Frames

The ultimate objective of basic molecular recognition research is to obtain a complete microscopic-level understanding of weak non-covalent interactions in ground and transition state complexes. The nature of these interactions has also been explored in a great variety of studies with small cyclophanes without receptor properties.

1.3.1 Transannular Interactions in Cyclophanes

As early as his first cyclophane paper in 1951, Cram discussed the possibility of a substituent on one ring of a monosubstituted [2.2]paracyclophane ([2.2]PCP) possessing a directive influence for electrophilic substitution in the second ring.[1] Such transannular substituent effects were subsequently observed, and Cram showed, in a very general way, that predominant substitution occurs *pseudogeminal* to the most basic position or substituent in the already substituted ring (equation 1.1).[15,16] These specific transannular substituent effects in electrophilic substitutions provide remarkable examples for functional group proximity and orientation effects on reaction rates and stereoselectivity.

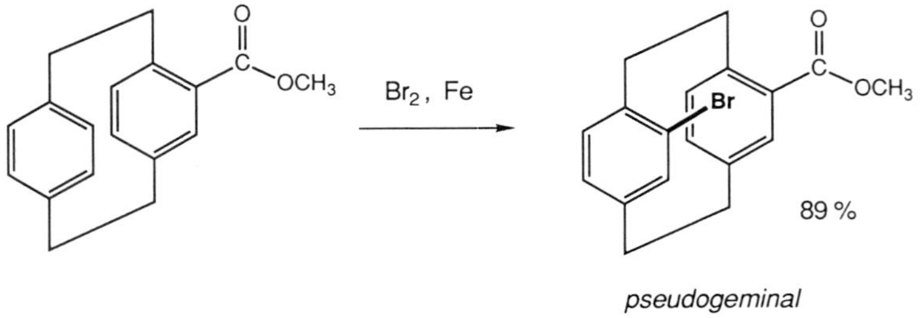

Also in 1951, Cram noted the abnormal electronic absorption spectrum of [2.2]PCP compared to an open-chain compound (Figure 1.1).[1] One of the characteristics of the cyclophane spectrum is the disappearance of fine structure and the appearance of a new long-wavelength transition around $\lambda_{max} = 300$ nm. The abnormalities in the electronic absorption spectra of

---

[15] H. J. Reich and D. J. Cram, *J. Am. Chem. Soc.*, 1969, **91**, 3505.
[16] H. J. Reich and D. J. Cram, *J. Am. Chem. Soc.*, 1969, **91**, 3534.

[*m.n*]paracyclophanes decrease with increasing length of the bridges: the spectra of [4.4]PCP and the nonmacrocyclic model compound 1,4-bis(*p*-ethylphenyl)butane closely resemble each other.[17] The peculiar electronic absorption spectra of [2.2]PCP along with the other smaller homologues in the series were explained by the distortions of their aromatic rings (Figure 1.2)[18,19] and by the rigidly enforced proximity of the two π-systems.[20] The relevance of transannular electronic interactions is supported by the differences between the spectra of [4.2]PCP and the open-chain compound. According to space-filling models, this cyclophane is fairly strain-free, and therefore the observed spectral changes cannot be associated with aromatic ring deformations. The spectroscopic investigations of the smaller [*m.n*]paracyclophanes were among the first to demonstrate particular electronic interactions between two chromophores as a result of specifically defined proximity and orientation.

1.3.2 Models for Intermolecular Excimers and Exciplexes

In the 1970s, the distance and orientation dependence of weak non-covalent interactions between two or more chromophores became the subject for a more systematic investigation. At that time, it was thought that it would be difficult to establish a well defined alignment of interacting groups in intermolecular complexes. Therefore, accurate information on the distance and orientation dependence of weak interactions would not be easily obtained by studying these systems. A precise arrangement of interacting chromophores would be better realized within a covalent molecular frame. The rigid skeleton of the smaller cyclophanes seemed ideal for a precise positioning of the chromophores and for intramolecularly analyzing weak non-covalent interactions. It was hoped that a comparison of the electronic interactions in intramolecular models to those in intermolecular complexes would provide useful information on the preferred distance and orientation of interacting groups in the latter.

Following this concept, Staab, Haenel, Misumi, and others systematically investigated the effect of orientation and distance on the electronic interactions of arene excimers.[21] Excimers are *exc*ited homod*imers* stabilized by exciton resonance ($AA^* \leftrightarrow {}^*AA$) and by charge-transfer configurations ($A^+A^- \leftrightarrow A^-A^+$).[22-24] Excimers are identified by their emission spectra: a

---

[17] D. J. Cram, N. L. Allinger, and H. Steinberg, *J. Am. Chem. Soc.*, 1954, **76**, 6132.
[18] H. Hope, J. Bernstein, and K. N. Trueblood, *Acta Crystallogr., Sect. B.*, 1972, **B28**, 1733.
[19] P. M. Keehn, in 'Cyclophanes', eds. P. M. Keehn and S. M. Rosenfeld, Academic Press, New York, 1983, Vol. 1, pp. 69.
[20] D. J. Cram, R. H. Bauer, N. L. Allinger, R. A. Reeves, W. J. Wechter, and E. Heilbronner, *J. Am. Chem. Soc.*, 1959, **81**, 5977.
[21] M. W. Haenel and D. Schweitzer in 'Polynuclear Aromatic Compounds', ed. L. B. Ebert, Advances in Chemistry Series Vol. 217, ACS, 1988, pp. 333.
[22] Th. Förster, *Angew. Chem.*, 1969, **81**, 364; *Angew. Chem. Int. Ed. Engl.*, 1969, **8**, 333.
[23] B. Stevens, *Adv. Photochem.*, 1971, **8**, 161.
[24] 'Organic Molecular Photophysics', ed. J. B. Birks, Wiley, New York, 1973 and 1975, Vols. 1 and 2.

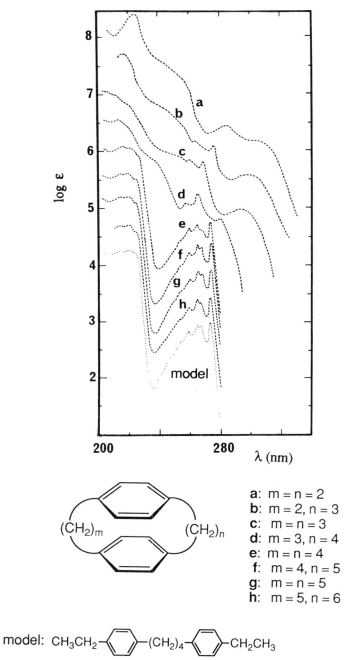

Figure 1.1 *Ultraviolet absorption spectra of the [m.n]paracyclophanes and an open-chain model in 95% ethanol. The curves, with the exception of the open-chain model at the bottom, have been displaced upward on the ordinate by 0.5 log ε unit increments from the curve immediately below (Reproduced with permission from reference 17. Copyright 1954 ACS.)*

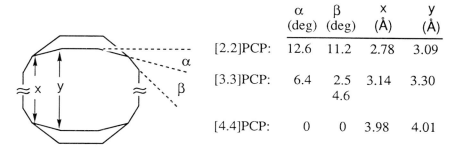

	α (deg)	β (deg)	x (Å)	y (Å)
[2.2]PCP:	12.6	11.2	2.78	3.09
[3.3]PCP:	6.4	2.5 / 4.6	3.14	3.30
[4.4]PCP:	0	0	3.98	4.01

Figure 1.2 *Schematic description of crystallographic data of [m.n]paracyclophanes.*[19] *The data for the [4.4]phane are those obtained for 6,9,16,19-tetramethoxy[4.4]paracyclophane*

singlet excimer shows a characteristic structureless band shifted to a longer wavelength compared to the corresponding monomer emission (Figure 1.3). This shift reflects the combination of the binding energy which stabilizes the complex with the repulsion energy experienced at the corresponding ground

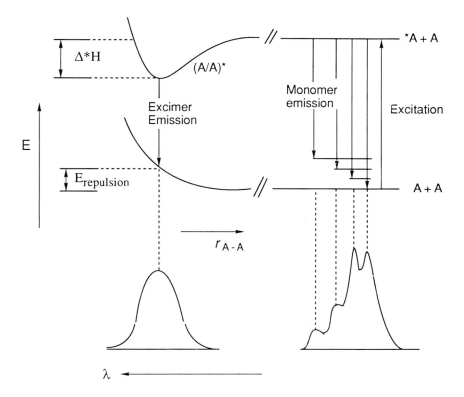

Figure 1.3 *Energetics of excimer formation*[33]

state configuration. The classical example for an arene excimer is the excited pyrene dimer first described by Förster.[22]

Scheme 1.1 shows three of the many different possible orientations of a naphthalene excimer. To model the $\pi-\pi$ interactions in intermolecular excimers, Haenel, Schweitzer, and coworkers prepared a variety of [2.2]-, [3.2]-, and [3.3]naphthalenophanes. Some of the [2.2]naphthalenophanes are shown in Scheme 1.1.[21] Following structural characterization, they studied the excited state properties of these compounds in glasses such as methylcyclohexane at 1.3 K. Most cyclophanes give a red-shifted and poorly structured fluorescence and phosphorescence spectrum similar to the characteristic emission of intermolecular excimers. Figure 1.4 shows the fluorescence and phosphorescence of chiral and achiral [2.2](2,6)naphthalenophanes compared to the monomer emission of 2,6-dimethylnaphthalene. The latter cannot form an intermolecular excimer when diluted in a rigid matrix. The emission from the achiral phane is shifted to a lower energy than the emission from the chiral isomer. The stronger red shift of the emission band is indicative of stronger $\pi-\pi$ interactions that stabilize the intramolecular excimer state. For intermolecular excimers, the degree of the fluorescence red shift relative to the monomer is proportional to the binding energy in the singlet excimer (Figure 1.3).[22,23] From their extensive investigations, Haenel and Schweitzer concluded that the orientation of interacting aromatic π-electron systems strongly influences the extent of the electronic interaction.[21] In the excited singlet and triplet states of the cyclophanes, the electronic interactions which give stability to the intramolecular excimers increase if the chromophore planes are parallel and reach a maximum when their six-membered rings are completely eclipsed. A similar geometry should also enable maximal stabilizing electronic interactions to occur between two chromophores in an intermolecular excimer.

The question of the distance dependence for excimer interactions was addressed by Staab and coworkers who prepared the [2.2]-, [3.3]-, and [4.4](2,7)pyrenophanes.[25-30] Table 1.1 shows their emission spectral data at $T = 1.3$ K in a 2-methyltetrahydrofuran matrix, contrasting the characteristic broad fluorescence from the intramolecular singlet excimer state in the pyrenophanes to the structured monomer emission of 2,7–dimethylpyrene. The magnitude of the red shift Δv is taken as a measure for the transannular interactions in the intramolecular singlet excimer state. The data in Table 1.1 show that the red shift of the fluorescence emission decreases with increasing length of the cyclophane bridges.

[25] H. Irngartinger, R. G. H. Kirrstetter, C. Krieger, H. Rodewald, and H. A. Staab, *Tetrahedron Lett.*, 1977, 1425.
[26] D. Schweitzer, K. H. Hausser, R. G. H. Kirrstetter, and H. A. Staab, *Z. Naturforsch., Teil A*, 1976, **31**, 1189.
[27] H. A. Staab, N. Riegler, F. Diederich, C. Krieger, and D. Schweitzer, *Chem. Ber.*, 1984, **117**, 246.
[28] T. Umemoto, S. Satani, Y. Sakata, and S. Misumi, *Tetrahedron Lett.*, 1975, 3159.
[29] T. Kawashima, T. Otsubo, Y. Sakata, and S. Misumi, *Tetrahedron Lett.*, 1978, 5115.
[30] R. H. Mitchell, R. J. Carruthers, and J. C. M. Zwinkels, *Tetrahedron Lett.*, 1976, 2585.

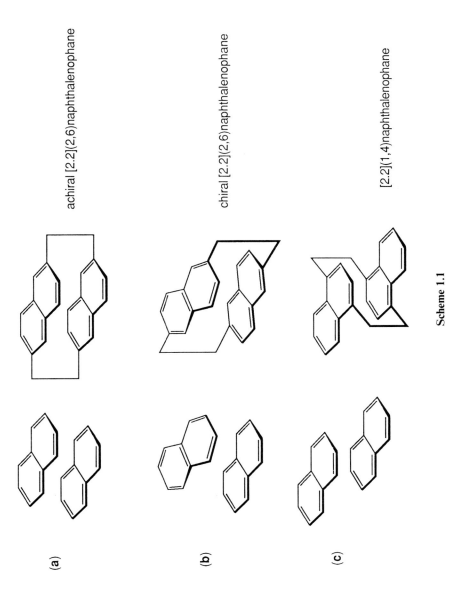

Scheme 1.1

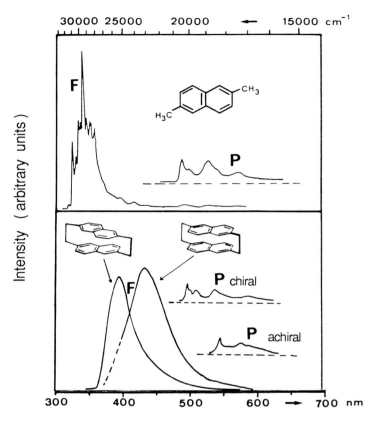

Figure 1.4 *Fluorescence (F) and phosphorescence (P) of 2,6-dimethylnaphthalene in n-octane and of chiral and achiral [2.2](2,6)naphthalenophanes in methylcyclohexane at 1.3 K*
(Reproduced with permission from reference 21. Copyright 1988 ACS.)

The pyrenophane study revealed some of the serious problems associated with modeling intermolecular interactions in small cyclophanes. In analogy to the [m.n]paracyclophane series, the smaller pyrenophanes are severely distorted. Table 1.2 shows relevant X-ray crystal structure data of the three pyrenophanes. Only the [4.4]phane is a strain-free molecule with approximately equal distances between the two planes measured at the edge and at the center. The pyrene moieties in the [2.2] and [3.3]cyclophanes are severely distorted from planarity and, most importantly, the lengths of the bridges do not directly correlate with the transannular distances. Distances between the centers of the two pyrenes in the [3.3] and [4.4]phanes are actually identical and smaller than in the [2.2]phane having two severely bent aromatic units (Table 1.2). Such chromophore distortions from planarity do not occur in intermolecular complexes. Also, the inter-chromophore distance does not correlate with

Table 1.1 *Fluorescence emission spectra (cm^{-1}) of the [n.n](2,7)pyrenophanes and 2,7-dimethylpyrene at 1.3 K in 2-methyltetrahydrofuran*[27]

Chromophore	ν_{max}	$\Delta\nu$
2,7-dimethylpyrene	25,850	
[2.2](2,7)pyrenophane	18,000	7,850
[3.3](2,7)pyrenophane	19,400	6,450
[4.4](2,7)pyrenophane	21,300	4,550

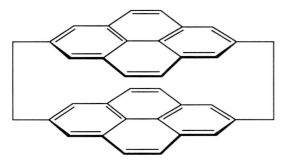

[2.2](2,7)Pyrenophane

Table 1.2 *Selected intraannular distances (Å) in the [n.n](2,7)pyrenophanes*[25,27]

[2.2]Phane	[3.3]Phane	[4.4]Phane
a: 2.79	3.18	3.68
b: —	3.13, 3.30	3.55, 3.61
c: 3.19	3.37, 3.39	3.67
d: —	3.48	3.48
e: 3.76	—	—

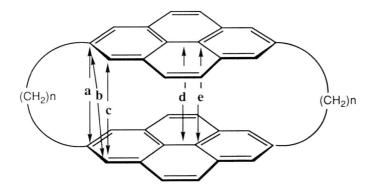

the length of the bridges, and the chromophore distortions considerably perturb the electronic and spectroscopic properties of the intramolecular model. Additionally, in all three phanes, the two pyrenes are not in a fully eclipsed arrangement to each other; lateral displacements perpendicular to the long pyrene axis occur to varying extents. Therefore, although the spectral properties (Table 1.1) indicate decreasing transannular interactions with increasing length of the pyrenophane bridges, a conclusion on the distance dependence of excimer interactions cannot be directly drawn from these studies.

Amine–arene exciplexes (*excited complexes*, originally called heteroexcimers), are stabilized by n–π interactions between the nonbonding nitrogen electron pair and aromatic π-electrons.[31-33] For exciplexes, the same energetic considerations (Figure 1.3) are valid as those seen for excimers formed by two chromophores of the same kind. It is assumed that the nonbonding nitrogen electron pair in arene–amine exciplexes is located directly over the aromatic π-systems. In naphthalene–amine exciplexes, the nitrogen has the choice of being positioned either above the ring or above the central naphthalene π-bond (Scheme 1.2). Haenel *et al.* prepared a series of naphthalenopyridinophanes and phanedienes (Scheme 1.2) to explore which orientation leads to the most stabilized intramolecular exciplex.[21,34,35] According to X-ray analysis, the pyridine lone pair in the phanedienes takes an almost perpendicular orientation above the naphthalene plane. Figure 1.5 suggests that the geometry with the nonbonding nitrogen electron pair located above the central π-bond is preferred. This conclusion holds if the degree of red shift of the exciplex emission compared to the monomer emission of 1,5-dimethylnaphthalene is taken as a measure for the stabilization of the intramolecular singlet exciplex state. Again, limitations of this cyclophane model should be pointed out. In an intermolecular exciplex, the intermolecular orbital interactions occur strictly through-space, whereas the intramolecular model has the potential for significant through-bond interactions between the two chromophores. These through-bond interactions have been increasingly studied in charge-transfer and photo-induced electron-transfer processes in bichromophoric molecules with covalently linked donor and acceptor moieties,[36,37] even though they remain poorly defined for intramolecular excimers and exciplexes. Through-bond interactions can considerably perturb the electronic and spectral properties of cyclophane models.

---

[31] 'The Exciplex', eds. M. Gordon and W. R. Ware, Academic Press, New York, 1975.
[32] R. S. Davidson, *Adv. Phys. Org. Chem.*, 1983, **19**, 1.
[33] S. L. Mattes and S. Farid, *Science (Washington, DC)*, 1984, **226**, 917.
[34] M. W. Haenel, B. Lintner, R. Benn, A. Rufinska, G. Schroth, C. Krüger, S. Hirsch, H. Irngartinger, and D. Schweitzer, *Chem. Ber.*, 1985, **118**, 4884.
[35] B. Lintner, D. Schweitzer, R. Benn, A. Rufinska, and M. W. Haenel, *Chem. Ber.*, 1985, **118**, 4907.
[36] P. Pasman, F. Rob, and J. W. Verhoeven, *J. Am. Chem. Soc.*, 1982, **104**, 5127.
[37] H. Oevering, M. N. Paddon-Row, M. Heppener, A. M. Oliver, E. Cotsaris, J. W. Verhoeven, and N. S. Hush, *J. Am. Chem. Soc.*, 1987, **109**, 3258.

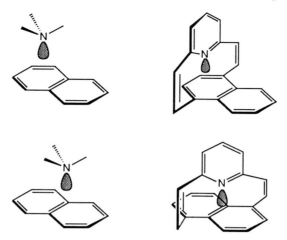

Scheme 1.2

1.3.3 Donor–Acceptor Cyclophanes as Models for Intermolecular Charge-transfer Complexes

With a large series of elegant donor–acceptor cyclophanes, Staab and coworkers modeled the orientation and distance dependence of charge-transfer interactions in charge-transfer complexes. Only a few examples from the paracyclophane series will be presented here.[38–46] Two *para*-disubstituted benzenes, one with two electron donor substituents and the other with two electron acceptor substituents, can interact in different orientations. Two distinguishable orientations, pseudogeminal with eclipsed substituents and pseudoortho with noneclipsing substituents, are shown in Scheme 1.3. These intermolecular complex orientations were modeled by donor–acceptor cyclophanes (Scheme 1.3). Figure 1.6 shows the strikingly different electronic absorption spectra of the pseudogeminal and pseudoortho [3.3]paracyclophane-quinhydrones.[44a] In dioxane, the pseudogeminal cyclophane-quinhydrone exhibits a very intense charge-transfer band [λ_{max} = 462 nm (ϵ = 3210)]. In contrast, the pseudoortho derivative exhibits a longer wavelength shoulder of much weaker intensity at λ_{max} = 500 nm (ϵ = 105) in addition to a stronger intensity band at λ_{max} = 356 nm (ϵ = 1904), both bands being assigned to a

---

[38] H. A. Staab and W. Rebafka, *Chem. Ber.*, 1977, **110**, 3333.
[39] H. A. Staab, C. P. Herz, and H.-E. Henke, *Chem. Ber.*, 1977, **110**, 3351.
[40] H. A. Staab and H. Haffner, *Chem. Ber.*, 1977, **110**, 3358.
[41] H. A. Staab and V. Taglieber, *Chem. Ber.*, 1977, **110**, 3366.
[42] H. A. Staab, A. Döhling, and C. Krieger, *Liebigs Ann. Chem.*, 1981, 1052.
[43] H. A. Staab, B. Starker, and C. Krieger, *Chem. Ber.*, 1983, **116**, 3831.
[44] (a) H. A. Staab, C. P. Herz, C. Krieger, and M. Rentea, *Chem. Ber.*, 1983, **116**, 3813; (b) H. Vogler, G. Ege, and H. A. Staab, *Tetrahedron*, 1975, **31**, 2441.
[45] H. A. Staab, G. Gabel, and C. Krieger, *Chem. Ber.*, 1983, **116**, 2827.
[46] H. A. Staab, G. H. Knaus, H.-E. Henke, and C. Krieger, *Chem. Ber.*, 1983, **116**, 2785.

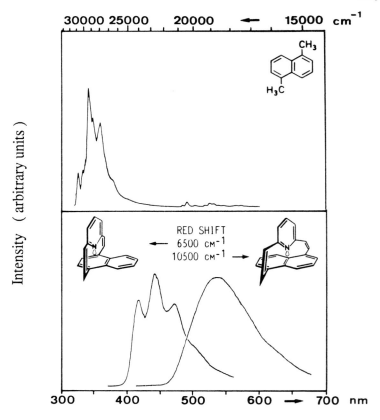

Figure 1.5 *Fluorescence spectra of naphthalenopyridinophane-dienes in n-octane at 1.3 K. The red shifts give the difference of the exciplex emission maximum and the monomer fluorescence of 1,5-dimethylnaphthalene (Reproduced with permission from reference 21. Copyright 1988 ACS.)*

charge-transfer transition.[44b] Comparable differences between the charge-transfer band intensities in the pseudogeminal and pseudoortho quinhydrones were also observed in the corresponding [2.2]phane series. A similar strong dependence on donor–acceptor orientation for charge-transfer absorptions was measured for other donor–acceptor [2.2]paracyclophanes (Scheme 1.3). In each case, the pseudogeminal derivative gave a longer wavelength charge-transfer band with a significantly higher extinction coefficient compared to the pseudoortho derivative. In all cyclophanes shown in Scheme 1.3, a charge-transfer occurs predominantly in the excited state and does not contribute significantly to the electronic ground state. Even in the donor–acceptor paracyclophane **12** with a N,N,N',N'-tetramethyl-p-phenylenediamine (TMPD) as a strong donor and a 7,7,8,8-tetracyanoquinodimethane (TCNQ)

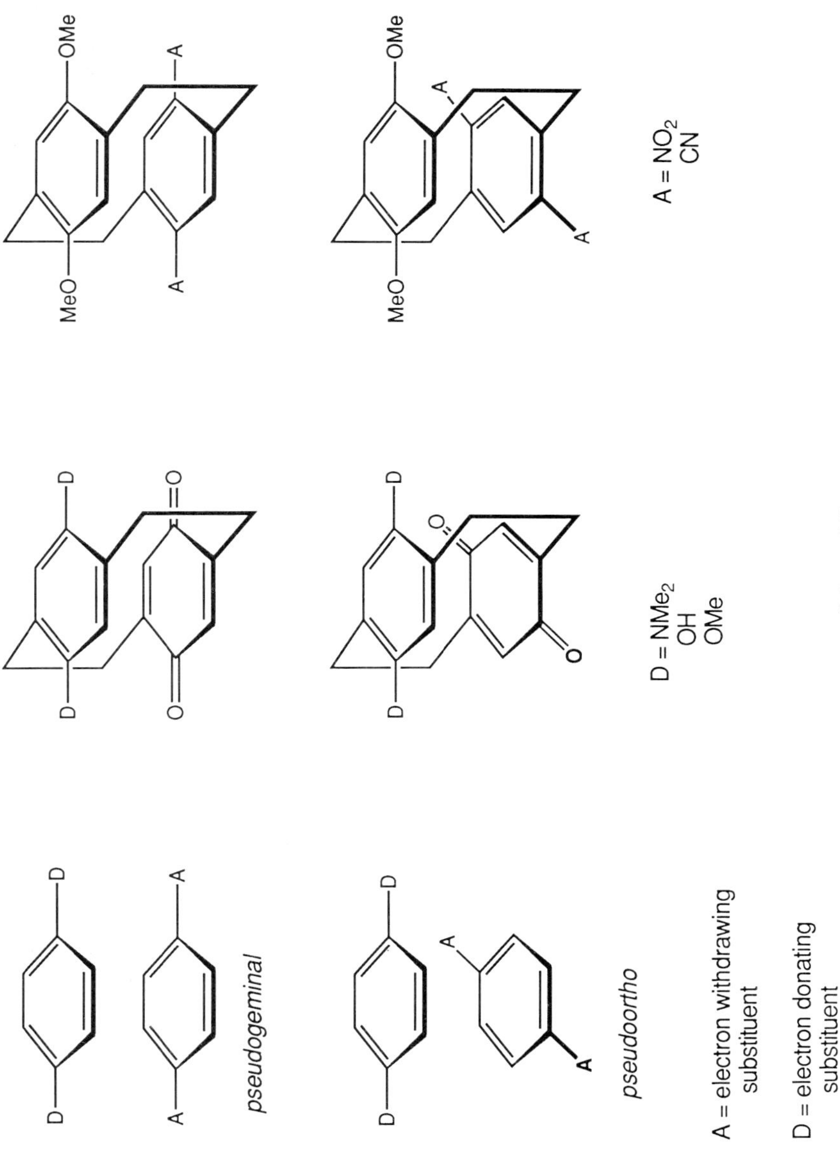

Scheme 1.3

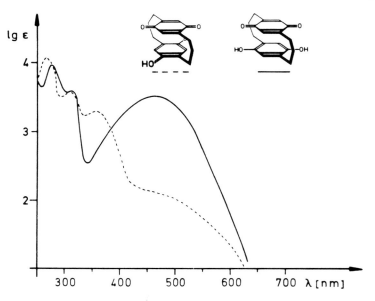

Figure 1.6 *Electronic absorption spectra of pseudogeminal and pseudoortho [3.3]paracyclophane-quinhydrones in dioxane (Reproduced with permission from reference 44a. Copyright 1984 VCH.)*

as a strong electron acceptor, a ground state charge-transfer that would form a diradical–zwitterionic species was not observed.[47]

Both in the Staab and the Misumi groups, the synthesis of multilayered cyclophanes for investigations of transannular and donor–acceptor interactions over larger chromophore distances ultimately led to spectacular molecular architecture as exemplified by compounds **13**[48] and **14**.[49]

1.4 Modeling Interactions and Reactions in Biological Systems: Stereochemical Control Imposed by Cyclophane Skeletons

Biological phenomena provide inspiration for much of the current work in chemical molecular recognition. This is equally true for many of the cyclophane studies in which binding does not play a role. The incorporation of biological building blocks, especially coenzymes and coenzyme models, into cyclophane skeletons for controlling the stereochemistry of biomimetic processes has been pursued with such vigor in the 1980s, that this chapter can

---

[47] H. A. Staab, R. Hinz, G. H. Knaus, and C. Krieger, *Chem. Ber.*, 1983, **116**, 2835.
[48] S. Misumi, in 'Cyclophanes', eds. P. M. Keehn and S. M. Rosenfeld, Academic Press, New York, 1983, Vol. 2, pp. 573.
[49] H. A. Staab and U. Zapf, *Angew. Chem.*, 1978, **90**, 807; *Angew. Chem. Int. Ed. Engl.*, 1978, **17**, 757.

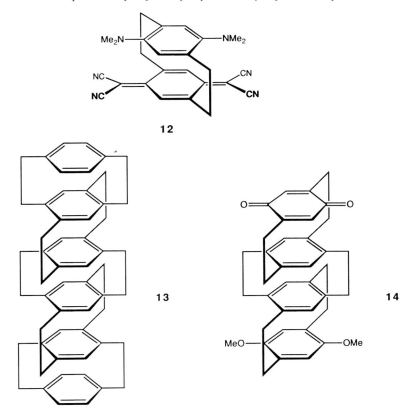

only provide a flavor of the many elegant investigations that have been performed.

1.4.1 Purinophanes

It is well known that the degree of stacking of nucleic acid bases affects their optical properties, and the electronic absorption intensity of a polynucleotide with n bases is generally less than n times the absorption of the free base. This decrease in absorption intensity is known as *hypochromism*.[50,51] For ordered macromolecular structures, hypochromic effects on the order of 10% to 50% are observed. A partial unwinding of double-stranded DNA reduces the stacking interactions between the bases and is detectable by a reduced hypochromism in the optical spectra. As models for stacking interactions of nucleic acid bases in DNA, Misumi and coworkers prepared a variety of

---

[50] A. M. Michelson, in 'Molecular Associations in Biology', ed., B. Pullman, Academic Press, New York, 1968, pp. 83.
[51] C. R. Cantor and P. R. Schimmel, 'Biophysical Chemistry', Freeman, New York, 1980, Part II, pp. 399.

doubly and triply bridged purinophanes, *e.g.* **15** and **16**.[52,53] In ethanolic and aqueous solutions, a strong hypochromism was observed for the longest wavelength band. For **15** having an average interplanar distance of 3.3 Å (X-ray), a hypochromic effect of 47% was calculated from the absorption spectra of dimer **15** compared to the monomeric base in water.[52] A similar evaluation afforded a hypochromism of 31% in 0.5 N HCl for **16** in which the two purines are fully eclipsed at an average interplanar distance of 3.7 Å. By comparing a series of purine dimers with varying stacking modes and interplanar distances, it was determined that the fully overlapped orientation of two purine rings at the closest interplanar distance gives the largest hypochromicity.

15

16

1.4.2 Flavinophanes

Staab and coworkers' concept of investigating chromophoric interactions as a function of distance and orientation in geometrically well defined cyclophane skeletons (Chapter 1.3) has also been applied to the modeling of π-stacking interactions involving coenzymes and coenzyme analogues. The enzyme glutathione reductase reduces the disulfide form of glutathione to the sulfhydryl form (equation 1.2). The enzyme uses two coenzymes, NADPH (dihydronicotinamide adenine dinucleotide phosphate) and FAD (flavin adenine dinucleotide) to achieve its function. At the enzyme's active site, electrons are transferred from NADPH to a tightly bound FAD, then to a disulfide bridge between two cysteine residues, and finally to a bound oxidized glutathione.[54] It was shown by X-ray crystallography[55] that the tricyclic isoalloxazine ring of bound FAD participates in π–π-stacking interactions. Prior to the binding of NADPH, a tyrosine ring (Tyr-114) is the partner of FAD in these interactions. In the course of the reaction, the tyrosine is replaced by the nicotinamide ring of NADP$^+$.

To model π–π-stacking interactions of flavins, Staab and coworkers prepared the chiral [3.3](3,10)isoalloxazinophane **17**[56] and the isoalloxazino-

[52] F. Hama, Y. Sakata, S. Misumi, M. Aida, and C. Nagata, *Tetrahedron Lett.*, 1982, **23**, 3061.
[53] K. Akahori, F. Hama, Y. Sakata, and S. Misumi, *Tetrahedron Lett.*, 1984, **25**, 2379.
[54] A. Meister and M. E. Anderson, *Ann. Rev. Biochem.*, 1983, **52**, 711.
[55] E. F. Pai and G. E. Schulz, *J. Biol. Chem.*, 1983, **258**, 1752.
[56] M. F. Zipplies and H. A. Staab, *Tetrahedron Lett.*, 1984, **25**, 1035.

γ-Glu—Cys—Gly
     |
     S
     |
     S
     |
γ-Glu—Cys—Gly
 + NADPH + H⁺ ⇌ 2 γ-Glu—Cys—Gly + NADP⁺
                           |
                           S
                           H

(1.2)

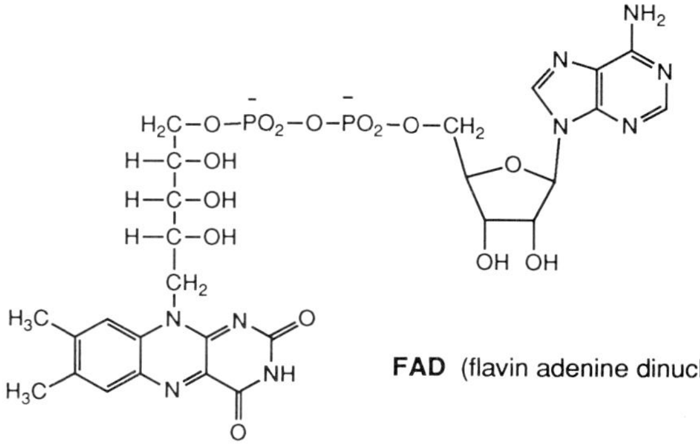

FAD (flavin adenine dinucleotide)

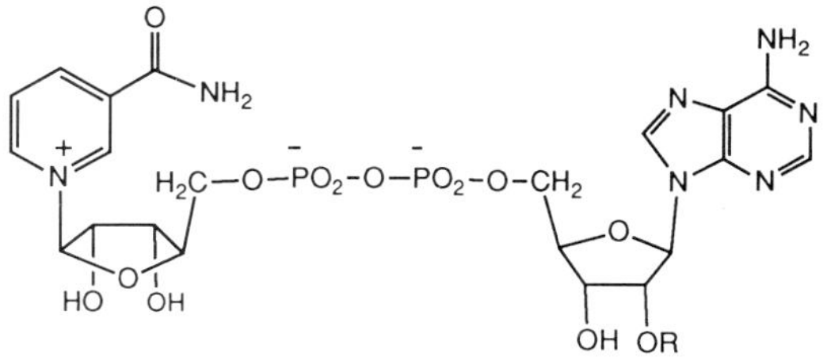

NAD⁺ R = H (nicotinamide adenine dinucleotide)

NADP⁺ R = PO_3^{--}

metacyclophane **18**.[57] The X-ray crystal structure of **18** shows that the overlap between benzene and isoalloxazine chromophores closely resembles the interactions between nicotinamide and flavin at the active site of glutathione reductase. The synthesis of **18** was developed as a preparatory step in the approach to a similar flavinophane in which the benzene ring is replaced by a nicotinamide.

Catalysts accelerate chemical reactions by reducing the free energy difference between ground states and transition states. Introducing strain into the ground state forcing it to become higher in energy and increasing its resemblance to the transition state provides one way of lowering the free energy of activation. This catalytic mechanism has been proposed to explain a

---

[57] M. F. Zipplies, C. Krieger, and H. A. Staab, *Tetrahedron Lett.*, 1983, **24**, 1925.

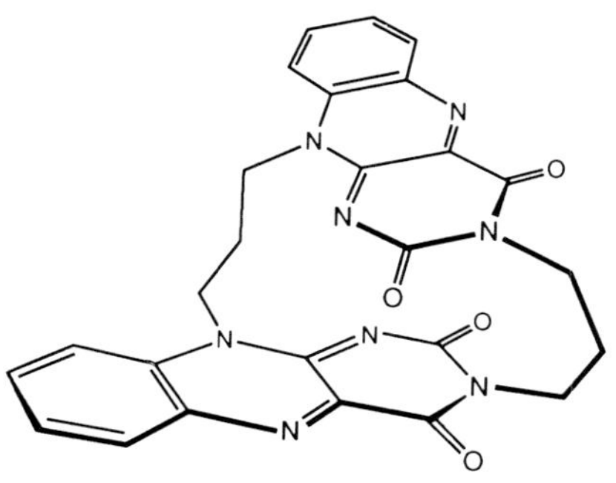

17 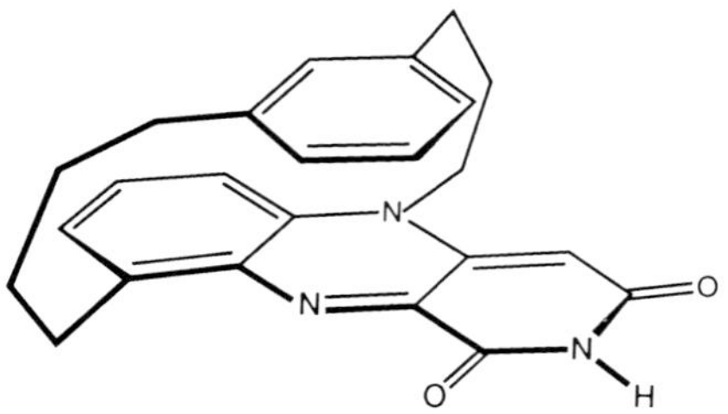 **18**

variety of enzymatic reactions.[58] Shinkai and coworkers prepared a series of flavinophanes Fl(*n*) and 5-deazaflavinophanes dFl(*n*) to investigate the relationship between the stability of their initial oxidized state and their reactivity in the oxidation of 1-benzyl-1,4-dihydronicotinamide (BNAH), used as a NADH model.[59]

Oxidized isoalloxazines are planar, whereas the 2e$^-$ reduced dihydroisoalloxazines, according to *X*-ray crystallography, adopt a butterfly shape by bending around the two nitrogens N-5 and N-10 of the central ring in the tricyclic system (equation 1.3).[60,61] This significant change in geometry suggests that processes which strain and therefore destabilize the planar oxidized flavin should lead to an enhanced reactivity in redox processes which produce the bent two-electron reduced state. A combination of spectroscopic studies, *X*-ray analysis, and computational geometry optimizations showed that the flavinophanes Fl(*n*) and the deazaflavinophanes dFl(*n*) become increasingly strained by shortening the alkyl strap bridging N-3 to the phenyl ring at N-10. Table 1.3 shows that the reactivity in the two series Fl(*n*) and dFl(*n*) is closely related to the strain in the oxidized phanes. Decreasing the length of the alkyl strap from $n=12$ to $n=6$ increases the second-order rate constant for the oxidation of 1-benzyl-1,4-dihydronicotinamide by a factor of 11 in the Fl-series and 16 in the dFl series. An attractive explanation for the increased rate seen for the shorter strapped derivatives is that the strained ground state more closely resembles the transition state leading to the bent reduced form, and therefore the energy of activation is lowered.

[58] A. Fersht, 'Enzyme Structure and Mechanism', 2nd ed., Freeman, New York, 1985.
[59] S. Shinkai, A. Kawase, T. Yamaguchi, O. Manabe, Y. Wada, F. Yoneda, Y. Ohta, and K. Nishimoto, *J. Am. Chem. Soc.*, 1989, **111**, 4928.
[60] C. Walsh, 'Enzymatic Reaction Mechanisms', Freeman, San Francisco, 1979.
[61] P. Kierkegaard, R. Norrestam, P.-E. Werner, I. Csöregh, M. Von Glehn, R. Karlsson, M. Leijonmark, O. Rönnquist, B. Stensland, O. Tillberg, and C. Torbjörnssen, in 'Flavins and Flavoproteins', ed. H. Kamin, University Park Press, Baltimore, 1971, Vol 3, pp. 1.

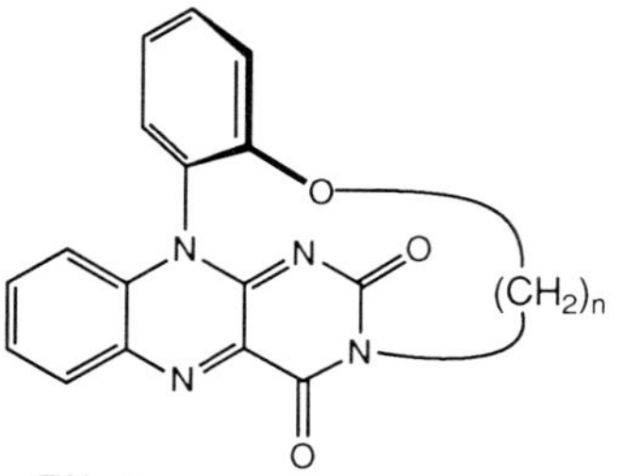

Fl(n) 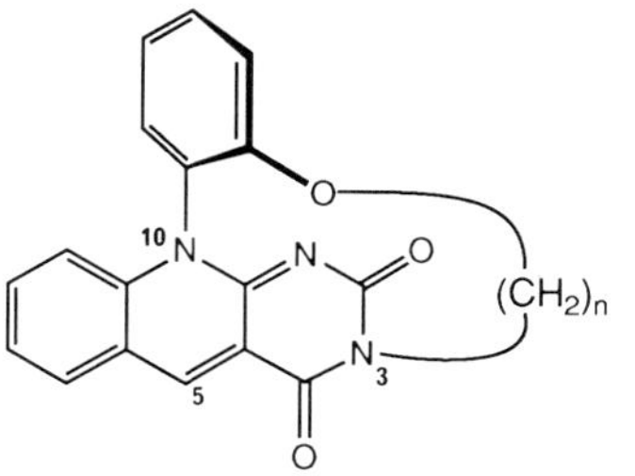   dFl(n)

n = 6, 7, 8, 10, and 12

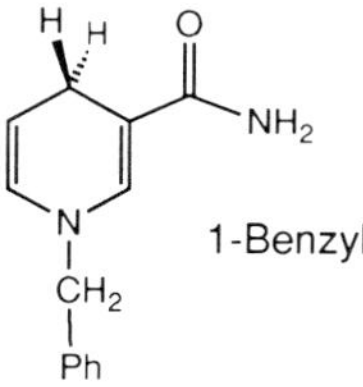

1-Benzyl-1,4-dihydronicotinamide
(BNAH)

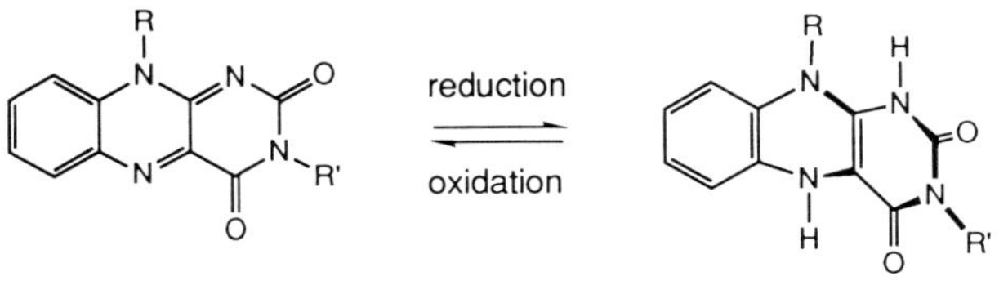

$$\text{reduction} \rightleftharpoons \text{oxidation} \qquad (1.3)$$

For dFl(6) and other members of the 5-deazaflavin series, a very interesting diastereo-differentiating[62] hydrogen transfer was observed.[63] The 5-deazaflavin chromophore (coenzyme F_{420}) resembles the nicotinamides in its shape and function as a two-electron, hydride transfer coenzyme. Walsh suggested that F_{420} could be considered as a nicotinamide in flavin's clothing.[64] With dFl(6), hydrogen transfer to the oxidized form and from the reduced form occurs exclusively at the axial C-5 position (Scheme 1.4). By ^{1}H NMR, it was shown that the reduced dFl(6) takes a boat-shaped conformation affording the diastereotopic, magnetically non-equivalent protons H_{ax} and H_{eq} at C-5 which were assigned by the nuclear Overhauser effect. Reduction with deuterated reducing agents, e.g. $NaBD_4$ in CD_3OD, afforded an NMR signal with the intensity of one proton only for H_{eq}, whereas the peak for H_{ax} had

---

[62] Y. Izumi and A. Tai, 'Stereo-Differentiating Reactions', Academic Press, New York, 1977.
[63] S. Shinkai, T. Yamaguchi, A. Kawase, O. Manabe, and R. M. Kellogg, *J. Am. Chem. Soc.*, 1989, **111**, 4935.
[64] C. Walsh, *Acc. Chem. Res.*, 1986, **19**, 216.

Table 1.3 *Second-order rate constants for the reduction of BNAH to 1-benzylnicotinamide in water/methanol (50.8:49.2, v/v), pH = 9.0 with 6 mM borate buffer*

Fl(n) n	k_{BNAH} ($L\ mol^{-1}\ s^{-1}$)	dFl(n) n	k_{BNAH} ($L\ mol^{-1}\ s^{-1}$)
6	24.2	6	1.00
7	13.0	7	0.349
8	6.54	8	0.244
10	2.72	10	0.0712
12	2.20	12	0.0625

almost completely disappeared. The incorporated deuterium almost exclusively occupied an axial position. When this deuterated reduced dFl$_{red}$ was reoxidized with N-methylacridinium iodide, a remaining deuterium content of $\approx 24\%$ was found in the reoxidized deazaflavin. Taking an experimentally determined primary H/D isotope effect of $k_H/k_D = 2.6$ for dFl$_{red}\rightarrow$dFl into account, the reactivity ratio of H$_{ax}$ *versus* H$_{eq}$ was calculated to be $(76/24) \times 2.6 = 8.6$. Thus, the total diastereo-differentiating ability of dFl(6) amounts to a factor of 8.6. The most plausible explanation for this axial preference is provided by stereoelectronic effects rather than by any steric effects caused by the alkyl strap. These effects are discussed below in Chapter 1.4.3.

1.4.3 Pyridinophanes

Prior to the deazaflavin studies by Shinkai *et al.*, Verhoeven and coworkers had proposed stereoelectronic effects to explain the high degree of diastereo-differentiating hydride transfer at bridged NADH models like **19**.[65] The degree of diastereo-differentiation for **19** is more than 90%, and hydride transfer to oxidized **19** and from reduced **19** occurs *via* the axial C-4 position. X-Ray crystallography and electronic absorption spectra showed that the strain exerted by the bridge leads to a boat-shaped distortion of the 1,4-dihydronicotinamide moiety which places the two C-4 hydrogens in diastereotopic, axial and equatorial, positions. Kinetic investigations of the hydride exchange processes also suggested that a partial boat shape of the pyridine ring occurs in the transition state. The authors attributed the diastereo-differentiation to stereoelectronic effects which favor the exchange *via* an 'axial' transition state (Figure 1.7). An axial transition state is more stabilized compared to the equatorial one through a π-type overlap, a kind of hyperconjugation, between the partially broken (or partially formed) C(4)-H bond and the neighboring π-system. Cyclophane **19** and related systems[65]

---

[65] F. Rob, H. J. van Ramesdonk, W. van Gerresheim, P. Bosma, J. J. Scheele, and J. W. Verhoeven, *J. Am. Chem. Soc.*, 1984, **106**, 3826.

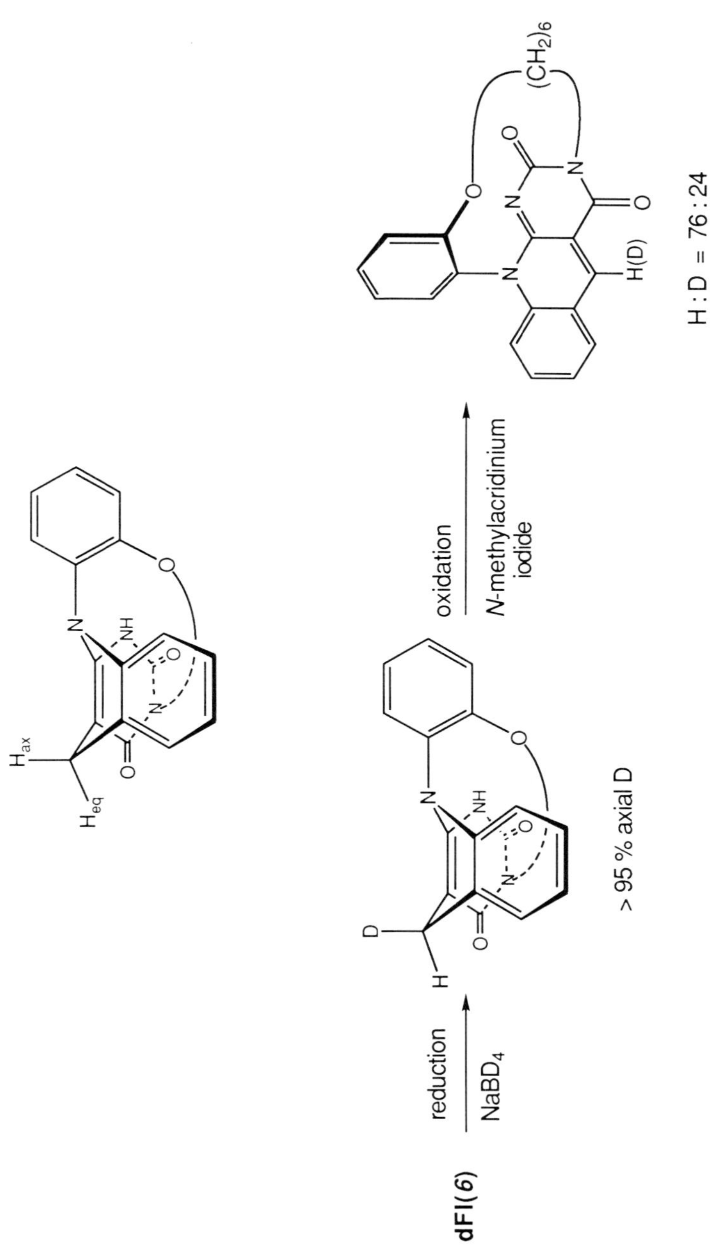
dFl(6)
reduction
NaBD4
> 95 % axial D
oxidation
N-methylacridinium
iodide
H : D = 76 : 24

Scheme 1.4

constituted the first examples of NADH models[66] capable of mimicking the diastereo-differentiating course of hydride exchange at pyridine dinucleotides under enzymatic conditions.[67]

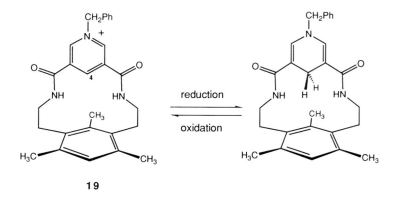

19

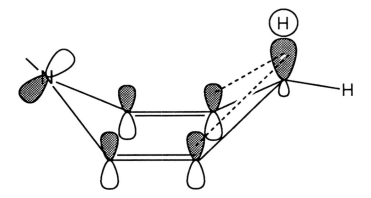

Figure 1.7 *Bonding π-type overlap between the HOMO of the π-system and the LUMO of the partial C–H bond in the axial transition state for hydride transfer at a distorted pyridine ring*[65]

The asymmetric reduction of carbonyl substrates by optically active NADH model compounds has received wide attention.[66,68] High enantioselectivity was obtained by Kellogg *et al.* in reductions by chiral dihydropyridine-cyclophanes with bridges containing (*S*)-valine residues, *e.g.* **20**.[68,69] Scheme

---

[66] S. Shinkai, in 'Enzyme Chemistry – Impact and Applications', 2nd ed., ed. C. J. Suckling, Chapman and Hall, London, 1990, pp. 50.
[67] N. J. Oppenheimer and N. O. Kaplan, *Bioorg. Chem.*, 1974, **3**, 141.
[68] R. M. Kellogg, *Angew. Chem.*, 1984, **96**, 769; R. M. Kellogg, *Angew. Chem. Int. Ed. Engl.*, 1984, **23**, 782.
[69] A. G. Talma, P. Jouin, J. G. De Vries, C. B. Troostwijk, G. H. W. Buning, J. K. Waninge, J. Visscher, and R. M. Kellogg, *J. Am. Chem. Soc.*, 1985, **107**, 3981.

1.5 shows the advantage of the optically active cyclophane **20** over the nonmacrocyclic control compound **21**. The reduction of ethyl phenylglyoxalate in the presence of Mg^{2+} ions in acetonitrile by the macrocycle yields the (S)-alcohol in 90% e.e., whereas the control affords the (R)-alcohol in only 18% e.e. The complexation of the highly electrophilic Mg^{2+} ions by the macrocycle plays an important role in organizing a productive ternary complex (Scheme 1.6) which presumably forms as the intermediate in the asymmetric reduction. Experimental support for the proposed coordination of the magnesium ion in this complex was provided by ^{13}C NMR. In the productive complex, the reactive substrate carbonyl is placed above the C-4 position of the 1,4-dihydropyridine. The phenyl ring of the substrate stacks cofacially with the dihydropyridine ring. For steric reasons, the glyoxalate residue turns away from the bulky valine isopropyl group and coordinates on the less hindered side together with the macrocyclic amide carbonyl to the Mg^{2+} ion. In such a ternary complex, hydride transfer occurs to the *re*-face of the carbonyl, yielding exclusively the (S)-product.

What is the role of the bridge in **20**? The comparison in Scheme 1.5 shows that the reaction with macrocycle **20** leads to preferential formation of a different enantiomer than the reaction with nonmacrocyclic **21**. In addition, the enantioselectivity is dramatically increased. It is known that the donor centers in the bridge do not participate in the Mg^{2+} coordination. Also, the replacement of the alkyl bridge by a polyether bridge – $(CH_2CH_2O)_n$ – does not significantly affect reactivity and stereoselectivity. It seems that the role of the bridge is to constrain and organize the macrocycle by enforcing the (S)-product generating conformations in the ternary productive complex. In these conformations, the bulky valine residues adjacent to the dihydropyridine are each blocking one side of the two C_2-symmetry-related faces of the cyclophane (Scheme 1.6).

Transhydrogenases[70] catalyze the hydride transfer between two face-to-face nicotinamide units having different oxidation states, NADH and NAD^+ Model studies by Staab and coworkers illustrate the importance of a correct geometry between hydride donor and acceptor in this biological redox process.[71,72] Starting from the corresponding [2.2](2,5)pyridinophanes, they prepared the semi-reduced derivatives **22** and **23** using 1,4-dihydro-3-(methoxycarbonyl)-1,2,5-trimethylpyridine as the reducing agent in aqueous methanol (Scheme 1.7). In their investigations of the intramolecular hydride transfer in these semi-reduced systems, they observed a large orientation dependence for the rate of mutual oxidation/reduction of the nicotinamide analogues. On the time scale of spin saturated transfer NMR experiments, dynamic intramolecular redox-equivalent exchange occurred only in **23** but not in **22**. Only in **23** do the donor and acceptor centers of the two rings have a correct spatial alignment for fast intramolecular hydride transfer.

---

[70] J. Rydström, J. B. Hoek, and L. Ernster, in 'The Enzymes', 3rd ed., ed. P. D. Boyer, Academic Press, New York, 1976, Vol. 13, pp. 51.
[71] H. A. Staab, H.-J. Hasselbach, and C. Krieger, *Liebigs Ann. Chem.*, 1986, 751.
[72] H.-J. Hasselbach, C. Krieger, M. Decker, and H. A. Staab, *Liebigs Ann. Chem.*, 1986, 765.

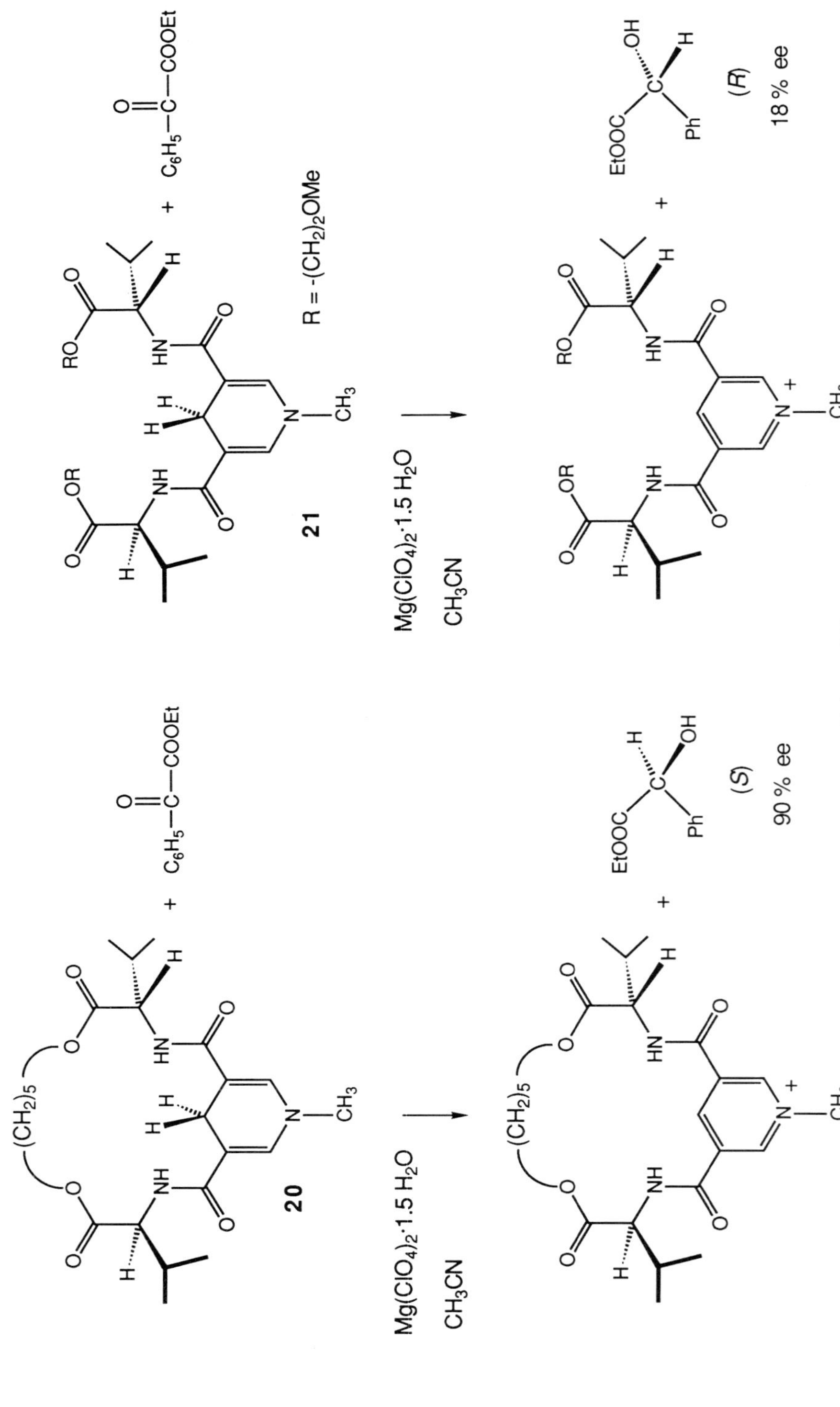

Scheme 1.5

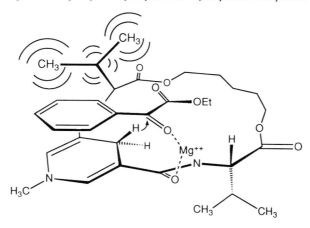

Scheme 1.6

Pyridoxal phosphate, oxidized Vitamin B_6, is another pyridine coenzyme derivative for which analogues have been incorporated into cyclophane structures to control the stereochemistry of coenzyme-mediated reactions.[60,73,74] Many of the transformations in amino acid metabolism that are catalyzed by pyridoxal phosphate-dependent enzymes can be mimicked by simple vitamin B_6 model systems.[75,76] Among the most interesting conversions is the asymmetric synthesis of α-amino acids through the enantioselective transamination of α-ketoacids. To model this process, Kuzuhara et al. constructed the chiral pyridoxamine-like pyridinophane **24**.[77-79] Starting from optically resolved **24** in methanol with 0.5 equivalents of Zn^{2+}, they obtained yields of 50-75% of optically active α-amino acids with enantiomeric excesses of 40-94% (Scheme 1.8). Reactions using (R)-**24a** gave the (S)-α-amino acid preferentially and similarly, (S)-**24a** gave (R)-amino acids (Table 1.4). The pyridoxal derivative **25a** was isolated from the reaction mixtures with chemical yields of 75-85% without loss of optical purity. When deuterated methanol (CD_3OD) was chosen as the solvent, the asymmetric synthesis led to α-deuterated α-amino acids.

The 1:0.5 ratio of **24a** to Zn^{2+} is crucial for a high enantioselectivity of the process. To rationalize this optimal stoichiometry and the stereochemical course of the transamination reaction, the authors proposed that the key protonation/deuteration reaction occurs via an octahedral Zn^{2+} chelate

[73] L. Davis and D. E. Metzler, in 'The Enzymes', 3rd ed., ed. P. D. Boyer, Academic Press, New York, 1972, Vol. 7, pp. 33.
[74] E. E. Snell and S. J. di Mari, in 'The Enzymes', 3rd ed., ed. P. D. Boyer, Academic Press, New York, 1970, Vol. 2, pp. 335.
[75] Y. Matsushima and A. E. Martell, J. Am. Chem. Soc., 1967, **89**, 1331.
[76] R. Breslow, Adv. Enzymol. Relat. Areas Mol. Biol., 1986, **58**, 1.
[77] Y. Tachibana, M. Ando, and H. Kuzuhara, Bull. Chem. Soc. Jpn., 1983, **56**, 3652.
[78] Y. Tachibana, M. Ando, and H. Kuzuhara, Chem. Lett., 1982, 1765.
[79] Y. Tachibana, M. Ando, and H. Kuzuhara, Chem. Lett., 1982, 1769.

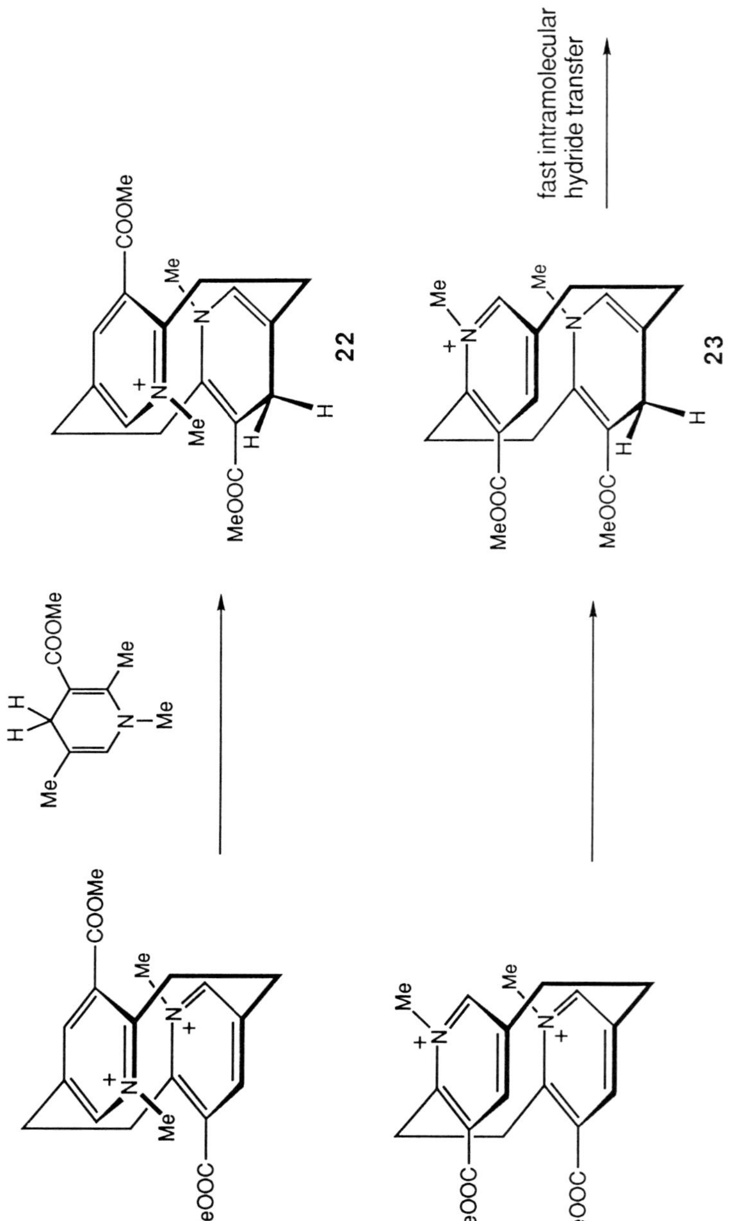

Scheme 1.7

formed by two carbanionic Schiff bases. The chelated complex **26a** derived from two molecules of (S)-**24a** is shown in Scheme 1.9. In this complex, the two bulky cyclophane straps are oriented away from each other. In addition, the bulky sulfur and neighboring alkyl groups of one coordinating Schiff base cover the *re*-face of the carbanion center of the second Schiff base, preventing protonation or deuteration by the solvent from this side. The (R)-amino acid is preferentially formed as a result of a kinetically controlled attack on the less shielded *si*-face of the carbanion.

$R = CH_2OH$, $X = OH$: Pyridoxine (Vitamine B_6)

$R = CHO$, $X = OPO_3^{2-}$: Pyridoxal phosphate (PLP)

$R = CH_2NH_3^+$, $X = OPO_3^{2-}$: Pyridoxamine (PMP)

Table 1.4 *Transamination reactions between the chiral pyridoxamine-cyclophane **24a** and α-ketoacids RCOCOOH*

R	**24a**	Solvent	Chem. Yield%	e.e.%	Product
$(CH_3)_2CH-CH_2-$	R	CH_3OH	68	96	S-leucine
	S	CH_3OH	66	95	R-leucine
	S	CD_3OD	66	94	R-leucine-α-d_1
$(CH_3)_2CH-$	R	CD_3OD	49	80	S-valine-α-d_1
CH_3-	R	CH_3OH	72	69	S-alanine
$Ph-CH_2-$	R	CH_3OH	60	61	S-phenylalanine
indolyl-CH_2-	S	CH_3OH	62	60	R-tryptophan

An interesting example of a biomimetic B_6-dependent aldolase reaction was observed by Kuzuhara et al.[80] for the system comprised of the Zn^{2+} chelate **26b** (R = H), formed by two carbanionic Schiff bases produced from glycine and (S)-**24b**.[81] This chelate was actually isolated in its neutral form, and, when combined with a very large excess of acetaldehyde, *allo*-threonine and threonine were obtained. After a 24 h reaction in methanol–water (1:1) at pH 10.23, these two β-hydroxy-amino acids were formed in a 1.7 : 1 ratio with high (S)-enantioselectivity at the α-amino acid carbon (88 and 74% e.e., respectively). Again, enantioface differentiation is responsible for the observed enantioselectivity (Scheme 1.9). The acetaldehyde is stereoselectively attacked

[80] H. Kuzuhara, N. Watanabe, and M. Ando, *J. Chem. Soc., Chem. Commun.*, 1987, 95.
[81] L. Schirch and M. Mason, *J. Biol. Chem.*, 1963, **238**, 1032.

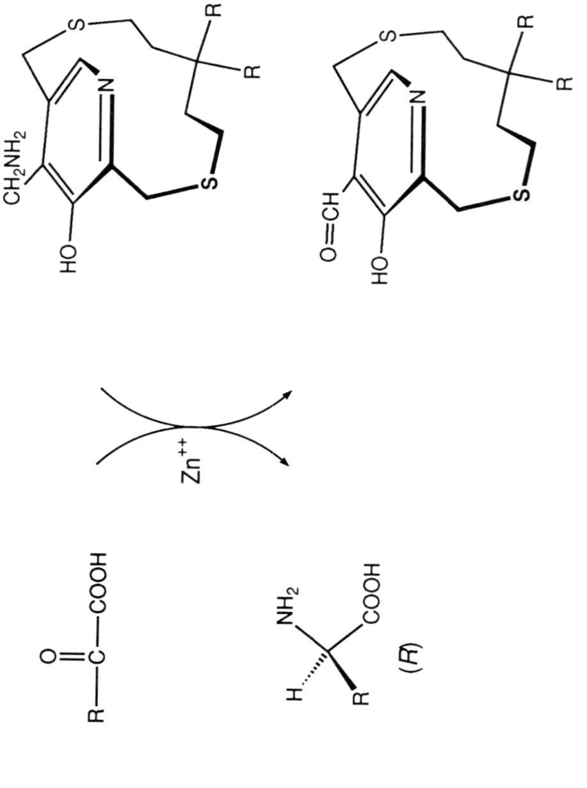

Scheme 1.8

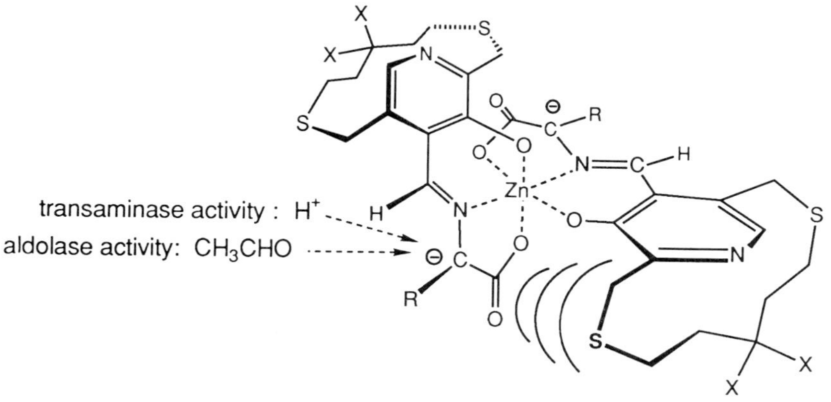

26a X = Me

26b X = H

Scheme 1.9

in a kinetically controlled process by one carbanionic Schiff base from the unprotected face opposite to the bulky cyclophane strap attached to the second coordinating Schiff base.

1.4.4 Porphyrinophanes

An extraordinary variety of porphyrin-cyclophanes have been prepared to model the function of metalloporphyrin-dependent proteins. Comprehensive review articles cover the synthesis and the study of bridged, capped, and strapped porphyrins[82-84] used to model (i) the oxygen transport and storage proteins haemoglobin and myoglobin,[85-88] (ii) the oxygen-reducing cytochrome C oxidase,[89,90] (iii) the hydrocarbon-oxidizing cytochrome P-450

---

[82] B. Morgan and D. Dolphin, *Struct. Bonding (Berlin)*, 1987, **64**, 115.
[83] J. E. Baldwin and P. Perlmutter, *Top. Curr. Chem.*, 1984, **121**, 181.
[84] I. Sutherland, in 'Cyclophanes', eds. P. M. Keehn, and S. M. Rosenfeld, Academic Press, New York, 1983, Vol. 2, pp. 679.
[85] M. F. Perutz, *Ann. Rev. Biochem.*, 1979, **48**, 327.
[86] T. G. Traylor, *Acc. Chem. Res.*, 1981, **14**, 102.
[87] M. Momenteau, *Pure Appl. Chem.*, 1986, **58**, 1493.
[88] E. Tsuchida and H. Nishide, *Top. Curr. Chem.*, 1986, **132**, 63.
[89] M. Wikstrom, M. Saraste, and T. Penttila, in 'The Enzymes of Biological Membranes', ed. A. N. Martonosi, Plenum, New York, 1985, Vol. 4, pp. 111.
[90] J. P. Collman, F. C. Anson, S. Bencosme, A. Chong, T. Collins, P. Denisevich, E. Evitt, T. Geiger, J. A. Ibers, G. Jameson, Y. Konai, C. Koval, K. Meier, P. Oakley, R. Pettman, E. Schmittou, and J. Sessler, in 'Organic Synthesis Today and Tomorrow', eds. B. M. Trost and C. R. Hutchinson, Pergamon Press, Oxford, 1981, pp. 29.

enzymes,[91-95] and (iv) the primary electron transfer processes at photosynthetic reaction centers.[96,97] This chapter can only highlight some of the diverse molecular architecture developed in these investigations.

The Fe(II) porphyrin-cyclophanes **27–30** are examples of haemoglobin and myoglobin models which were constructed to study the structural requirements for reversible dioxygen binding, the criteria for differential binding of dioxygen and carbon monoxide, and the mechanisms of these processes. In the natural systems, the imidazole of a histidine acts as the fifth axial ligand for the iron(II). On the other porphyrin face, the surrounding protein forms a hydrophobic environment in which dioxygen is reversibly bound to iron(II) as the sixth axial ligand.[85] In the synthetic iron(II) porphyrinates, free or bridged imidazoles or pyridines serve as axial ligands. The arene-caps in **27–29** and the C_{10}-strap in **30** mimic the hydrophobic dioxygen binding site of the natural systems. In addition, capping is crucial for preventing the irreversible oxidation of the iron(II) upon dioxygen binding. Simple planar iron(II) porphyrinates react rapidly and irreversibly with dioxygen yielding in most cases a μ-oxo dimer, L-PFe(III)-O-Fe(III)P-L (L = ligand, P = porphyrin). In **27–30**, the space above the metalloporphyrin surface on the arene-capped side is only large enough to accept small molecules, *e.g.* O_2 or CO, as sixth axial ligands (Scheme 1.10). An imidazole derivative cannot be accommodated at these sterically shielded axial coordination sites which therefore remain accessible for dioxygen binding. This was of particular importance for the dioxygen binding studies with **27**[83] and **28**[86] which were executed in solutions containing a large excess of imidazole. Alternatively to the bridging in **27–30**, the steric encumbrance at one porphyrin face, which is required for reversible dioxygen binding, was also successfully achieved in the so-called *picket-fence porphyrins*.[98] These compounds, prepared by Collman *et al.*, are tetra(*meso*-phenyl)porphyrins with one face being shielded by bulky substituents attached to the four *meso*-phenyl rings.

All four porphyrin-cyclophanes shown are reversible dioxygen carriers at room temperature in organic solvents such as toluene. The models **29**[99] and **30**[87] with an intramolecular axial base are truly mononuclear dioxygen carriers with distinct advantages over the models **27** and **28** without a covalently attached base.[87] The covalent fixation of the base to the bridge on one side of the porphyrin ensures the presence of a five-coordinate Fe(II) in studies with **29** and **30**. In contrast, an external ligand needs to be added in

---

[91] 'Cytochrome P-450', ed. P. R. Ortiz de Montellano, Plenum, New York, 1986.
[92] S. D. Black and M. J. Coon, *Adv. Enzymol. Relat. Areas Mol. Biol.*, 1987, **60**, 35.
[93] D. Mansuy, *Pure Appl. Chem.*, 1987, **59**, 759.
[94] T. C. Bruice, in 'Design of Enzymes and Enzyme Models', Proceedings of the Robert A. Welch Foundation Conferences on Chemical Research XXXI, Houston, 1987, pp. 37.
[95] J. T. Groves, *J. Chem. Educ.*, 1985, **62**, 928.
[96] J. Deisenhofer and H. Michel, *Angew. Chem.*, 1989, **101**, 872; *Angew. Chem. Int. Ed. Engl.*, 1989, **28**, 829.
[97] R. Huber, *Angew. Chem.*, 1989, **101**, 849; R. Huber, *Angew. Chem. Int. Ed. Engl.*, 1989, **28**, 848.
[98] J. P. Collman, *Acc. Chem. Res.*, 1977, **10**, 265.
[99] A. R. Battersby, S. A. J. Bartholomew, and J. Nitta, *J. Chem. Soc., Chem. Commun.*, 1983, 1291.

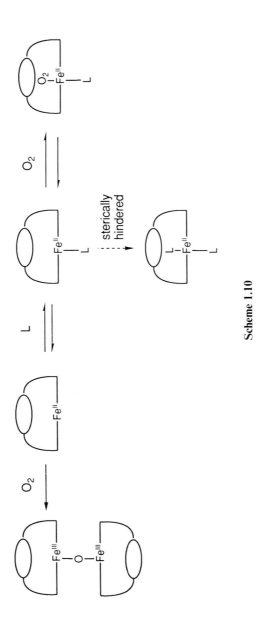

Scheme 1.10

27

28

29

30

31

excess to the solutions of **27** and **28** in order to ensure fivefold coordination, required for dioxygen binding. The doubly bridged dioxygenated species formed by **29** and **30** possess considerable stability against irreversible oxidation. The dioxygen adduct of **29** has a half-lifetime $t_{1/2}$ of *ca.* 24 hours in DMF at 20 °C,[99] whereas a lifetime of approximately one day was reported for the dioxygen complex of **30** in dry toluene under 1 atm of oxygen.[87] On the other hand, the dioxygen adducts of the noncovalently ligated systems **27** and **28** are less stable; their lifetimes depend on the nature and concentration of the axial base. In addition, irreversible μ-oxo dimer formation at the open porphyrin face (Scheme 1.10) competes with axial base and with dioxygen binding.

Reversible dioxygen binding is not limited to bridged porphyrin-cyclophanes. Busch *et al.* described the iron(II) cyclophane **31** in which the bridge at one face of the tetraazamacrocycle takes on a role similar to that of the caps and straps in the porphyrin derivatives **27**–**30**.[100] Pyridine or 1-methylimidazole coordinate axially on the macrocycle side opposite to the bridge. The narrow void created by the bridge prevents axial ligation of a second base, but offers a suitable hydrophobic space for dioxygen binding. In acetone/water or acetone/pyridine at −35 °C, dioxygen is reversibly bound to iron(II) in this void. At room temperature, the O_2 complexes are unstable and rapidly produce iron(III) containing species.

Cytochrome c oxidase is essential for O_2 reduction in aerobic systems and catalyzes the transfer of electrons from reduced cytochrome c to molecular oxygen with the formation of water (equation 1.4).[89] At the active site, the enzyme employs four metals, two Cu(I) and two heme-Fe(II) ions, and dioxygen is bound between one Fe(II) and one Cu(I) center. To model the mode of action of cytochrome c oxidase and to generate the binding of dioxygen between two metal ions in a μ-peroxo dimer-type configuration (Me^{n+}–O–O–Me^{n+}), Collman and coworkers prepared bis(metalloporphyrin)cyclophanes, *e.g.* the cobalt derivative **32**.[90] The porphyrins adopt a cofacial orientation giving sufficient space for intercalative dioxygen binding between the two metal centers. In collaboration with Anson *et al.*, it was shown by rotating-disk voltammetry that **32** is an excellent catalyst for the 4-electron reduction of dioxygen.[101,102] This catalyst, applied to a graphite disk electrode, operates at rates near those of the enzyme cytochrome c oxidase. Oxygen reduction commences at a disk current of +0.8 V and reaches a maximum at +0.6 V. The 2-electron reduction leading to H_2O_2 was efficiently suppressed and accounted for only <4% of the formed product. Scheme 1.11 shows the mechanism proposed for the reduction of oxygen by the bis(metalloporphyrin)cyclophane **32**.

---

[100] N. Herron and D. H. Busch, *J. Am. Chem. Soc.*, 1981, **103**, 1236.
[101] J. P. Collman, F. C. Anson, C. E. Barnes, C. S. Bencosme, T. Geiger, E. R. Evitt, R. P. Kreh, K. Meier, and R. B. Pettman, *J. Am. Chem. Soc.*, 1983, **105**, 2694.
[102] R. C. Durand, Jr., C. S. Bencosme, J. P. Collman, and F. C. Anson, *J. Am. Chem. Soc.*, 1983, **105**, 2710.

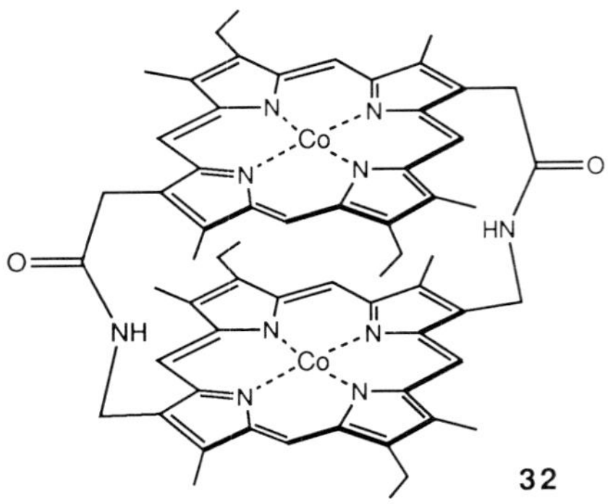

32

$$O_2 + 4e^- + 4H^+ \rightleftharpoons 2H_2O \tag{1.4}$$

Cytochrome P-450 enzymes represent an extraordinarily versatile class of biological oxidation catalysts.[91,92] In their function as monooxygenases, they are responsible for the metabolism of endogenous as well as exogenous lipophilic substrates. Prominent examples are the hydroxylation of steroids, the epoxidation of unsaturated fatty acids, and the hydroxylation of polycyclic aromatic hydrocarbons. A large body of biochemical studies[91,92] and X-ray crystal structure analyses[103] have revealed some highly characteristic features of these enzymes. (i) An iron porphyrin forms one side of the oxygen and substrate binding site which is deeply buried in the protein and, therefore, has a pronounced hydrophobic character. (ii) A cysteinyl ligand, located on the porphyrin side opposite to the binding site, acts as a sixth axial ligand for the heme-bound iron. For monooxygenase activity, molecular oxygen is bound by the heme-iron(II) complex (Scheme 1.12). In a sequence of redox processes, the bound oxygen is cleaved with the formation of water and a high-valent iron-oxo-complex, believed to be Porph$^{+\cdot}$Fe(IV)=O. This iron-oxo-complex subsequently transfers its oxygen atom to the complexed substrate. In model studies, the natural process of oxygen activation is normally circumvented, with the catalytically active iron-oxo-intermediate being generated by direct conversion of iron(III) porphyrinates with oxygen-transfer agents like iodosylbenzene (Scheme 1.12).

Simple Fe(III) or Mn(III) porphyrinates as well as bridged derivatives mimic many of the transformations catalyzed by P-450 cytochromes. Olefin epoxidations and alkane hydroxylations are catalyzed with high turnover and remarkable regio- and stereoselectivity. Bridged derivatives have been particularly successful in asymmetric oxidation processes. Mansuy and coworkers prepared the chiral iron(III) porphyrinate **33** with L-phenylalanine units incorporated into the bridges.[93] In the presence of this catalyst, they

---

[103] R. Raag and T. L. Poulos, *Biochemistry*, 1989, **28**, 917.

From Ansa Compounds to [2.2]Paracyclophane to Cyclophane Complexes 39

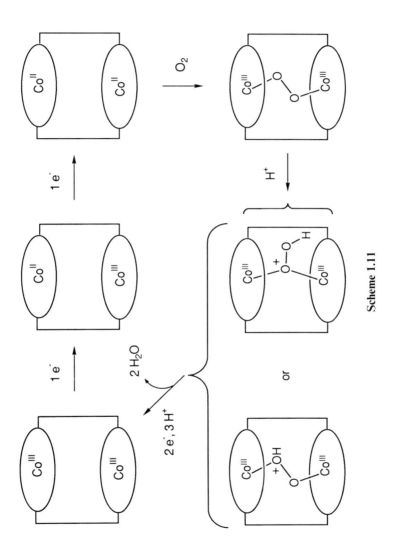

Scheme 1.11

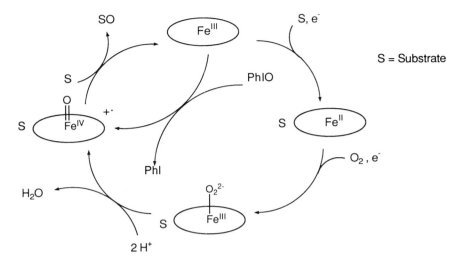

Scheme 1.12

obtained enantiomeric excesses up to 50% in the asymmetric epoxidation of styrenes (equation 1.5). Groves and Viski prepared the chiral iron porphyrinate **34** with the space atop the metal center on both porphyrin faces being shaped by the major groove (Chapter 4.5) of the 1,1′-binaphthyl unit.[104] In hydroxylation reactions with optically pure (*S*,*S*)-**34** as catalyst, ethylbenzene afforded a 40% yield of 1-phenylethanol with a 71:29 ratio of the *R* to *S* enantiomers (42% e.e.) (equation 1.6). Detailed mechanistic investigations of **34** elegantly showed how supramolecular complexation imposes stereoselectivity on stepwise free-radical reactions.

The question remains open to what extent the axial thiolate ligand in the natural P-450 cytochromes assists in dioxygen cleavage, stabilization of the higher-valent iron-oxo-intermediate, and promotion of the oxygen transfer to a bound substrate.[105,106] To investigate the role of the axial thiolate ligand in the enzymatic mechanism, Woggon *et al.*[107] prepared the doubly bridged porphyrin **35**. A sterically fixed thiophenol ligand with a buried thiol group is aligned for axial coordination after insertion of a metal center into the porphyrin. Upon treatment of the free ligand **35** in a direct metal ion insertion reaction with $FeBr_2$/lutidine in THF/benzene, the Fe(III) complex **36** was formed. This result is remarkable considering the presence of the intramolecular redox pair Fe(III)/RSH, and demonstrates that steric encumbrance of the

---

[104] J. T. Groves and P. Viski, *J. Am. Chem. Soc.*, 1989, **111**, 8537; *J. Org. Chem.*, 1990, **55**, 3628.
[105] A. R. Battersby, W. Howson, and A. D. Hamilton, *J. Chem. Soc., Chem. Commun.*, 1982, 1266.
[106] J. P. Collman and S. E. Groh, *J. Am. Chem. Soc.*, 1982, **104**, 1391.
[107] B. Stäubli, H. Fretz, U. Piantini, and W.-D. Woggon, *Helv. Chim. Acta*, 1987, **70**, 1173.

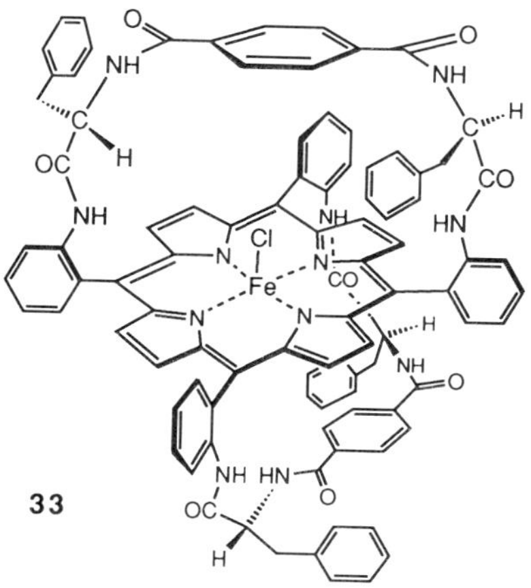

33

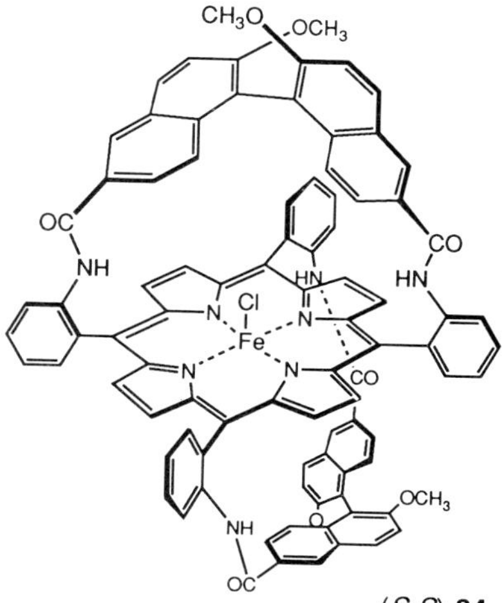

(S,S)-**34**

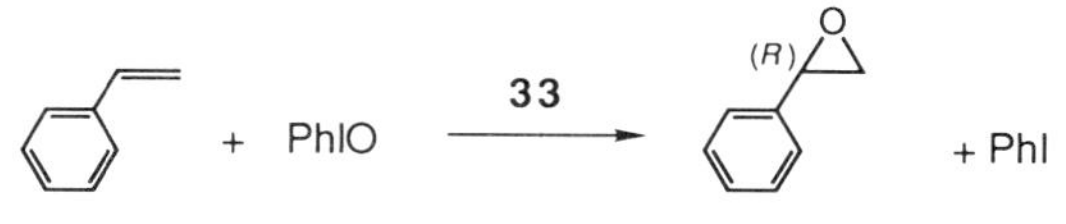

styrene + PhIO →[**33**] (R)-styrene oxide + PhI (1.5)
50 % e.e.

1-phenylethane + PhIO →[(S,S)-**34**] (R)-1-phenylethanol + PhI (1.6)
41 % e.e.

axial thiophenol prevents it from undergoing disulfide formation in a redox process. On the stage of the Fe(II) derivative **37**, formed upon reduction with sodium dithionite, the severely hindered thiophenol ligand could be deprotonated in toluene with KH in the presence of 18-crown-6, and subsequent exposure to CO led to the hexacoordinated CO-complex **38**. In the electronic absorption spectrum, this complex shows a split *Soret* band with peaks of equal intensities at 403 and 457 nm. Such a split *Soret* band is highly characteristic of the CO-complexes of the natural P-450 cytochromes.[91]

The last chapter in the biomimetic chemistry of porphyrinophanes describes the modeling of the electron transfer reactions occurring between porphyrins and quinones at photosynthetic reaction sites.[96,97] Among the many porphyrin–quinone cyclophanes, prepared to investigate the criteria for fast, efficient, and stable electron transfer,[82] compounds **39** and **40**, reported from the Staab group, possess the best defined overlap geometry between the two

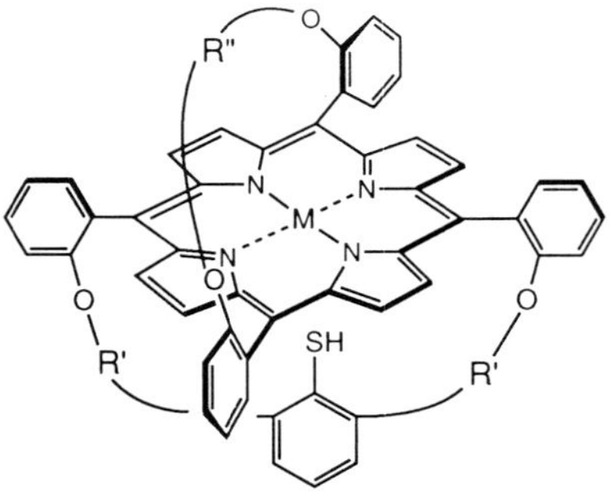

R' = -(CH$_2$)$_3$-
R" = -(CH$_2$)$_{11}$-

35 M = 2H,
36 M = Fe(III)
37 M = Fe(II)

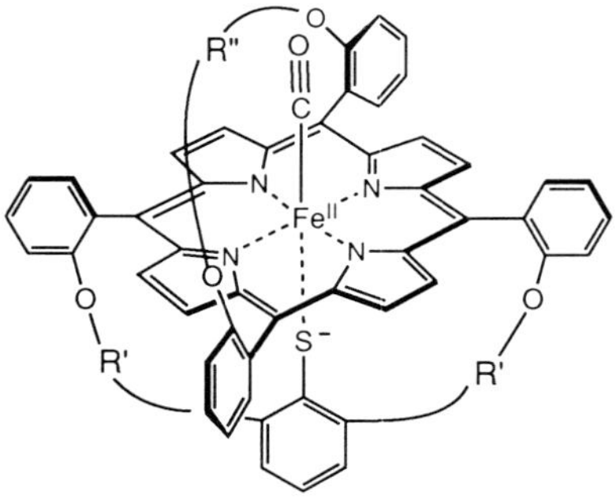

38

chromophores.[108,109] The X-ray crystal structure of **39** revealed that the porphyrin and benzoquinone moieties are in a parallel arrangement and are exactly centered. The transannular distance between the porphyrin and each of the two benzoquinones is 3.42 Å. For both cyclophanes, a very strong quenching of the porphyrin fluorescence occurs due to efficient intramolecular electron transfer from porphyrin to quinone. If the *p*-benzoquinone units in **39** are replaced by *p*-dimethoxybenzene units, only a small degree of fluorescence quenching is observed.

In cyclophane **41**, described by Mauzerall, Lindsey *et al.*,[110] the interplanar separation between porphyrin and quinone varies between 6.5 and 8.5 Å which is about twice as large as in **39** and **40**. At this larger interchromophoric distance, the quenching of the fluorescence emission is not as extensive as found for **39** or **40**. Cyclophane **41** is much more flexible, and fluorescence lifetime measurements indicate the presence of two preferred conformers.

Photophysical measurements showed that in all three cyclophanes **39–41**, electron transfer from the photoexcited porphyrin to the quinone occurs

[108] C. Krieger, J. Weiser, and H. A. Staab, *Tetrahedron Lett.*, 1985, **26**, 6055.
[109] J. Weiser and H. A. Staab, *Tetrahedron Lett.*, 1985, **26**, 6059.

rapidly *via* through-space nonadiabatic electron tunneling.[110,111] Fluorescence lifetime data gave electron transfer times between 0.5 and 15 ns for the two conformers of **41**. Following flash photolysis of **41** in polar solvents, a transient band at 415 nm was observed which is assigned to a charge-separated ZnP$^+$ Q$^-$ state. In contrast to the zinc derivative **41**, the free base porphyrin–quinone shows no electron transfer interactions. Electron transfer times for **39** and **40**, which are on the order of 50 ps to 1 ns, are faster than for **41** due to the shorter porphyrin–quinone transannular distances.

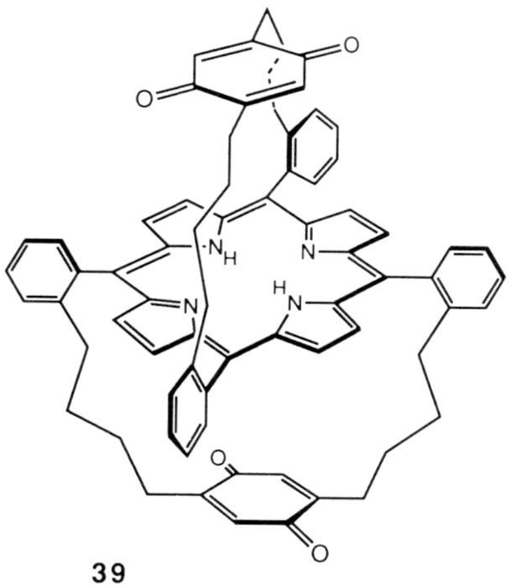

39

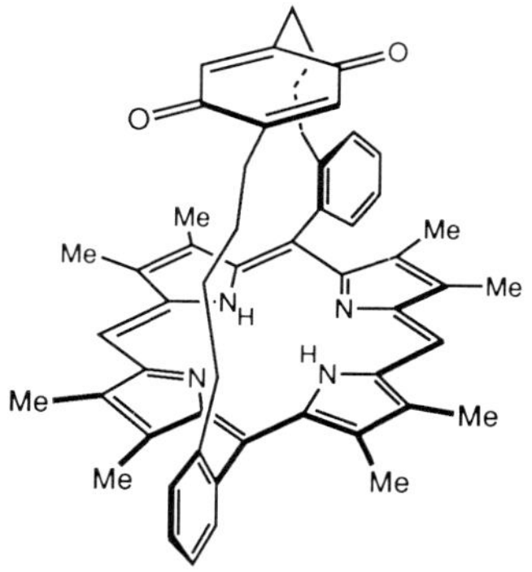

40

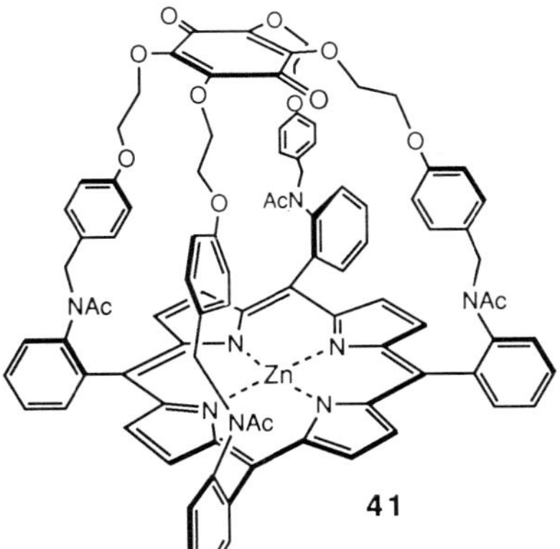

41

[110] J. S. Lindsey, J. K. Delaney, D. C. Mauzerall, and H. Linschitz, *J. Am. Chem. Soc.*, 1988, **110**, 3610.
[111] D. Mauzerall, J. Weiser, and H. Staab, *Tetrahedron*, 1989, **45**, 4807.

1.5 Cyclophane Complexation: From π–π to Inclusion Complexes

1.5.1 Charge-transfer Complexes of Tetracyanoethylene in Solution

In the 1950s, intermolecular charge-transfer complexes formed in solution between stacked electron-rich and electron-deficient aromatic rings became an intensely investigated subject.[112,113] Many of the methods currently used in the quantitative analysis of the stability of synthetic host–guest complexes were developed in these studies.

Shortly after the synthesis of tetracyanoethylene (TNCE) was reported[114] and its potential to form charge-transfer complexes with aromatic hydrocarbons described,[115] Cram et al. investigated the π–π complexation between TCNE and the series of [m.n]paracyclophanes.[116-118] The 1:1 complexes formed by the homologous series of cyclophanes provided a 'rainbowlike series of deep and beautiful colors, ranging from yellow to deep purple'.[7a] These colors resulted from long-wavelength bands assigned to charge-transfer absorptions. However, as previously mentioned in Chapter 1.3.3, a significant stabilization of the complexes through a charge-transfer from the cyclophane donor to the TNCE acceptor $(D + A \leftrightarrow D^+ + A^-)$ does not take place in the ground state but rather in the excited state $(D + A \leftrightarrow D^+ + A^-)^*$. Intermolecular electron donor–acceptor (EDA) complexes formed by stacking π-systems with different electron affinities (different ionization potentials) are predominantly stabilized by a combination of weak intermolecular forces such as electrostatic interactions, polarization interactions, and dispersion energy.[119] It is common to name EDA complexes which show a distinctive long-wavelength transition in the electronic absorption spectra, not present in the spectra of the individual components, as charge-transfer complexes. This name does not normally implicate a significant ground state stabilization of the complexes through charge-transfer interactions.

In Chapter 1.3.1, it was shown that transannular interactions are responsible for the characteristic electronic absorption spectra of the smaller [m.n]cyclophanes and for the directional substituent effects observed in electrophilic aromatic substitution reactions of these compounds. Transannular interactions are also responsible for the differences in energy between the

---

[112] G. Briegleb, 'Elektronen-Donator-Acceptor-Komplexe', Springer, Berlin, 1961.
[113] R. Foster, 'Organic Charge-Transfer-Complexes', Academic Press, London, 1969.
[114] T. L. Cairns, R. A. Carboni, D. D. Coffman, V. A. Engelhardt, R. E. Heckert, E. L. Little, E. G. McGeer, B. C. McKusick, W. J. Middleton, R. M. Scribner, C. W. Theobald, and H. E. Winberg, *J. Am. Chem. Soc.*, 1958, **80**, 2775.
[115] R. C. Merrifield and W. D. Phillips, *J. Am. Chem. Soc.*, 1958, **80**, 2778.
[116] D. J. Cram and R. H. Bauer, *J. Am. Chem. Soc.*, 1959, **81**, 5971.
[117] L. A. Singer and D. J. Cram, *J. Am. Chem. Soc.*, 1963, **85**, 1080.
[118] M. Sheehan and D. J. Cram, *J. Am. Chem. Soc.*, 1969, **91**, 3553.
[119] K. Morokuma, *Acc. Chem. Res.*, 1977, **10**, 294.

Table 1.5 *Correlation between the complex stability constants K ($L\ mol^{-1}$) and the maxima of the charge-transfer bands in the visible absorption spectra of TCNE-cyclophane complexes in CH_2Cl_2, 298 K*

Hydrocarbon	$K\ (L\ mol^{-1})$	λ_{max} (nm)
[3.3]PCP	79 ± 12	599
[4.3]PCP	52 ± 2	538
[2.2]PCP	42	521
[4.4]PCP	36 ± 1	476
[6.6]PCP	13 + 1	490
1,3-Bis(4-ethylphenyl)-propane	13 ± 2	477
4-Ethyl[2.2]PCP	52	540
4-Acetyl[2.2]PCP	24.5	496
4-Cyano[2.2]PCP	8	475

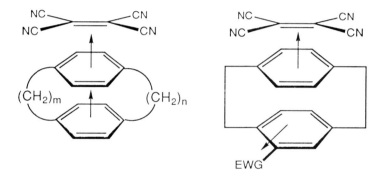

(EWG = electron withdrawing group)

$$\text{TCNE} + \text{Cyclophane} \underset{}{\overset{K}{\rightleftharpoons}} \pi\text{-}\pi\text{-Complex} \qquad (1.7)$$
$$\text{A} \qquad \text{D} \qquad\qquad\qquad \text{DA}$$

$$\text{Complex stability constant}\ \ K(L\ mol^{-1}) = \frac{DA}{A \cdot D}$$

charge-transfer transitions of the various TCNE-[m.n]paracyclophane complexes. Table 1.5 shows the charge-transfer absorption maxima together with the stability constants K ($L\ mol^{-1}$) for the 1:1 complexes formed (equation 1.7).[118]

The absorption maximum changes to higher energies in the series [3.3]PCP > [4.3]PCP > [2.2]PCP > [4.4]PCP > [6.6]PCP > (p-$CH_3C_6H_4CH_2$)$_2$. Similarly, in the same sequence the stability constant K of the complexes decreases. Except for the position of [2.2]PCP, the transannular distance in the cyclophanes increases in this sequence. The exceptional position of [2.2]PCP presumably results from its severely distorted geometry. With decreasing intraannular distance in the cyclophane, the non-bound benzene ring further releases its electron density into the ring that stacks with TNCE which increases its π-base strength or electron donor potential. With increasing donor strength, the complexes become more stable (increasing K values) and the charge-transfer transitions occur at lower energy. In agreement with these explanations, substitution of one cyclophane ring with electron-withdrawing substituents leads to a decrease in stability of the complexes and to higher energy charge-transfer transitions (Table 1.5). The change in transition energy correlates with the Hammett substituent constant, σ_m, of the electron withdrawing substituent. In the complexes of mono-acceptor-substituted cyclophanes, it is assumed that the more π-basic, unsubstituted ring binds to TCNE. Similar substituent effects on the transition energy were also found in TCNE-complexes formed by donor–acceptor cyclophanes in which the donor ring binds to TCNE.[120] A recent X-ray and theoretical study has confirmed the π–π-stacking geometries of the TCNE–cyclophane complexes.[121]

1.5.2 Cyclophane Clathrates

The first evidence for the inclusion of a guest within the intramolecular cavity of a host was reported in 1938 by Freudenberg for the cyclodextrin–iodine adduct.[122] In contrast, the first descriptions of clathrate formation, the inclusion of organic and inorganic guests in the interstices of the crystal lattices of compounds such as hydroquinone, urea, and cholic acid, date back to the middle of the nineteenth century.[123-125]

With cyclophanes, two modes of solid state complex formation are possible. If they possess a spacious organized cavity, solid state host–guest complexes can form in which suitably sized organic guests are included into the intramolecular cavity of the macrocycle. This type of solid state inclusion complexation will not be presented in this chapter. Rather, cyclophanes capable of incorporating a guest into their intramolecular cavity in the solid state along with their ability to form inclusion complexes in solution will be discussed in the following chapters on molecular complexation in the liquid phase. Alternatively, crystallization can lead to large interstitial spaces between cyclophane molecules in the crystal lattice into which guests are incorporated. The orthocyclophanes tri-*o*-thymotide (TOT, **42**),[126] cyclotriveratrylene (**43**) and cyclotricatechylene (**44**),[127] and tetraphenylene (**45**)[128]

[120] L. Schanne and H. A. Staab, *Tetrahedron Lett.*, 1984, **25**, 1721.
[121] A. Renault, C. Cohen-Addad, J. Lajzerowicz, E. Canadell, and O. Eisenstein, *Mol. Cryst. Liq. Cryst.*, 1988, **164**, 179.
[122] K. Freudenberg and M. Meyer-Delius, *Ber. Dtsch. Chem. Ges.*, 1938, **71**, 1596.
[123] F. Wöhler, *Ann. Chem. Pharm.*, 1849, **69**, 294.
[124] A. Clemm, *Ann. Chem. Pharm.*, 1859, **110**, 345.
[125] R. Anschütz, *Ber. Dtsch. Chem. Ges.*, 1892, **25**, 3512.
[126] R. Gerdil, *Top. Curr. Chem.*, 1987, **140**, 71.
[127] A. Collet, *Tetrahedron*, 1987, **43**, 5725.
[128] T. C. W. Mak and H. N. C. Wong, *Top. Curr. Chem.*, 1987, **140**, 141.

have been extensively utilized to shape crystal lattices with interstices available for guest incorporation. This work has been reviewed in detail, and this chapter only intends to illustrate some of the potential and applications of clathrate formation.[129]

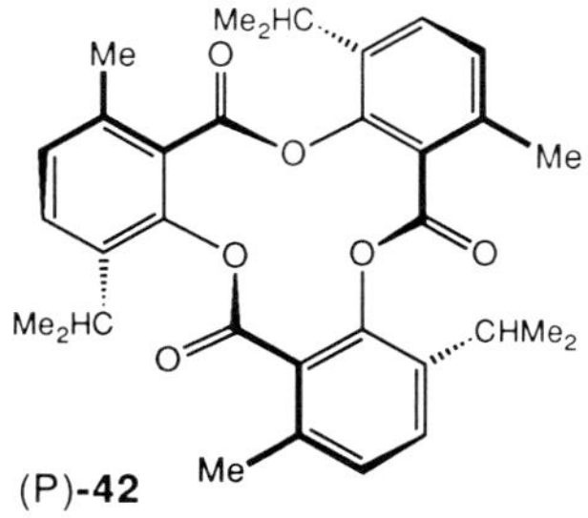

(P)-**42**

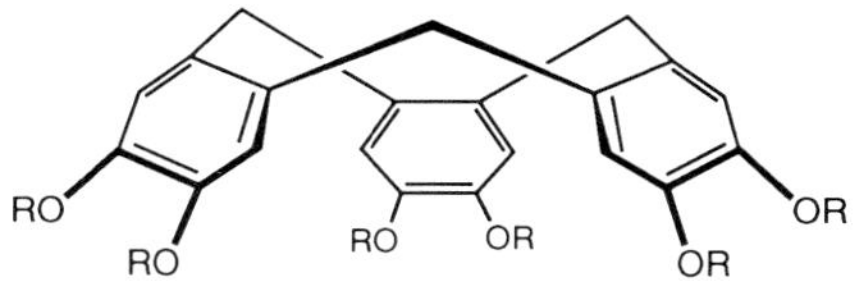

43 R = Me
44 R = H

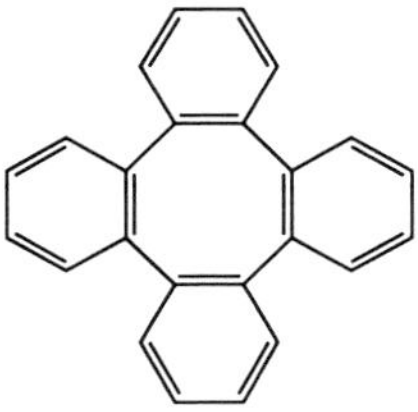

45

The clathrate-forming properties of tri-*o*-thymotide (TOT, **42**) have been known since the early 1950s. In solution, this flexible orthocyclophane exists predominantly in two rapidly interconverting, propellor-shaped chiral conformations having C_3-symmetry.[126] Pure TOT crystallizes into the propellor form. The crystal lattice is achiral being formed by equal amounts of both conformational enantiomers *P* (plus) and *M* (minus)-**42**. Co-crystallization with a great variety of guests leads to clathrate formation. The stability of these clathrates is very high. Depending on the shape of the guest, two different clathrate forms are obtained: compact guests form cage-type clathrates, whereas long-chain-like molecules prefer to form channel-type clathrates. The cage or channel interstices occupied by the guests are shaped by the TOT packing pattern in the crystal. The occurrence of two different forms considerably expands the spectrum of guests which form clathrates with TOT.

With racemic guests, spontaneous resolution very often occurs giving single crystals of TOT in either *P* or *M* configuration with one guest enantiomer preferentially enclosed.[130] Table 1.6 shows examples of racemate resolutions

---

[129] E. Weber, *Top. Curr. Chem.*, 1987, **140**, 1.
[130] D. Worsch and F. Vögtle, *Top. Curr. Chem.*, 1987, **140**, 22.

Table 1.6 *Resolution of racemates using (P)-(+)-tri-o-thymotide (TOT, 42)*[130]

Guest	e.e.%	Configuration	Type of Clathrate
2-Chlorobutane	32–45	(S)-(+)	cage
3-Bromooctane	4	(S)-(+)	channel
trans-2,3-Dimethyloxirane	47	(S,S)-(−)	cage
Methylsulfinic acid methyl ester	14–15	(R)-(+)	cage
Ethylmethylsulfoxide	40–80	(R)-(+)	cage

using (P)-(+)-TOT. In sequences with similar guests, strong correlations were observed between the absolute configuration of TOT in the clathrate single crystal and the configuration of the preferentially included guest enantiomer.[131] Such correlations might be useful for assigning absolute configurations to new optically active guests having comparable geometry.

An interesting cage-controlled reaction was described by Jefford *et al.*[132] The prochiral guest molecule (Z)-2-methoxybut-2-ene **46** reacts with singlet oxygen to yield the hydroperoxide **47** (equation 1.8). If this reaction occurs in a separated cage-type clathrate formed by (P)-TOT and **46**, (+)-**47** is obtained in high enantiomeric excess (equation 1.9). Similarly, the photooxygenation of the (M)-TOT clathrate yields (−)-**47** (equation 1.10). Singlet oxygen permeates the crystal lattice, and the asymmetric environment of the cage transfers chirality to the hydroperoxide under formation.

Another interesting aspect of TOT clathrate chemistry is that strained species can be stabilized in the matrix formed by this host.[133] The X-ray crystal structure of a TOT–chlorocyclohexane clathrate revealed that only chair and boat conformations with *axial*-Cl had been included in the cage-type interstices.

A major direction in organic materials research is the generation of nonlinear optical properties, *e.g.* second-harmonic generation, through control of bulk dipolar alignment.[134] Clathrate formation appears to be a general technique for achieving a desired dipolar alignment.[135] In solution, dipoles prefer to aggregate in a face-to-face, antiparallel arrangement, which minimizes electrostatic repulsion. In a narrow channel clathrate, dipoles are forced by the host lattice to arrange themselves in a linear, head-to-tail sequence (Scheme 1.13). This alignment generates a bulk dipole necessary for nonlinear optical properties. Besides other materials such as β-cyclodextrin, thiourea, and deoxycholic acid, Eaton *et al.* found that TOT is capable of forming polar, noncentrosymmetric structures that are efficient second-

---

[131] R. Arad-Yellin, B. S. Green, M. Knossow, and G. Tsoucaris, *J. Am. Chem. Soc.*, 1983, **105**, 4561.
[132] R. Gerdil, G. Barchietto, and C. W. Jefford, *J. Am. Chem. Soc.*, 1984, **106**, 8004.
[133] R. Gerdil and E. Frew, *J. Incl. Phenom.*, 1985, **3**, 335.
[134] D. J. Williams, *Angew. Chem.*, 1984, **96**, 637; *Angew. Chem. Int. Ed. Engl.*, 1984, **23**, 690.
[135] D. F. Eaton, *Tetrahedron*, 1987, **43**, 1551.

$$\text{CH}_3\text{O}\diagup\text{C}=\text{C}\diagdown\text{CH}_3 \quad \xrightarrow{{}^1\text{O}_2} \quad \text{CH}_3\text{O}-\overset{\overset{\text{OH}}{|}}{\underset{\underset{\text{CH}_3}{|}}{\text{C}}}-\text{C}(=\text{CH}_2)\text{H}$$

$$\mathbf{46} \qquad\qquad \mathbf{47} \quad 100\% \qquad (1.8)$$

$$(P)\text{-}(+)\text{-TOT/} \; \mathbf{46} \xrightarrow{{}^1\text{O}_2} (+)\text{-}\mathbf{47} \; [\alpha]^{20}_{546} = +120° \pm 20° \qquad (1.9)$$

$$(M)\text{-}(-)\text{-TOT/} \; \mathbf{46} \xrightarrow{{}^1\text{O}_2} (-)\text{-}\mathbf{47} \; [\alpha]^{20}_{546} = -146° \pm 20° \qquad (1.10)$$

harmonic generation (SHG) materials.[136] Among the TOT-clathrates with SHG-capabilities are those formed from p-(N,N-dimethylamino)cinnamaldehyde and p-(N,N-dimethylamino)benzonitrile.

Cyclotriveratrylene (CTV, **43**) forms clathrates that are not as stable as the solid TOT inclusion compounds.[127] Heating under vacuum readily liberates the guests that are enclosed in channels between stacks of CTV. These guests include benzene, chloroform, acetone, ethanol, tetrahydrofuran, methyl ethyl ketone, among others. The host–guest inclusion stoichiometry can vary from 1:0.1 to 1:3. More stable clathrates form between cyclotricatechylene and small molecules capable of hydrogen-bonding. Guest molecules like N,N-dimethylformamide, dimethylsulfoxide, or 2-propanol located in interstices

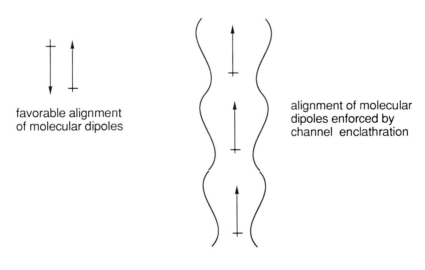

Scheme 1.13

[136] D. F. Eaton, A. G. Anderson, W. Tam, and Y. Wang, *J. Am. Chem. Soc.*, 1987, **109**, 1886.

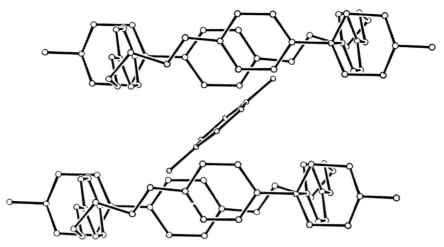

Figure 1.8 *Molecular structure of the 1:1 clathrate between **49** and p-xylene in a view parallel to the least-squares plane of the host (Reproduced with permission from reference 139. Copyright 1985 VCH.)*

interact with the host lattice not only through van der Waals interactions. Hydrogen-bonding between the phenolic OH groups of the host lattice and the guest provides additional stabilization to these clathrate structures. Such additional stabilizing interaction is absent in the CTV clathrates.

The clathration cavity in crystalline tetraphenylene is spheroidal with a diameter of 7.2–7.4 Å, the size approximately of a carbon tetrachloride molecule.[128] Carbon tetrachloride, tetrahydrofuran, benzene, acetone, 2-bromopropane are among the small molecules which, according to X-ray crystallography, form defined clathrates.

Many other cyclophanes form stable clathrates, *e.g.* **48** which includes stoichiometric amounts of solvents such as DMF, DMSO, or *o*-dichlorobenzene.[137] Even the incorporation of guests into lattice interstices has often been observed in crystalline complexes of cyclophanes which possess extended cavities and are capable of inclusion complexation in solution.[138,139] In the 1:1 complex formed by the tetraoxa[6.1.6.1]paracyclophane **49** and *p*-xylene, the aromatic rings of the guests are not included into the sufficiently large cavity.[139] Rather, they are sandwiched between two host molecules in stacks formed by **49** (Figure 1.8).

At the end of this chapter, it should be noted that clathrate chemistry, as a subdiscipline of organic solid state and materials chemistry,[140] has grown

---

[137] F. Bottino, S. Pappalardo, and G. Ronsisvalle, *Gazz. Chim. Ital.*, 1981, **111**, 437.
[138] D. J. Cram, S. Karbach, H-E. Kim, C. B. Knobler, E. F. Maverick, J. L. Ericson, and R. C. Helgeson, *J. Am. Chem. Soc.*, 1988, **110**, 2229.
[139] C. Krieger and F. Diederich, *Chem. Ber.*, 1985, **118**, 3620.
[140] L. Addadi, Z. Berkovitch-Yellin, I. Weissbuch, J. van Mil, L. J. W. Shimon, M. Lahav, and L. Leiserowitz, *Angew. Chem.*, 1985, **97**, 476; *Angew. Chem. Int. Ed. Engl.*, 1985, **24**, 466.

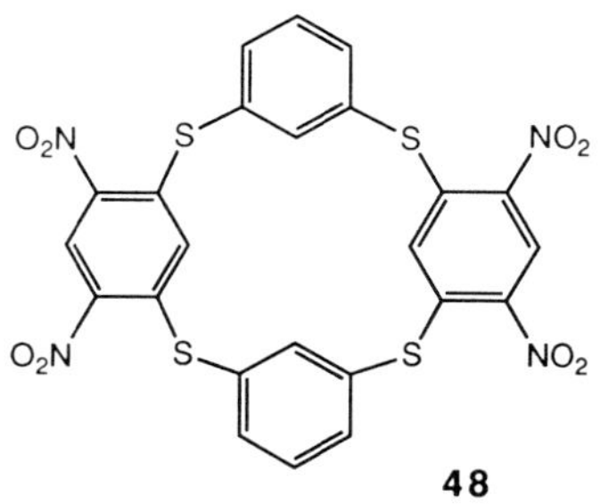

48

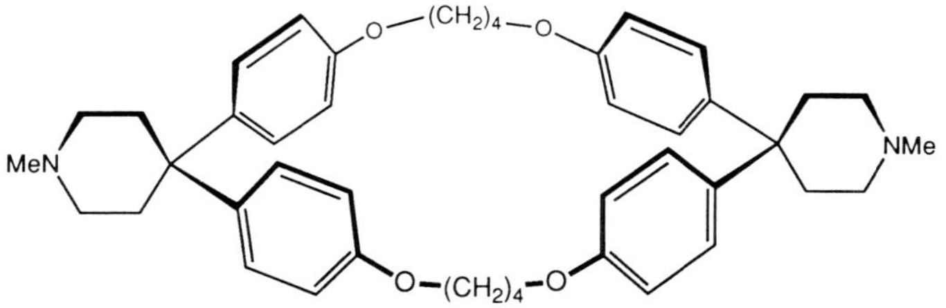

49

dramatically. Rational design methods are now being applied to the invention of novel clathrate-forming hosts.[141,142] Interests include controlling, *via* clathrate formation, bulk dipolar alignment to generate nonlinear optical properties which was previously described. Also, great progress has been made in the application of enclathration for bulk, even on an industrial scale, separation processes of constitutional and stereoisomers, in particular optical isomers.[143] The exploitation of the rigorous stereochemical control provided by rigid, precisely defined chiral host lattices will increase in solid state asymmetric synthesis.[144]

[141] E. Weber and M. Czugler, *Top. Curr. Chem.*, 1988, **149**, 45.
[142] M. C. Etter, *Acc. Chem. Res.*, 1990, **23**, 120.
[143] F. Toda, *Top. Curr. Chem.*, 1987, **140**, 43.
[144] F. Toda, *Top. Curr. Chem.*, 1988, **149**, 211.

CHAPTER 2
Inclusion Complexes of Neutral Molecules in Aqueous Solution

2.1 Introduction

In 1980, Koga *et al.*[1] reported the first unambiguous evidence for stoichiometric inclusion of an apolar guest into the cavity of a synthetic cyclophane host both in aqueous solution and in the solid state. Chapter 2 analyzes the progress that has been made since this milestone contribution in the development of cyclophanes that complex apolar organic molecules in aqueous solution. These complexes are held together predominantly by van der Waals interactions, with dispersion interactions dominating, and solvent-specific forces. A detailed analysis of the role of solvent in molecular complexation is presented later in Chapter 7. After a short outline of historic developments, Chapter 2 describes the structural characteristics of cyclophanes that were specifically designed for apolar substrate binding. Subsequently, the discussion focuses upon the complexation performance of these hosts with different classes of apolar solutes. Complexes of apolar solutes bearing ionic functional groups are only included in Chapter 2 if these ionic groups do not contribute significantly to the strength of the association. Complexes of charged organic guests that are stabilized by a major attractive Coulombic term will be presented in Chapter 4.

Most cyclophane hosts have been prepared to complex apolar solutes in water rather than in organic media. Complexation in water has always attracted special interest since it can directly model molecular recognition events in biological systems. In addition, earlier developments had shown that apolar complexation would be stronger in aqueous solutions compared to organic solvents. Prior investigations on inclusion complexation, chiral recognition, and catalysis by cyclodextrins, initiated by F. Cramer about 1950, were conducted in water.[2-4] Furthermore, extensive work on the partitioning

---

[1] K. Odashima, A. Itai, Y. Iitaka, and K. Koga, *J. Am. Chem. Soc.*, 1980, **102**, 2504.
[2] F. Cramer, 'Einschlußverbindungen', Springer, Berlin, 1954.
[3] M. L. Bender and M. Komiyama, 'Cyclodextrin Chemistry', Springer, Berlin, 1978.
[4] J. Szejtli, 'Cyclodextrin Technology', Kluwer Academic Publishers: Dordrecht, 1988.

of apolar solutes between water and organic solvents or the gas phase as well as investigations into the formation of micelles and membranes had suggested a specifically strong driving force for the complexation of apolar molecules in aqueous solution.[5]

2.2 The First Cyclophanes for the Inclusion of Apolar Substrates

In 1955, Stetter and Roos first recognized the potential of cyclophanes to form inclusion complexes, and prepared the macrocycles **1–3** derived from benzidine.[6] They described the formation of stable crystalline 1:1 complexes with benzene and dioxane when the macrocycles **2** and **3** but not the smaller macroring **1** were recrystallized from these solvents. Until 1982, these complexes were thought to be of the cavity inclusion type. At that time, Hilgenfeld and Saenger[7] showed by X-ray crystallography that the largest macrocycle **3** and benzene do not form an intramolecular cavity inclusion complex, but rather, a clathrate in which benzene is accommodated between host molecules in the crystal lattice. The structural data also shows that the nonplanar biphenyl units in **3** partially fill the cavity which reduces the space available for guest inclusion. A more detailed discussion on the importance of binding site preorganization prior to complexation will be given in Chapter 2.3. The formation of a 1:1 solid state complex with dioxane was also reported in 1960 by Faust and Pallas[8] for **4** in which the benzidine units of the earlier cyclophanes had been replaced by corresponding 4,4'-diamino-1,1'-binaphthyl units. Although X-ray evidence is not available, it must be assumed that dioxane is located in crystal lattice interstices rather than in the cavity of **4**. If the binaphthyls assume energetically favorable conformations with dihedral angles about their chirality axis of $\approx 60–120°$, their naphthalene rings can easily fill a possible cavity binding site.

1 n = 2
2 n = 3
3 n = 4

4

[5] C. Tanford, 'The Hydrophobic Effect: Formation of Micelles and Biological Membranes', 2nd ed., Wiley, New York, 1980.
[6] H. Stetter and E.-E. Roos, *Chem. Ber.*, 1955, **88**, 1390.
[7] R. Hilgenfeld and W. Saenger, *Angew. Chem.*, 1982, **94**, 788; *Angew. Chem. Int. Ed. Engl.*, 1982, **21**, 787; *Angew. Chem. Suppl.*, 1982, 1690.
[8] G. Faust and M. Pallas, *J. Prakt. Chem.*, 1960, **11**, 146.

Around 1970, the groups of Inazu and Tabushi began pursuing the development of large cyclophanes with suitably sized cavities allowing guest inclusion. In addition to tetraoxa-analogues of **1–3**,[9] Inazu *et al.* prepared the fully carbocyclic macro-rings **5**[10] and **6**.[11] Tabushi *et al.* described the synthesis of the paracyclophanes **7–9**.[12,13] In the [2.2.2] and the [2.2.2.2] derivatives **7** and **8**, the benzene rings substantially prefer the face-to-face conformation shown in Scheme 2.1. The enclathration or inclusion complexation of guest molecules by macrocycles **5–9** was not reported.

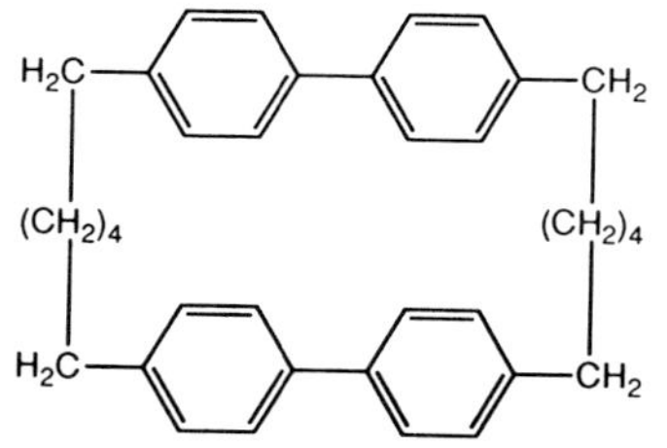

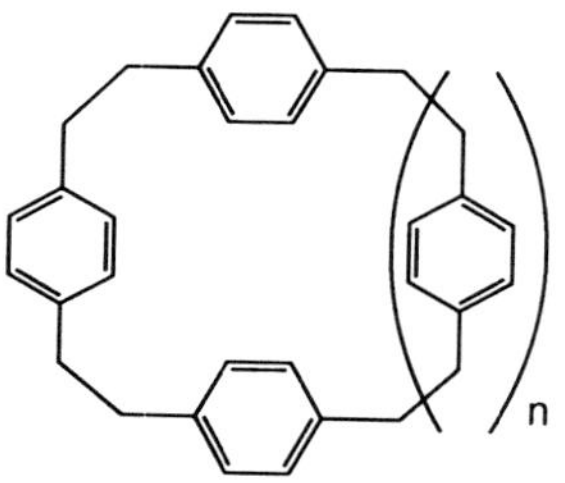

7 n = 0
8 n = 1
9 n = 2

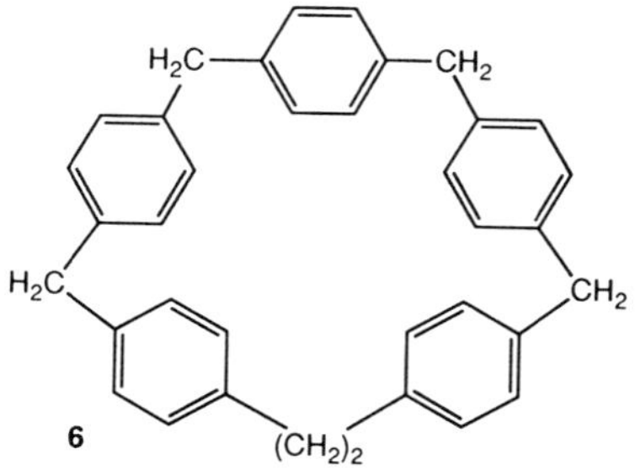

In 1971, Inazu *et al.* published the synthesis of the tetraaza[3.3.3.3]paracyclophane **10** which, when crystallized out of benzene or dioxane, afforded crystals including one equivalent of solvent molecule.[14] Later, it was found by *X*-ray crystallography that the dioxane molecules in the 1:1 complex are not located in the intramolecular cavity.[15] The cyclophanes pack one above the other

---

[9] J. Nishikido, T. Inazu, and T. Yoshino, *Bull. Chem. Soc. Jpn.*, 1973, **46**, 263.
[10] R. Nagano, J. Nishikido, T. Inazu, and T. Yoshino, *Bull. Chem. Soc. Jpn.*, 1973, **46**, 653.
[11] T. Inazu and T. Yoshino, *Bull. Chem. Soc. Jpn.*, 1968, **41**, 652.
[12] I. Tabushi, H. Yamada, K. Matsushita, Z. Yoshida, H. Kuroda, and R. Oda, *Tetrahedron*, 1972, **28**, 3381.
[13] I. Tabushi and K. Yamamura, *Top. Curr. Chem.*, 1983, **113**, 145.
[14] Y. Urushigawa, T. Inazu, and T. Yoshino, *Bull. Chem. Soc.*, 1971, **44**, 2546.
[15] S. J. Abbott, A. G. M. Barrett, C. R. A. Godfrey, S. B. Kalindjian, G. W. Simpson, and D. J. Williams, *J. Chem. Soc., Chem. Commun.*, 1982, 796.

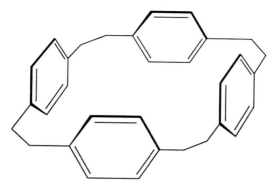

8 face-to-face conformation

Scheme 2.1

producing continuous channels, and the dioxane molecules are included in secondary cavities between two stacking cyclophanes in these channels. However, X-ray crystal structures of 1:1 complexes between **10** and chloroform, dichloromethane, acetonitrile, or carbon dioxide showed the location of these guests in the intramolecular cavity of **10**.[16,17] Figure 2.1 shows different perspectives of the **10**·CH_2Cl_2 complex.[16] The cavity takes the shape of a square molecular box; the space available for guest inclusion is 4.6–6.4 Å in width and *ca.* 6 Å in depth.

For complexation in aqueous solution, Tabushi *et al.* prepared the water-soluble [3.3.3.3]paracyclophanes **11**[18] and **12**.[19,20] CPK models show that the distance between opposite benzene rings in the face-to-face conformation of **11** is ≈ 7 Å, large enough to accommodate a substrate's phenyl or naphthyl moiety. Cyclophane **11** is readily water-soluble at pH ≈ 7, and fluorescence binding titrations, evaluated according to Benesi and Hildebrand,[21] demonstrated the formation of a 1:1 complex with 8-phenylamino-1-naphthalenesulfonate (1,8-ANS) with an association constant K_a of 1.6×10^3 L mol^{-1} at room temperature (equation 2.1).

Whereas CPK model examinations indicate that the cavity in **11** is large enough for the inclusion of benzene and naphthalene substrates, the information on the cavity size gained from models of the quaternary tetraaza[3.3.3.3]paracyclophane **12** is not as clear. Although kinetic analysis

---

[16] I. Tabushi, K. Yamamura, H. Nonoguchi, K. Hirotsu, and T. Higuchi, *J. Am. Chem. Soc.*, 1984, **106**, 2621.
[17] K. Hirotsu, S. Kamitori, T. Higuchi, I. Tabushi, K. Yamamura, and H. Nonoguchi, *J. Incl. Phenom.*, 1984, **2**, 215.
[18] I. Tabushi, H. Sasaki, and Y. Kuroda, *J. Am. Chem. Soc.*, 1976, **98**, 5727.
[19] I. Tabushi, Y. Kimura, and K. Yamamura, *J. Am. Chem. Soc.*, 1978, **100**, 1304.
[20] I. Tabushi, Y. Kimura, and K. Yamamura, *J. Am. Chem. Soc.*, 1981, **103**, 6486.
[21] H. A. Benesi and J. H. Hildebrand, *J. Am. Chem. Soc.*, 1949, **71**, 2703.

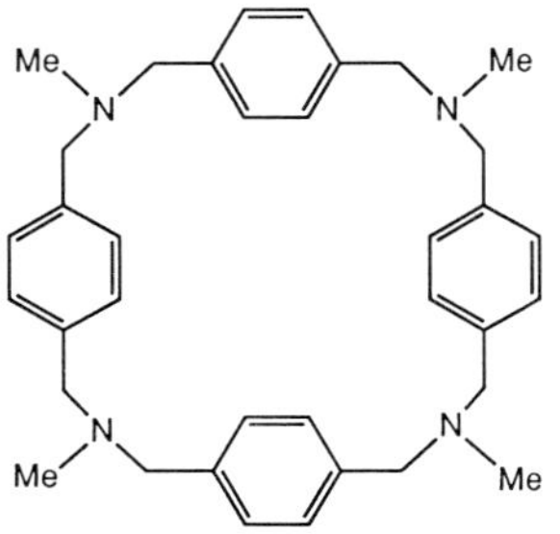

10

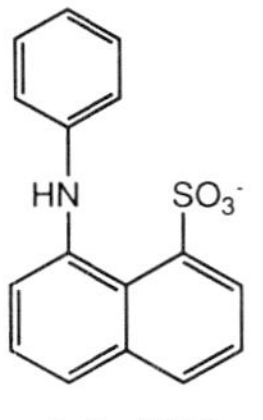

1,8-ANS

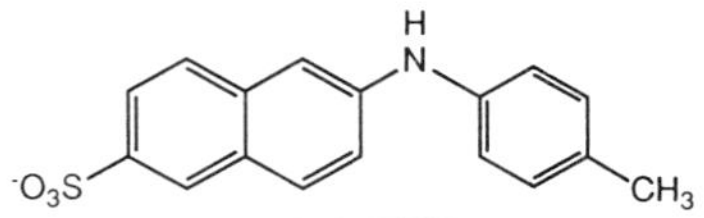

2,6-TNS

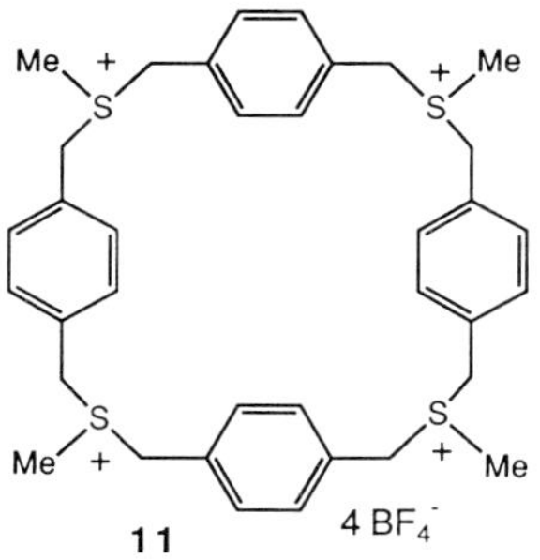

11

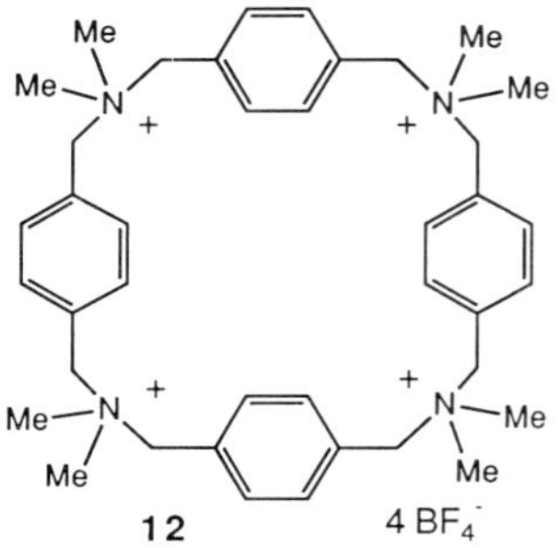

12

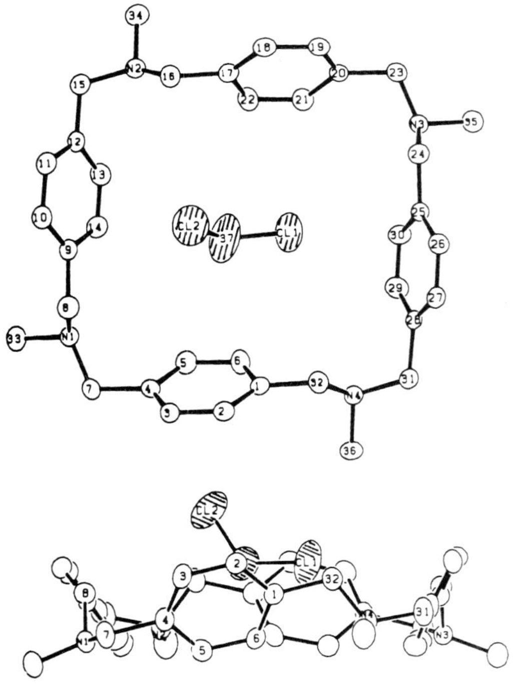

Figure 2.1 *Top and edge views of the X-ray crystal structure of the* ***10***$\cdot CH_2Cl_2$ *inclusion complex*
(Reproduced with permission from reference 16. Copyright 1984 ACS.)

$$\text{Host} + \text{Guest} \underset{}{\overset{K_a}{\rightleftharpoons}} \text{Host-Guest Complex} \quad (2.1)$$
$$H \quad G \qquad\qquad\qquad\qquad HG$$

Association constant K_a (L mol^{-1}) = $\dfrac{HG}{H \cdot G}$ (H, G, HG = concentrations of free host and guest and of the host-guest complex)

suggests that considerable binding of hydroxynaphthoic acids by **12** occurs in aqueous solution (K_a's around 10^3 L mol^{-1}),[20] a full inclusion complexation of these substrates does not seem very probable. Compared to the binding site in **11**, the cavity size in all-face **12** is significantly reduced as a result of the introduction of the smaller quaternary nitrogen atoms and their additional methyl groups. A CPK model of an inclusion complex with an aromatic guest can only be constructed by inducing strain. Therefore, this author proposes nesting complexation (Scheme 2.2) or apolar surface interactions as possible binding modes in the complexes formed between **12** and apolar solutes. Such weaker host–guest interactions are in agreement with the small association constants (K_a's around 100 L mol^{-1}) measured by Lepropre and Fastrez for complexes formed by tetraaza[3.3.3.3]paracyclophane derivatives, which are similar to **10** and **12**, with 1,8-ANS or 6-(4-methylphenyl)amino-2-naphthalenesulfonate (2,6-TNS) as guests.[22]

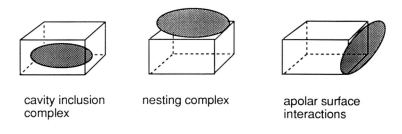

cavity inclusion complex   nesting complex   apolar surface interactions

Scheme 2.2

Since 1974, Murakami *et al.* reported the synthesis of a variety of functionalized cyclophanes as hydrolase models, *e.g.* **13**–**16**.[23,24] The cavities of these macrocycles are either too small or not sufficiently preorganized for inclusion complexation of organic substrates. The accelerations of the hydrolyses of long-chain *p*-nitrophenyl esters in aqueous solutions containing these cyclophanes presumably originate from micellar effects rather than from catalysis in structurally defined stoichiometric inclusion complexes.

[22] G. Lepropre and J. Fastrez, *J. Incl. Phenom.*, 1987, **5**, 157.
[23] Y. Murakami, J. Sunamoto, and K. Kano, *Bull. Chem. Soc. Jpn.*, 1974, **47**, 1238.
[24] Y. Murakami, *Top. Curr. Chem.*, 1983, **115**, 107.

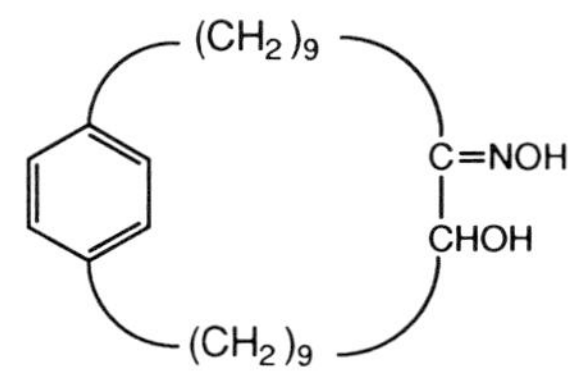

13

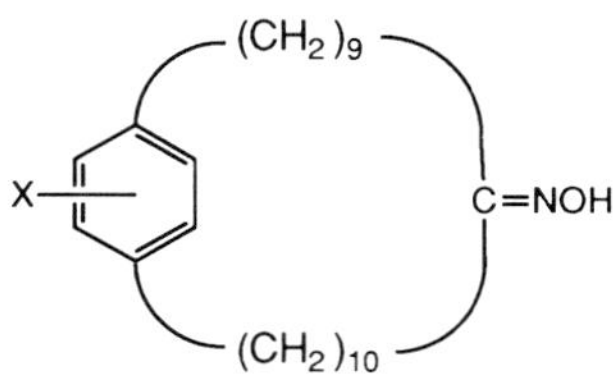

14 X = COOH
CH$_2$N$^+$(CH$_3$)$_3$Cl$^-$
CONHCH$_2$CH$_2$-(imidazole)

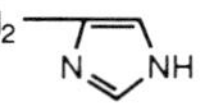

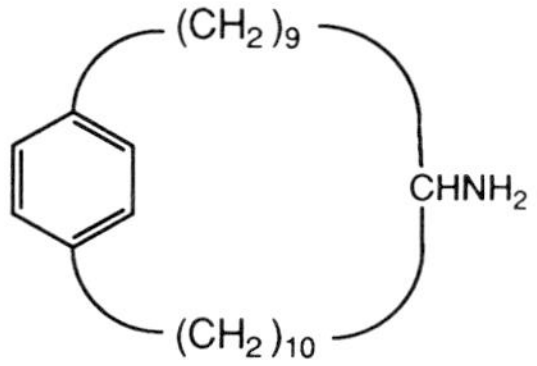

15

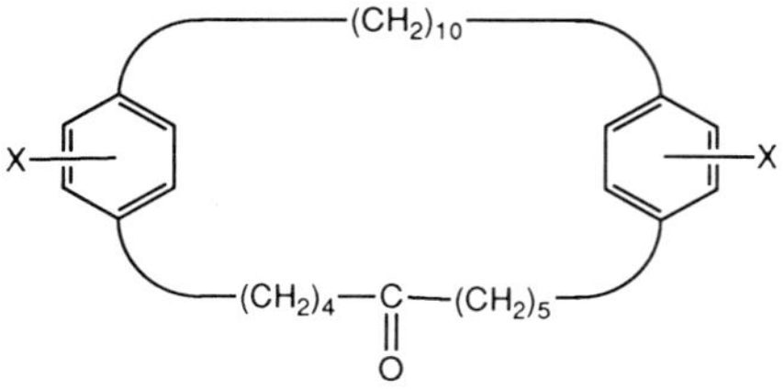

16 X = COOH
CH$_2$N$^+$(CH$_3$)$_3$Cl$^-$
CO(CH$_2$)$_3$COOH

Around 1980, Whitlock et al. prepared a series of naphthalenophanes with rigid diacetylene spacers, e.g. **17**[25] and **18**.[26] Optical as well as ^{1}H NMR spectroscopy provided some experimental evidence for the inclusion complexation of aromatic substrates, e.g. 1,8-ANS and 2-naphthalenesulfonate, by these macrocycles in aqueous solution. Most importantly, these investigations document very well the problems associated with unambiguously assigning cavity inclusion geometries to complexes formed by receptors with large apolar surfaces like the naphthalene rings in **17** and **18**. These hosts are capable of undergoing strong exo-cavity π—π-stacking interactions with aromatic substrates (Figure 2.2). In addition, external π—π-stacking interactions can lead to self-association of the hosts. All these equilibria compete with stoichiometric host–guest inclusion complexation and therefore, they interfere with the accurate determination of inclusion thermodynamics. The work by Whitlock suggests that extended aromatic cavity walls in cyclophane hosts should be sterically protected from outside π—π-stacking to obtain exclusive cavity inclusion (for related findings, see Chapter 8.3).[27]

---

[25] S. P. Adams and H. W. Whitlock, *J. Am. Chem. Soc.*, 1982, **104**, 1602.
[26] E. T. Jarvi and H. W. Whitlock, *J. Am. Chem. Soc.*, 1982, **104**, 7196.
[27] E. M. Seward, R. B. Hopkins, W. Sauerer, S.-W. Tam, and F. Diederich, *J. Am. Chem. Soc.*, 1990, **112**, 1783.

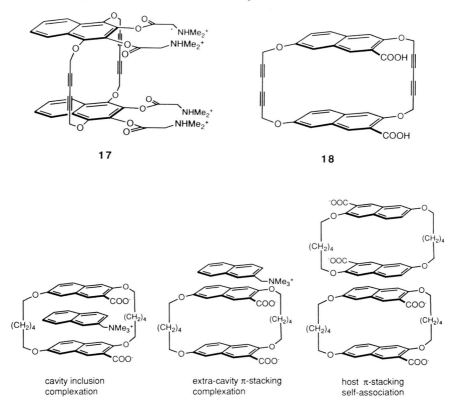

Figure 2.2 *Competing association modes in solutions of hosts and guests with large apolar surfaces capable of undergoing intra- and extra-cavity π-stacking interactions*

In 1980, Koga *et al.* reported the first unambiguous evidence for stoichiometric inclusion of an apolar guest in the cavity of a synthetic cyclophane host in both aqueous solution and in the solid state.[1] By ^{1}H NMR spectroscopy, they demonstrated the formation of a 1:1 complex ($K_a = 2.8 \times 10^3$ L mol^{-1}, $T = 298$K) between tetraprotonated **19** and 2,7-naphthalenediol in dilute hydrochloric acid at pH < 2.[28] Figure 2.3 shows the specific and individual complexation-induced changes in the proton chemical shifts of both binding partners that are observed in the spectrum of a complex solution with [host] = 0.05 M and [guest] = 0.025 M. Due to fast complexation–decomplexation rates on the ^{1}H NMR time scale ('at fast exchange'), the signals of the host (guest) appear as a weighted average of the chemical shifts of free host (guest) and host (guest) bound in each possible orientation. The observed changes in chemical shift are best explained by differential

[28] K. Odashima, A. Itai, Y. Iitaka, Y. Arata, and K. Koga, *Tetrahedron Lett.*, 1980, **21**, 4347.

orientations of host and guest protons towards the shielding and deshielding regions of surrounding aromatic rings in the complex. The particular upfield shifts measured for solutions of the **19**·2,7-naphthalenediol complex support a favorable axial-type guest inclusion (Figure 2.3, see Chapter 2.6.2). In this complex geometry, the protons 3,6-H of the guest are directed outside the binding site and encounter only a small amount of shielding by the aromatic rings of the host. These protons are only moderately shifted upfield. On the other hand, guest protons 1,8-H and 4,5-H are pointing into the shielding regions of the host and are therefore moved strongly upfield.

One of the merits of Koga's work is the establishment of ^{1}H NMR spectroscopy as *the* method of choice for studying cyclophane inclusion complexation in solution. By this method, extensive information is obtained about the structures of the complexes. In many cases, both the thermodynamics and kinetics of complexation can be evaluated. Since the inclusion of an organic guest into a cyclophane cavity is convincingly demonstrated by ^{1}H NMR spectroscopy, this should be the first analytical method used to investigate solution complexation of a new cyclophane.

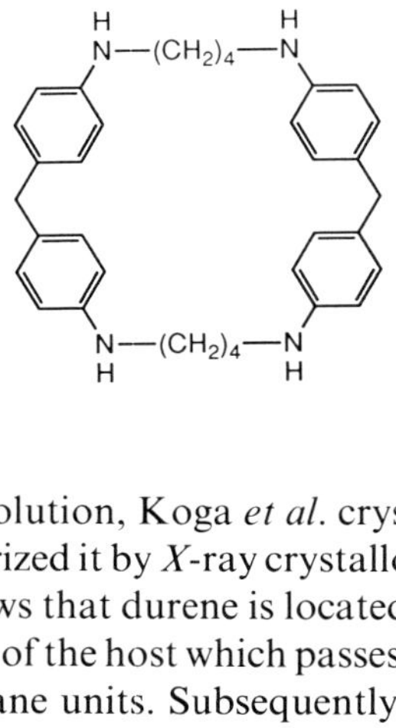

From acidic aqueous solution, Koga *et al.* crystallized a 1:1 complex of **19** with durene and characterized it by X-ray crystallography as a cavity inclusion complex.[1] Figure 2.4 shows that durene is located in a plane perpendicular to the mean molecular plane of the host which passes through the central carbons of the two diphenylmethane units. Subsequently, the communications given by Koga *et al.* in 1980[1,28] started a vigorous development of cyclophanes which could bind organic molecules in aqueous solution.[29-31]

2.3 Structures of Cyclophane Receptors for Apolar Inclusion Complexation

The design of efficient cyclophane receptors is guided by two principles that are valid throughout the entire molecular recognition field: (i) the principle of

---

[29] K. Odashima and K. Koga, in 'Cyclophanes', eds. P. M. Keehn and S. M. Rosenfeld, Academic Press, New York, 1983, Vol. 2, pp. 629.
[30] F. Franke and F. Vögtle, *Top. Curr. Chem.*, 1986, **132**, 135.
[31] F. Diederich, *Angew. Chem.*, 1988, **100**, 372; *Angew. Chem. Int. Ed. Engl.*, 1988, **27**, 362.

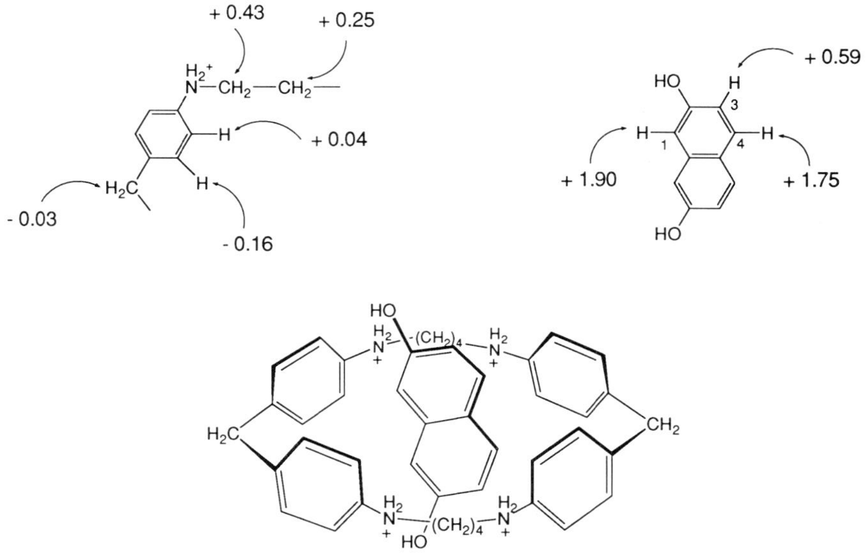

Inclusion Complexes of Neutral Molecules in Aqueous Solution

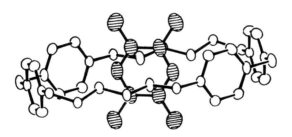

Figure 2.3 *Complexation-induced changes of the ^1H NMR resonances of **19** and 2,7-naphthalenediol in a DCl–D_2O solution with [host] = 0.05 M and [guest] = 0.025 M, pD 1.2, T = 301 K; (+) = upfield shift. These chemical shift changes are taken as support for the axial-type inclusion geometry shown*

Figure 2.4 *X-ray crystal structure of the complex formed by **19**·4HCl and durene (Reproduced with permission from reference 1. Copyright 1980 ACS.)*

stereoelectronic complementarity between host and guest and (ii) the principle of preorganization of a binding site prior to complexation. The former essentially represents a modern formulation of the lock-and-key principle of Emil Fischer.[32] The latter, demonstrated by Cram *et al.*,[33] states that preorganization of a host prior to complexation is an essential factor in controlling association strength. If a host's binding site is not organized, a

---

[32] E. Fischer, *Ber. Dtsch. Chem. Ges.*, 1894, **27**, 2985.
[33] D. J. Cram, *Angew. Chem.*, 1986, **98**, 1041; *Angew. Chem. Int. Ed. Engl.*, 1986, **25**, 1039.

structural reorganization must occur upon complexation of the guest. This reorganization can cost part of or all of the binding free energy.

Cyclophanes capable of forming stable inclusion complexes with apolar organic molecules or residues in water possess open cavities shaped by lipophilic walls. So far, apolar inclusion complexation by nonpreorganized binders has not been reported. The aromatic rings in cyclophane hosts are multifunctional. Their rigidity is crucial for organizing the binding site. Complex stability arises from specific interactions between the highly polarizable aromatic rings of the cyclophane and the encapsulated guest (Chapter 4). In addition, they provide suitable sites for the introduction of functional groups in catalytically active systems ('catalytic cyclophanes', Chapter 8). The aromatic rings in most cases are part of structurally well defined spacer units, e.g. the diphenylmethane unit in **19** or the Tröger's base unit in **23**. Usually, the opening of the spacer defines the width of the binding site. Diphenylmethane units, as in **19**, define cavity shapes that are ideal for flat aromatic guests (Figure 2.4). The distance between the two nitrogens at one diphenylmethane unit in **19** or between the two oxygens at one diphenylmethane unit in cyclophanes **25–35** varies between ≈ 8.5–9.1 Å. To generate cavity binding sites, two or more spacers are connected by bridges having adjustable chain length. The length of the bridges between the spacers allows the cavity to be tailor-fitted for guests of a specific size. Molecular-dispersed water-solubility of the receptors is achieved by introducing a suitable number of charged centers. These are either located in the cavity periphery or at more remote distance from the binding site.

Four protonated or quaternary ammonium centers at the cavity's periphery provide high molecular-dispersed water-solubility for the tetraaza[n.1.n.1] paracyclophanes, e.g. **19** and **20**, prepared and studied by Koga et al.[29,34] and others.[35a,36] Protonated ammonium centers in the cavity periphery were also used by Vögtle et al.[37a] to solubilize macrobicyclic receptors such as **21**. In another cyclophane (**22**) from the same group,[37b] water-solubility is provided by attaching carboxylate residues to the bridges between two triphenylmethane units.[35b] Wilcox introduced derivatives of Tröger's Base as rigid chiral spacers into macrocycles like **23**.[38] Similar to diphenylmethane units, spacers derived from Tröger's base give flat open binding sites for aromatic rings. Computer-aided molecular modeling studies using the MACROMODEL modeling

---

[34] T. Soga, K. Odashima, and K. Koga, *Tetrahedron Lett.*, 1980, **21**, 4351; K. Odashima, A. Itai, Y. Iitaka, and K. Koga, *J. Org. Chem.*, 1985, **50**, 4478.

[35] (a) J. Winkler, E. Coutouli-Argyropoulou, R. Leppkes, and R. Breslow, *J. Am. Chem. Soc.*, 1983, **105**, 7198; (b) D. O'Krongly, S. R. Denmeade, M. Y. Chiang, and R. Breslow, *J. Am. Chem. Soc.*, 1985, **107**, 5544.

[36] H. J. Schneider, K. Philippi, and J. Pöhlmann, *Angew. Chem.*, 1984, **96**, 907; *Angew. Chem. Int. Ed. Engl.*, 1984, **23**, 908.

[37] (a) F. Vögtle, W. M. Müller, U. Werner, and H.-W. Losensky, *Angew. Chem.*, 1987, **99**, 930; *Angew. Chem. Int. Ed. Engl.*, 1987, **26**, 901; (b) T. Merz, H. Wirtz, and F. Vögtle, *Angew. Chem.*, 1986, **98**, 549; *Angew. Chem. Int. Ed. Engl.*, 1986, **25**, 567.

[38] M. D. Cowart, I. Sucholeiki, R. R. Bukownik, and C. S. Wilcox, *J. Am. Chem. Soc.*, 1988, **110**, 6204.

Inclusion Complexes of Neutral Molecules in Aqueous Solution 63

program[39] show that the cavity in **23** is, to a large extent, preformed prior to guest inclusion. Hünig et al.[40] and the group of Stoddart[41] independently prepared cyclophanes derived from paraquat (1,1'-dimethyl-4,4'-bipyridinium), e.g. **24**. Figure 2.5 shows the X-ray crystal structure of **24**·4PF$_6$ as the acetonitrile solvate.[41] The cavity in **24** provides a tight space for incorporation of flat aromatic substrates between the two paraquat units. As discussed in Chapter 7, the solvated charged centers in the periphery of the binding sites in **19–24** reduce the driving forces for complexation that result from the desolvation of apolar surfaces. However, the ionic centers may also stabilize complexes of aromatic substrates by participating in effective pole–dipole interactions (Chapter 4.3).[42,43]

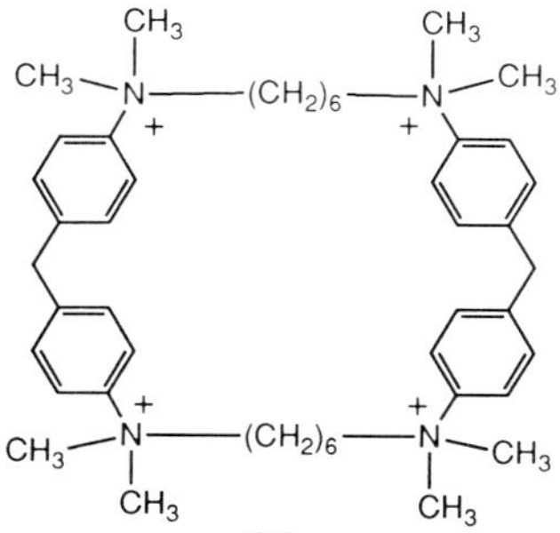

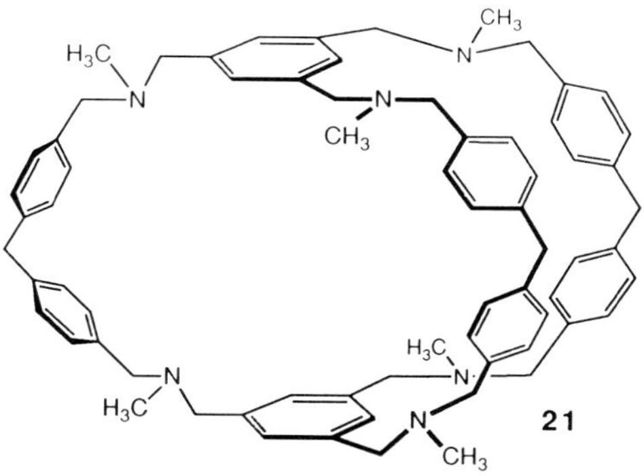

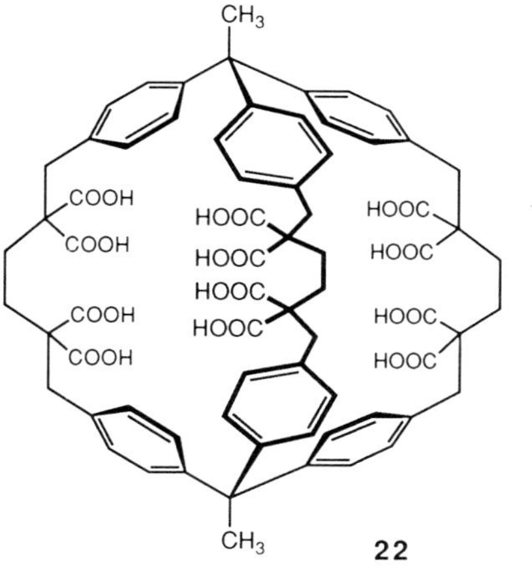

---

[39] MACROMODEL molecular modeling program, Professor W. C. Still, Columbia University, New York.
[40] M. Bühner, W. Geuder, W.-K. Gries, S. Hünig, M. Koch, and T. Poll, *Angew. Chem.*, 1988, **100**, 1611; *Angew. Chem. Int. Ed. Engl.*, 1988, **27**, 1553.
[41] B. Odell, M. V. Reddington, A. M. Z. Slawin, N. Spencer, J. F. Stoddart, and D. J. Williams, *Angew. Chem.*, 1988, **100**, 1605; *Angew. Chem. Int. Ed. Engl.*, 1988, **27**, 1547.
[42] M. A. Petti, T. J. Shepodd, R. E. Barrans, Jr., and D. A. Dougherty, *J. Am. Chem. Soc.*, 1988, **110**, 6825.
[43] H.-J. Schneider and T. Blatter, *Angew. Chem.*, 1988, **100**, 1211; *Angew. Chem. Int. Ed. Engl.*, 1988, **27**, 1163.

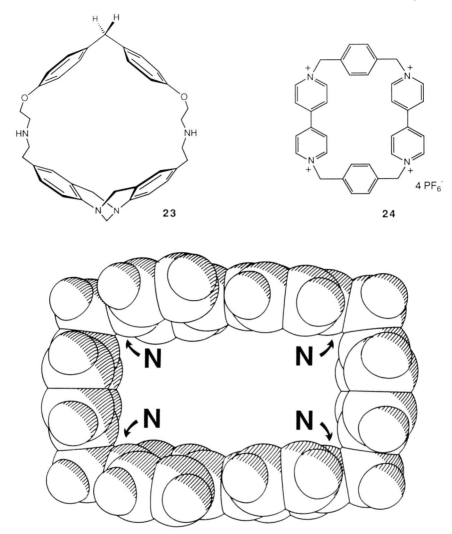

Figure 2.5 *X-ray crystal structure of* **24** *in the solvate* **24**·$4PF_6^-$·$3CH_3CN$
(Reproduced with permission from reference 41. Copyright 1988 VCH.)

In the second type of water-soluble cyclophane hosts, the ionic centers providing water-solubility are located remote from the cavity. Therefore, the hydrophobic character of the binding sites in these hosts is not perturbed by strongly hydrated charged centers.[44] Thus, they allow meaningful investigations into the nature and strength of apolar host–guest binding interactions, and into the role desolvation of apolar surfaces plays as a driving force for

[44] F. Diederich and K. Dick, *Tetrahedron Lett.*, 1982, **23**, 3167.

complexation. The quaternary ammonium centers in **25**–**33**, that provide water-solubility at neutral pH, are introduced in a spiro-type arrangement.[31] In the series **25**–**28**, the length of the α,ω-dioxaalkane bridges between the two diphenylmethane spacers is varied to accommodate substrates of different size and shape. Compounds **29**–**32** have respectively two, four, six, and eight methyl groups attached to the host. As a result, the cavity becomes successively deeper and more hydrophobic. To increase the molecular-dispersed water-solubility (Chapter 2.4), spiro piperidinium rings attached to the bridges in **33** locate two additional ionic centers away from the cavity. Macrobicyclic systems like **34** and **35** provide a tighter and more complete guest encapsulation than the corresponding monocyclic derivatives.

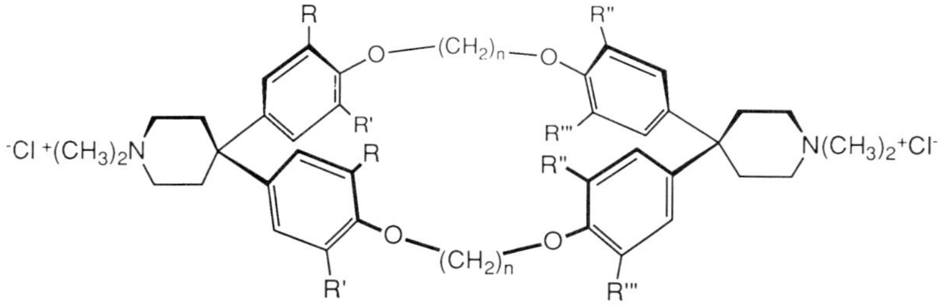

R-R'''	n		R	R'	R''	R'''	n
25	H	2	29	CH$_3$	H	H	H	4
26	H	3	30	CH$_3$	CH$_3$	H	H	4
27	H	4	31	CH$_3$	CH$_3$	CH$_3$	H	4
28	H	5	32	CH$_3$	CH$_3$	CH$_3$	CH$_3$	4

The high degree of cavity preorganization provided by diphenylmethane units and the considerable distance between the piperidine nitrogens and the cavity is clearly visible from the X-ray crystal structures obtained from the bis(N-methylpiperidine) precursor to the bisquaternary host **27** (compound **49** in Chapter 1.5.2).[31,45] In the 1:1 clathrate obtained by recrystallization from toluene, the disordered toluene molecules are located within the interstitial spaces of the crystal lattice (Figure 2.6). From wet benzene, a crystalline 1:2:1 (host–benzene–water) complex was obtained. One of the benzene rings is perfectly enclosed within the intramolecular cavity of the host (Figure 2.7). The second benzene ring and the water molecule are located within interstitial spaces of the crystal lattice.

In **36**, a representative for a series of cyclophanes prepared by Dougherty *et al.*, 9,10-bridged 9,10-dihydroanthracenes locate carboxylates remote of the cavity.[42] These rigidly structured spacers lead to a highly preorganized

---

[45] C. Krieger and F. Diederich, *Chem. Ber.*, 1985, **118**, 3620.

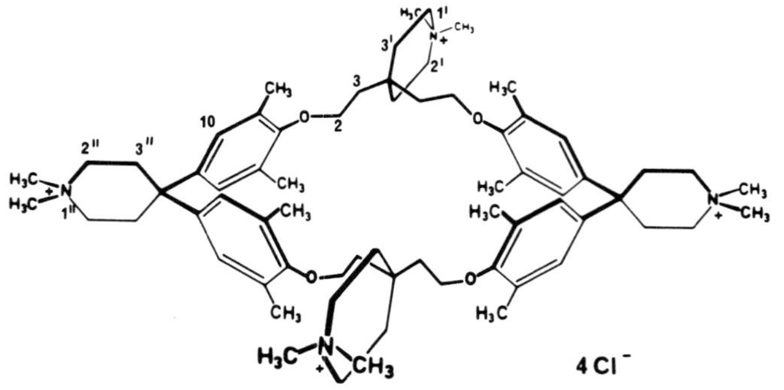

33

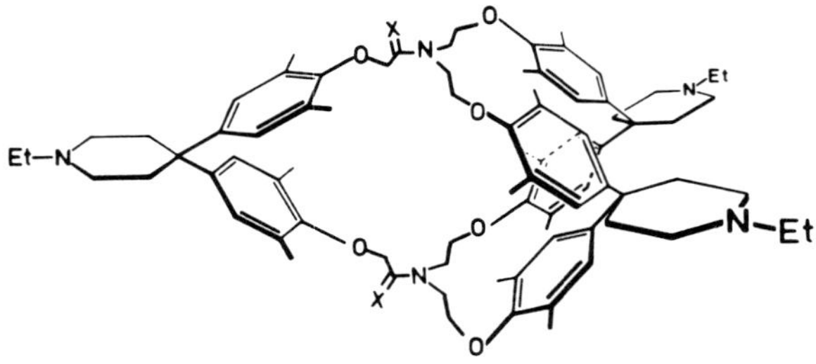

34 X = O
35 X = 2H

binding site. Computer models generated with the program BIOGRAPH[46] show that the spacers in **36** create a wider cavity space than the diphenylmethane units in the cyclophanes **19** or **25–35**. Adamantane derivatives are fully encapsulated into **36** but not in bis(diphenylmethane) hosts. Collet *et al.* introduced the cyclotriveratrylene unit to shape spherical hosts, *e.g.* **37**.[47] Its spherical binding site is seen in the X-ray crystal structure of a CH_2Cl_2 complex shown in Figure 2.8.[47] In **38**, structurally related to **37**, the special alignment of the solvated carboxylates prevents them from occupying the cavity.[48]

2.4 Aggregation Behavior of Cyclophanes in Aqueous Solution

A serious problem arises in the complexation between apolar hosts and guests in aqueous solution. It involves the potential of the host and/or guest

---

[46] BIOGRAPH: Biodesign Inc., Pasadena, CA.
[47] A. Collet, *Tetrahedron*, 1987, **43**, 5725; J. Canceill, M. Cesario, A. Collet, J. Guilhem, and C. Pascard, *J. Chem. Soc., Chem. Commun.*, 1985, 361.
[48] J. Canceill, L. Lacombe, and A. Collet, *J. Chem. Soc., Chem. Commun.*, 1987, 219.

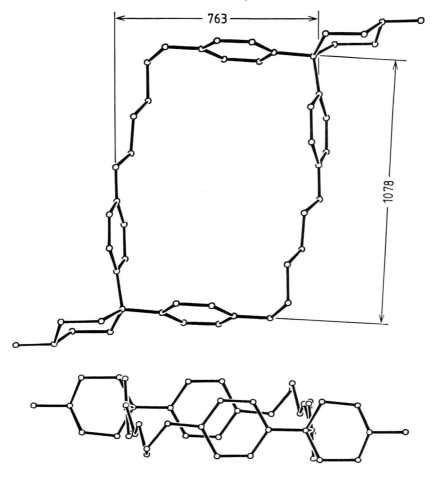

Figure 2.6 *View onto the least-squares plane and view parallel to the least-squares plane of the bis(N-methylpiperidine) precursor to 27 in a 1:1 toluene clathrate. Only the cyclophane is shown and not the toluene molecules which are located in a disordered way in the crystal lattice. Distances shown are in pm*
(Reproduced with permission from reference 31. Copyright 1988 VCH.)

molecules for self-aggregation which leads to the formation of higher molecular weight aggregates. The same hydrophobic interactions that act as a driving force for stoichiometric host–guest complexation can also lead to the segregated association of host and/or guest molecules and to the formation of mixed aggregates with poorly defined orientation of host and guest, similar to micellar systems. Therefore, the formation of stoichiometric complexes needs to be studied in a concentration range where the complexation equilibrium is not perturbed by additional equilibria resulting from aggregation. Hence, the

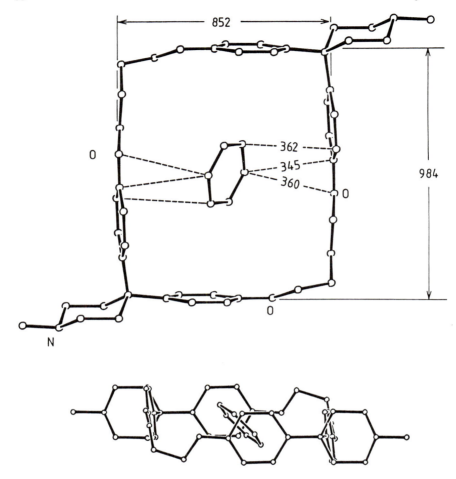

Figure 2.7. *View on the least-squares plane and view parallel to the least-squares plane of the bis(N-methylpiperidine) precursor to **27** in a 1:2 benzene monohydrate. Only the host and the benzene ring in the intramolecular cavity are shown; an additional benzene ring and a water molecule are located in the crystal lattice. Distances shown are in pm (Reproduced with permission from reference 45. Copyright 1985 VCH.)*

aggregation behavior of both host and guest should be known. A study of stoichiometric complexation in aqueous solution with a new cyclophane, designed to provide significant thermodynamic and kinetic data, should always begin by investigating the aggregation behavior of the host.

The critical aggregation concentration (CAC) of cyclophane hosts, below which the complexation is studied, can be determined easily by ^{1}H NMR spectroscopy.[49] Other methods that have been used are light scattering[44] and

---

[49] F. Diederich and K. Dick, *J. Am. Chem. Soc.*, 1984, **106**, 8024.

Inclusion Complexes of Neutral Molecules in Aqueous Solution

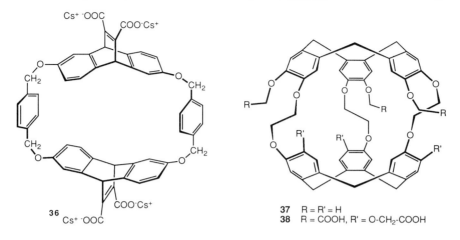

37 R = R' = H
38 R = COOH, R' = O-CH$_2$-COOH

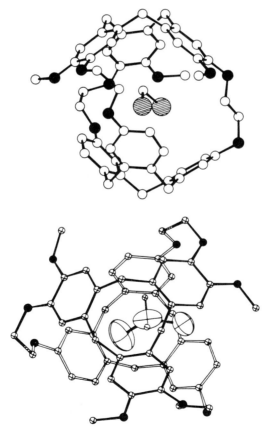

Figure 2.8 *X-ray crystal structure of the 37·CH$_2$Cl$_2$ complex in views perpendicular to the pseudo-C$_3$-axis and looking down the C$_3$-axis (Reproduced with permission from reference 47. Copyright 1985 Chemical Society London.)*

surface tension measurements.[24,50] Figure 2.9 shows the chemical shifts of the protons of **33** plotted as a function of the host concentration in aqueous solution. The well-defined discontinuity in the curves, approximately at the same concentration, defines the ^{1}H NMR CAC of this compound. Below this value, the host is present in a molecular-dispersed form, and the chemical shifts are independent of concentration. Above the CAC, the chemical shifts become dependent upon host concentration since the anisotropic regions of the aromatic rings influence the signal positions of the aggregating hosts.

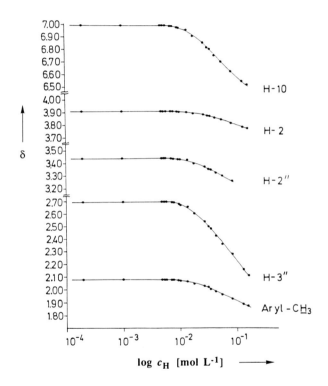

Figure 2.9 *Chemical shifts (δ-values) of the protons of **33** as a function of concentration (c_H) in D_2O [360 MHz, 303 K, ext. standard: sodium 2,2,3,3–tetradeuterio-3-(trimethylsilyl)propionate (TSP)] (Reproduced with permission from reference 31. Copyright 1988 VCH.)*

For molecular-dispersed water-solubility in concentration ranges suitable for ^{1}H NMR studies, the relationship between lipophilicity and hydrophilicity for the cyclophanes needs to be well balanced. Table 2.1 shows some examples which should help in designing this important balance into new hosts. In the paracyclophanes **12** and **19**, the four ionic charges aligned in the periphery of the cavity effectively break the hydrophobic surfaces and, correspondingly, these compounds have high CACs.[50] Many of the cyclophanes with two spiro piperidinium ions, e.g. **27–32**, show low CACs which severely limits the analysis of the receptor properties of these compounds. A large reduction in aggregation tendency occurs upon introduction of two additional spiro piperidinium rings into **33**. The CAC of **33** is more than two orders of magnitude higher than the CAC of the comparable host **32** which has only two ionic centers (Table 2.1).

Table 2.1 *Examples for critical aggregation concentrations (CACs) of cyclophane hosts*

Host	CAC $[mol\ L^{-1}]$	Reference
19[a]	$> 8 \times 10^{-2}$	50
25[b]	2.5×10^{-3}	31
26[b]	2.5×10^{-3}	31
27[b]	1.6×10^{-4}	31
32[b]	$\leq 2 \times 10^{-5}$	31
33[b]	7.5×10^{-3}	49
34[c]	1.0×10^{-3}	64
35[c]	5.0×10^{-3}	64
36[d]	$\approx 1 \times 10^{-3}$	60
39[b]	$\approx 1 \times 10^{-2}$	51
40[b]	$\approx 1 \times 10^{-2}$	51
42[b]	$\approx 1 \times 10^{-1}$	40
43[b]	$\approx 1 \times 10^{-1}$	40

[a] 0.25 M HCl–H_2O; [b] D_2O; [c] 0.5 M KD_2PO_4; [d] Aqueous borate buffer, pD $\approx$ 9.

The introduction of additional charged centers into host **33** to increase the CAC led to a much longer synthetic route. Whereas compounds **25–32** are prepared in seven-step syntheses, a fourteen-step route was required for **33**. A more facile way to obtain a more favorable hydrophobicity/hydrophilicity balance was found in the octamethoxy hosts **39** and **40**, available in seven-step syntheses.[51] The introduction of methoxy groups at the aromatic rings greatly increases the molecular-dispersed water-solubility of these compounds compared to the corresponding hosts with methyl substituents, e.g. **32** (Table 2.1). With **39** and **40**, no aggregation was observed by ^{1}H NMR up to concentrations of 10^{-2} mol L^{-1}. It is a reasonable assumption that other polar substituents aligned around apolar cyclophane cavities should also reduce the aggregation tendency.

To gain insight into binding and aggregation equilibria above the CAC of a cyclophane, Janzen *et al.* studied the interaction between **39** and the α-substituted benzyl *tert*-butyl nitroxide spin probes **41a–e** in water by electron

[50] I. Takahashi, K. Odashima, and K. Koga, *Chem. Pharm. Bull.*, 1985, **33**, 3571.
[51] S. B. Ferguson, E. M. Seward, F. Diederich, E. M. Sanford, A. Chou, P. Inocencio-Szweda, and C. B. Knobler, *J. Org. Chem.*, 1988, **53**, 5593.

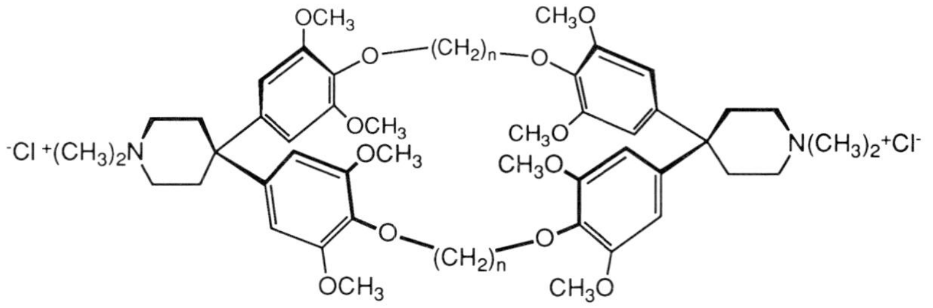

39 n = 3
40 n = 4

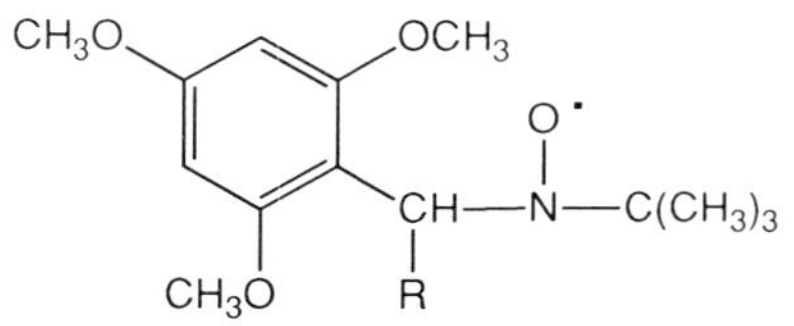

41 a R = Ph
  b R = Et
  c R = n-hexyl
  d R = cyclopentyl
  e R = cyclohexyl

spin resonance.[52] They had previously shown that the inclusion complexation of nitroxide spin probes by cyclodextrins affects the nitrogen and β-hydrogen hyperfine splitting constants (hfsc).[53,54] These changes are a result of the different polarity of the environment which the probe encounters in the complex compared to the bulk solution. In addition, the population of the different spin probe conformers presumably changes upon binding which also affects the relative magnitudes of the hfsc's. Since complexation–decomplexation equilibria are slow on the ESR time scale, the changes in hfsc's produce a new spectrum of the spin probe along with the spectrum of free probe remaining in solution (Figure 2.10). These spectral changes can be evaluated in binding titrations to provide the thermodynamic parameters for the stoichiometric binding event. The 2,4,6-trimethoxyphenyl ring and the *tert*-butyl group in **41** are too bulky for incorporation into the cavity of **39**, and thus any spectral changes must result from the inclusion of the residue R. Below the CAC of **39** (0.01 M), binding was only observed for **41a** (R = Ph) (Figure 2.10), and spectral simulations afforded $K_a = 290 \pm 40$ L mol^{-1} as the association constant for the 1:1 complex formed at 293 K. No new ESR spectra were

---

[52] E. G. Janzen, Y. Kotake, F. N. Diederich, and E. M. Sanford, *J. Org. Chem.*, 1989, **54**, 5421.
[53] Y. Kotake and E. G. Janzen, *J. Am. Chem. Soc.*, 1989, **111**, 2066.
[54] Y. Kotake and E. G. Janzen, *J. Am. Chem. Soc.*, 1989, **111**, 5138.

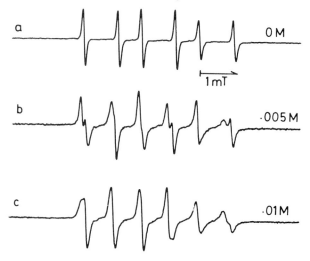

Figure 2.10 *ESR spectra at 293K in water: (a) of pure phenyl probe **41a**, (b) of **41a** in the presence of 0.005 M **39**, and (c) of **41a** in the presence of 0.01 M **39**. Observed hfscs are the following: free probe $a_N = 1.642$ mT, $a_H = 0.978$ mT; included probe, $a_N = 1.602$ mT, $a_H = 0.752$ mT (Reproduced with permission from reference 52. Copyright 1989 ACS.)*

obtained in the presence of the cyclophane for the spin probes **41b–e** with R = alkyl. Since no inclusion of alkyl groups was found, the association constants for complexes of these groups must be less than 1 L mol^{-1}.

Around the CAC of **39**, an almost pure spectrum of the **39·41a** complex was obtained (Fig 2.10). When the concentration of **39** in the solution of the phenyl probe **41a** was further increased above its CAC, no additional change was observed in the hfsc's of the probe. The polarity of the environment around the probe is not influenced by aggregate formation because the probe stays in the cavity irrespective of the aggregation state of the host. On the other hand, probe **41d** with R = cyclopentyl did not show any sign of complex formation below the CAC of **39**. However above the CAC, the N-hfsc started to decrease (from 1.62 mT at [**39**] = 0.01 M to 1.59 mT at [**39**] = 0.08 M). The smaller hfsc means that the probe experiences a more hydrophobic environment. Also, spectral line broadening was observed indicating that the molecular tumbling motion becomes restricted above the CAC. These findings indicate that probe **41d** starts to be sequestered into the aggregate that is formed by host **39** at its CAC. Upon increasing the concentration of **39** above the CAC, the decrease of the N-hfsc continues until all probe molecules are in the aggregate. This experiment by Janzen *et al.* clearly shows how important it is to determine the CAC prior to the analysis of host–guest interactions. Above the CAC, it is not possible by ^{1}H NMR or any other spectral analysis to decide whether spectral changes observed in host–guest solutions result from molecular complexation or from aggregation.

If aggregation in water or aqueous buffers is a problem, complexation studies are best executed in alcohol–water mixtures with an alcohol content of ≥40% (v/v).[55] Methanol is the best cosolvent to break water structure and to prevent aggregation.[56,57]

2.5 Complexes of Polycyclic Aromatic Hydrocarbons

Polycyclic aromatic hydrocarbons (PAH) are extremely insoluble in water. Cyclophanes like **21**,[37a] **33**,[58] **34**,[59] **36**,[60] **42**,[40] and **43**[40] form stable complexes with PAHs, *e.g.* perylene, pyrene, anthracene, and phenanthrene. The amount of hydrocarbon that can be solubilized in water is dramatically increased by host–guest complexation. As an example, the solubility of pyrene at 293 K in water is only 8×10^{-7} mol L^{-1}. However, with a 5.5×10^{-3} molar aqueous solution of **33** using solid–liquid extractions, a complex solution can be prepared containing a total pyrene concentration of 2.8×10^{-3} mol L^{-1}. Solid–liquid[31,49] and liquid–liquid extractions[61,62] are the methods of choice for the preparation of aqueous solutions containing complexes of PAHs. In the solid–liquid extraction procedure, an excess of finely powdered solid guest is extracted at constant temperature with an aqueous solution of the host using vigorous shaking and stirring or exposure to ultrasonic irradiation until the extraction equilibrium is established.[49] In liquid–liquid extractions, the guests are extracted from a *n*-hexane solution into an aqueous solution containing the host. Both methods allow for the determination of association constants of

[55] R. Dharanipragada, S. B. Ferguson, and F. Diederich, *J. Am. Chem. Soc.*, 1988, **110**, 1679.
[56] Y. Murakami, Y. Aoyama, and M. Kida, *J. Chem. Soc., Perkin Trans. 2*, 1977, 1947–1952.
[57] W. D. Harkins, R. Mittelmann, M. L. Corrin, *J. Phys. Colloid Chem.*, 1949, **53**, 1350.
[58] F. Diederich and K. Dick, *Angew. Chem.*, 1983, **95**, 730; *Angew. Chem. Int. Ed. Engl.*, 1983, **22**, 715; *Angew. Chem. Suppl.*, 1983, 957.
[59] F. Diederich and K. Dick, *Angew. Chem.*, 1984, **96**, 789; *Angew. Chem. Int. Ed. Engl.*, 1984, **23**, 810.
[60] M. A. Petti, T. J. Shepodd, and D. A. Dougherty, *Tetrahedron Lett.*, 1986, **27**, 807.
[61] H. K. Frensdorff, *J. Am. Chem. Soc.*, 1971, **93**, 4684.
[62] K. E. Koenig, G. M. Lein, P. Stuckler, T. Kaneda, and D. J. Cram, *J. Am. Chem. Soc.*, 1979, **101**, 3553.

the complexes; the liquid–liquid extraction procedure is more accurate and reproducible.[49]

Table 2.2 shows the stabilities of arene complexes formed by **33** in aqueous solution.[49] Perylene has the highest stereoelectronic complementarity to the binding site and forms the most stable complex. Upon binding, the large apolar guest surface becomes favorably desolvated and subsequently provides numerous van der Waals contacts in the complex (Chapter 7). The geometric complementarity between **33** and smaller arenes is less favorable and, expectedly, the complex stability decreases. Perylene binds better than pyrene and fluoranthene, and a strong decrease in complex stability is observed with the smaller guests biphenyl, azulene, naphthalene, and durene.

Table 2.2 *Association constants (K_a) and free energies of complexation ($-\Delta G°$) for the 1:1 complexes formed between cyclophanes and polycyclic aromatic hydrocarbons in aqueous solutions*

Guest	K_a [L^{-1} mol]	$-\Delta G°$ [kcal mol^{-1}]
(a) Complexes of host **33** in water ($T = 293$–295 K)[31]		
Perylene	1.6×10^7	9.6
Fluoranthene	1.8×10^6	8.4
Pyrene	1.8×10^6	8.4
Biphenyl	2.2×10^4	5.8
Azulene	2.1×10^4	5.8
Naphthalene	1.2×10^4	5.5
Durene	1.9×10^3	4.4
(b) Complexes of host **34** in 0.5 M KH_2PO_4, pH = 4.7 ($T = 293$–295 K)[31]		
Pyrene	3.1×10^6	8.7
Naphthalene	1.6×10^4	5.6
(c) Complex of host **36** in borate buffer, pH ≈ 9.0 ($T = 298$ K)[60]		
Pyrene	2×10^6	8.6

As mentioned in Chapter 2.2, detailed information about the spatial relationship between the two binding partners in arene–cyclophane complexes is obtained by ^{1}H NMR spectroscopy.[63] In the complexes, the diatropic ring currents of both host and guest mutually influence the positions of their proton resonances, and selective complexation shifts are observed. The extent of these complexation shifts depends on the distance and orientation of the two binding partners in the complex.

The large PAHs are preferentially located in the cavity of **33** within a specific plane which is perpendicular to the mean plane of the host and passes through the two spiro carbon atoms of the two diphenylmethane units.[63] Aromatic guest incorporation into this plane is easily recognized in the ^{1}H NMR spectrum by characteristic upfield and downfield shift patterns of the host

[63] F. Diederich and D. Griebel, *J. Am. Chem. Soc.*, 1984, **106**, 8037.

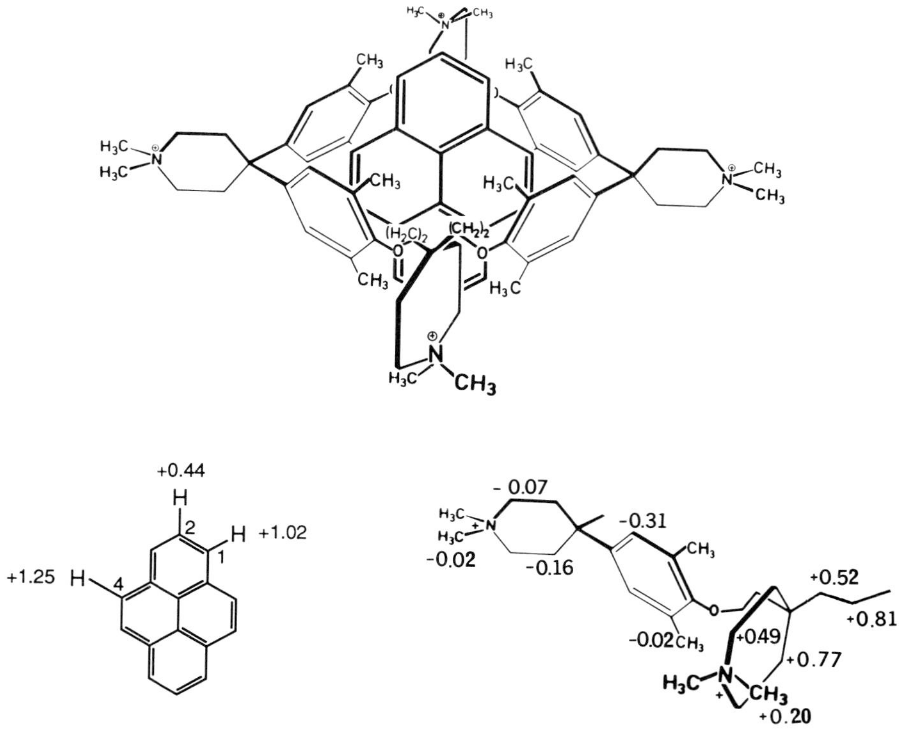

Figure 2.11 *Schematic drawing of the highly favored geometry of the **33**·pyrene complex in aqueous solution. The piperidinium rings of the bridges can also be oriented to the same cavity side. The geometry shown is supported by the characteristic complexation-induced shifts ($\Delta\delta$ values, $+$ = upfield shift) of host and guest resonances in the 360 MHz ^{1}H NMR spectrum of a solution of the complex in D_2O with $[33] = 5.5 \times 10^{-3}$ mol L^{-1} and pyrene $= 2.8 \times 10^{-3}$ mol L^{-1}*

signals as well as by specific upfield shifts of the guest signals. Figure 2.11 shows the complexation-induced shifts of the host and guest resonances in an aqueous solution of the **33**·pyrene complex, and the preferred orientation of bound pyrene as deduced from these characteristic signal shifts. The protons 2,7-H of the complexed guest are positioned outside the cavity, and therefore their triplet exhibits the smallest chemical shift change upon complexation. The signals for the host protons located within the plane of the aromatic guest move downfield, whereas the signals of protons oriented about perpendicular to the pyrene plane move upfield. As depicted in Figure 2.11, the piperidinium rings attached to the aliphatic bridges envelop the enclosed guest. This generates additional favorable van der Waals contacts and shielding of the apolar guest from the solvent.

In the cavities of the macrobicycles **21**[37a] and **34**,[59,64] which are water-soluble in the protonated form, PAHs are located in the plane passing through the central sp^3-carbon atoms of the three diphenylmethane units (Figure 2.12). These specific inclusion geometries are also supported by characteristic ^{1}H NMR complexation shifts. Upfield shifts are observed for the host resonances in the bridges connecting the three diphenylmethane units, whereas the signals of the aromatic protons at these spacers move downfield. The cavity of **21** is wider than the binding site in **34**. Pyrene shows the best fit in the cavity of **34** and, accordingly, the larger cyclophane **21** incorporates more spacious arenes like triphenylene, benzo[*def*]chrysene (1,2-benzpyrene), and benzo[*ghi*]perylene (1,12-benzperylene).

Host **21** forms a more stable complex with phenanthrene than with anthracene. This binding selectivity, in addition to the higher water-solubility of phenanthrene, was exploited for the separation of phenanthrene from anthracene by solid–liquid extraction.[37a] Furthermore, it was possible to separate, through complex formation with **21**, partially hydrogenated arenes from the parent arenes, *e.g.* hexahydropyrene from pyrene and dodecahydro-triphenylene from triphenylene. In liquid–liquid extractions, only the stronger binding parent arenes are transferred from the *n*-hexane phase into the aqueous solution. The poorer binding partially hydrogenated guests which, in addition, possess the smaller distribution constants K_{dist} ($= G_{H_2O}/G_{n\text{-hexane}}$) are not extracted. Such remarkable selectivity in a molecular recognition event could well be utilized in novel separation processes and technologies which make use of recyclable host and solvent (see also Chapter 8.2).

The pyrene complexes of the macromonocyclic cyclophanes **33** and **36** and the macrobicyclic system **34** possess very similar stability (Table 2.2). Pyrene complexes of considerable stability ($K_a \approx 10^5$–10^6 L mol^{-1}) are also formed with hosts **42** and **43**[40]; they are more stable than the corresponding complexes of phenanthrene and anthracene. Characteristic complexation-induced shifts in the ^{1}H NMR spectra indicate that pyrene is sandwiched between the paraquat units of **42** and the 1,2-bis(pyridinium)ethene units of **43**. Figure 2.13 shows the geometry of the **42**·pyrene complex as modeled by a MM2 force field[65] computational study. The calculated orientation of the guest is supported by the observed ^{1}H NMR complexation shifts. In a solution of the **42**·pyrene complex with almost nearly all of the pyrene in the bound state, the resonances of 2,7-H exhibit the largest upfield shift ($\Delta\delta = +1.73$). This is in agreement with the preferred orientation of these protons into the shielding regions of the diphenylmethane units of the host. With values of $\Delta\delta = +1.37$ (1,3,6,8-H) and $+0.94$ (4,5,9,10-H), the upfield shifts of the two other proton groups in the complexed guest are smaller.

In spite of the high association strength between the larger arenes and **33** in D$_2$O, complexation–decomplexation equilibria at room temperature are

---

[64] F. Diederich, K. Dick, and D. Griebel, *J. Am. Chem. Soc.*, 1986, **108**, 2273.
[65] U. Burkert and N. L. Allinger, 'Molecular Mechanics', American Chemical Society, Washington, DC, 1982.

78   Chapter 2

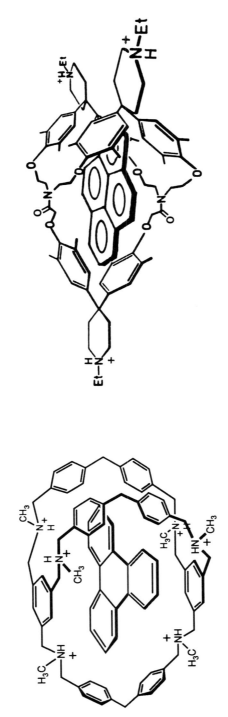

Figure 2.12 *Schematic drawing of the inclusion geometries of the **21**·triphenylene and **34**·pyrene complexes which are highly favored according to the analysis of ^1H NMR complexation shifts*

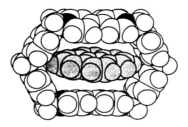

Figure 2.13 *MM2-optimized geometry of the **42**·pyrene complex which is also supported by characteristic ^1H NMR complexation shifts (Reproduced with permission from reference 40. Copyright 1988 VCH.)*

rapid on the 360 MHz ^{1}H NMR time scale. For the dissociation of the **33**·pyrene complex, the first-order decomplexation rate constant was determined, at the coalescence temperature, as $k_{-1} = 495$ s^{-1} ($T = 273$ K). With an association constant $K_a = k_1/k_{-1}$ of 1.8×10^6 L mol^{-1} ($T = 293$ K, Table 2.2), the second-order complexation rate constant can be estimated as approximately $k_1 \approx 9 \times 10^8$ s^{-1} L mol^{-1}. Hence, the formation of highly stable arene complexes in water is close to diffusion controlled. This fast complexation rate is not unexpected since the binding sites in [n.1.n.1]paracyclophane hosts are sterically accessible. In addition, the desolvation of apolar binding sites and apolar arenes during the complexation step apparently requires only a small activation energy. This is in contrast to the much higher activation energies required for the desolvation of cations and their polar binding sites.[66,67]

Slower exchange kinetics are observed if the aromatic guests bear polar substituents capable of forming strong hydrogen bonds to the solvent (–CONH$_2$, –NH$_2$, –COOH, –OH). The decomplexation rates for a variety of complexes between naphthalene derivatives having polar functionalities and host **33** in water were either on the ^{1}H NMR time scale or even slower. At association constants around $K_a \approx 10^4$ L mol^{-1} or less, the complexation steps must be significantly slower than diffusion controlled for the dissociation of these complexes to be observed on the ^{1}H NMR time scale. The slower complexation of arenes with polar, hydrogen-bonding substituents reflects the higher energies of activation needed to partially desolvate the substituents upon penetration of the guest into the cyclophane cavity. Nevertheless, complexation equilibria generally are fast enough to make cyclophanes with apolar binding sites promising candidates for further transformation into catalytically active systems (Chapter 8). Many supramolecular reactions would need very large catalytic rate accelerations before decomplexation rates become rate-determining.

Despite these preliminary kinetic data from ^{1}H NMR measurements, the kinetics of the complexation between cyclophanes and organic molecules have

[66] J.-M. Lehn, *Struct. Bonding (Berlin)*, 1973, **16**, 1.
[67] R. M. Izatt, J. S. Bradshaw, S. A. Nielsen, J. D. Lamb, J. J. Christensen, and D. Sen, *Chem. Rev.*, 1985, **85**, 271.

been poorly investigated. The application of faster relaxation methods, *e.g.* ultrasonic relaxation, seems very desirable.[67,68] From such studies, interesting information on the energies needed to desolvate neutral polar substituents should be obtained.

2.6 Complexes of Naphthalene Derivatives

2.6.1 Determination of Stoichiometry and Stability of Solution Complexes

In contrast to polycyclic aromatic hydrocarbons, many functionalized naphthalene and most benzene derivatives possess sufficient water solubility to allow stoichiometry and stability of their complexes with cyclophanes to be determined from binding titrations in homogeneous aqueous solution. Besides ^{1}H NMR titrations,[69] optical titrations are most commonly applied for this analysis.[31,70] In binding titrations, the total guest concentration is usually a constant, and the change of its property X is monitored as a function of increasing host concentration. Titrations using a constant host concentration are less frequently executed. In ^{1}H NMR studies, the property X normally is the chemical shift of a guest proton. In optical spectroscopy, X is the intensity of an electronic absorption or emission band of the guest. Titration curves (see Figure 8.4) at constant guest concentration are obtained by plotting the change of X as a function of increasing host concentration. From these curves, the stoichiometry of the complexes and the K_a values are calculated using a variety of graphic or computational curve-fitting methods. Defined experimental boundary conditions need to be observed during titrations to obtain significant association constants.[70-72] These boundary conditions are valid, independent of the method chosen to calculate K_a. According to Deranleau[71] and Person,[72] meaningful titrations have a concentration range in which about 20–80% saturation binding of the minor component, held at fixed concentration, is observed. As a guideline for choosing suitable concentration ranges, the amount of the major component should be varied during the titration starting from the concentration corresponding to approximately half the dissociation constant K_d ($= 1/K_a$) of the complex to about a twentyfold higher concentration ($\approx 10\ K_d$). If suitable concentrations of the two binding partners are within the ^{1}H NMR sensitivity range ($c \approx \geq 1 \times 10^{-4}$ mol L^{-1}), ^{1}H NMR titrations are normally the method of choice. They are superior to optical studies being less sensitive to small impurities and provide useful structural information on the complex. At lower concentration ranges, optical binding titrations are applied.

[68] G. W. Liesegang and E. M. Eyring in 'Synthetic Multidentate Macrocyclic Compounds', eds. R. M. Izatt and J. J. Christensen, Academic Press, New York, 1978, pp. 245.
[69] R. Foster and C. A. Fyfe, *Prog. Nucl. Magn. Reson. Spectrosc.*, 1969, **4**, 1.
[70] K. A. Conners, 'Binding Constants, The Measurement of Molecular Complex Stability', Wiley, New York, 1987.
[71] D. A. Deranleau, *J. Am. Chem. Soc.*, 1969, **91**, 4050.
[72] W. B. Person, *J. Am. Chem. Soc.*, 1965, **87**, 167.

For a graphical analysis of titration curves using the methods of Benesi and Hildebrand,[21] Scatchard,[73] or Scott,[74] the experimental boundary condition $H \gg G$ needs to be observed during the titration. This is best done by choosing over the entire titration range a total host concentration H_0 which is much larger (≥ 10 times) than the total guest concentration G_0. No boundary condition with respect to the ratio of the concentrations of the two binding partners needs to be obeyed in titrations that are evaluated by computer-assisted, nonlinear least-squares curve fitting. Several nonlinear regression methods have been described,[38,42,75,76] and programs for microcomputers are available.[77-79] In most studies, stoichiometries of the complexes have been determined directly from the titration curves[63] or from the best fit in the nonlinear regression analysis; an additional useful method employs the analysis of Job plots.[80]

Other methods have been applied to determine the stability of cyclophane complexes. In fluorescence titrations, the competitive inhibition of the binding of fluorescing guests with known complexation properties by non-fluorescing inhibitors was evaluated to determine the association strength of the inhibitors.[49,81] Batch or titration calorimetry can also provide complex stability constants and represents the most accurate method for the determination of complexation enthalpies $\Delta H°$ and heat capacity changes $\Delta C_p° = (\partial \Delta H°/\partial T)_p$.[82] Heat capacity changes have been measured for a variety of apolar association processes particularly in aqueous environments (Chapter 7.4.3).[5] In many studies, the association constants of 1:1 complexes were determined independently by several of the previously discussed methods, including solid–liquid and liquid–liquid extraction procedures described in Chapter 2.5.[31,49] Good agreement was found between the values generated by different methods.

2.6.2 Complexation of Neutral Naphthalene Derivatives

The complexation of neutral naphthalene derivatives in aqueous solution has been studied with a variety of cyclophanes.[28,49,63,76,81,83,84] Table 2.3 shows the stability of some characteristic complexes. It is seen that complexes formed

---

[73] G. Scatchard, *Ann. N. Y. Acad. Sci.*, 1949, **51**, 660.
[74] R. L. Scott, *Recl. Trav. Chim. Pays-Bas*, 1956, **75**, 787.
[75] S. B. Ferguson, Ph. D. Thesis, University of California at Los Angeles, 1989.
[76] H.-J. Schneider, R. Kramer, S. Simova, and U. Schneider, *J. Am. Chem. Soc.*, 1988, **110**, 6442.
[77] 'Computational Methods for the Determination of Formation Constants', ed. D. J. Leggett, Plenum, New York, 1985.
[78] R. J. Leatherbarrow, 'Enzfitter', Elsevier-BIOSOFT, 68 Hills Road, Cambridge CB2 1LA, UK.
[79] A. E. Martell and R. J. Motekaitis, 'Determination and Use of Stability Constants', VCH, Weinheim, 1988.
[80] A. Job, *Ann. Chim.*, 1928, **9**, 113.
[81] K. Odashima, T. Soga, and K. Koga, *Tetrahedron Lett.*, 1981, **22**, 5311.
[82] M. R. Eftink, M. L. Andy, K. Bystrom, H. D. Perlmutter, and D. S. Kristol, *J. Am. Chem. Soc.*, 1989, **111**, 6765.
[83] F. Vögtle, W. M. Müller, U. Werner, and J. Franke, *Naturwissenschaften*, 1985, **72**, 155.
[84] J. Franke and F. Vögtle, *Angew. Chem.*, 1985, **97**, 224; *Angew. Chem. Int. Ed. Engl.*, 1985, **24**, 219.

by receptors without charged centers in the periphery of the binding site are more stable than the complexes of hosts having charged ammonium centers aligning their cavity. The study of host **33** with naphthalene derivatives investigated the influence that the nature and the position of guest substituents have on structure and stability of the complexes in aqueous solution.[49,63] Cyclophane **33** forms complexes of similar stability with naphthalene and its derivatives bearing non-ionic substituents, *e.g.* –OH, –NMe$_2$, –Me ($K_a \approx 10^4$ L mol^{-1}, $\Delta G° \approx -5.4$ kcal mol^{-1}, Table 2.3), if these substituents extend out of the cavity into the aqueous solution. This orientation is favored for steric reasons and, especially in the case of polar or ionic groups, for favorable solvation. Depending on the nature of the substituents, the guests, especially the disubstituted naphthalenes, have a highly preferred orientation in the cavity of **33** (Figure 2.14). Naphthalene itself does not seem to prefer a specific orientation; Figure 2.14 shows its equatorial inclusion. Both 1,3- and 2,3-naphthalenediols prefer an axial orientation, 2,6- and 2,7-naphthalenediols a pseudoaxial orientation, and 1,5-disubstituted naphthalenes a pseudoequatorial position in the cavity of **33**. These structural assignments are based on characteristic upfield shift patterns of the guest resonances in the ^{1}H NMR spectra. It is apparent from Figure 2.14 that protons 1,4,5,8-H of an axially enclosed guest should exhibit specifically large upfield complexation shifts. Similarly, a large upfield shift can be expected for the resonances of 4,8-H in a pseudoaxial inclusion and for the resonances of 3,7-H in a pseudoequatorial inclusion. Experimentally, these complexation shifts are observed in the ^{1}H NMR spectra. Figure 2.15 compares the spectrum of the **33**·1,3-naphthalenediol complex to the spectra of the two isolated binding partners; the differential upfield shifts of the guest protons in the complex solution are clearly visible. The nomenclature used in Figure 2.14 to describe the different orientations of guests in the cavity was initially proposed by Koga *et al.*[29]

The geometries derived from ^{1}H NMR studies for the complexes of [*n*.1.*n*.1]paracyclophanes and naphthalene derivatives are also supported by computer modeling and X-ray crystallography. MM2 calculations revealed a preference for the pseudoaxial inclusion of 2,6-naphthalenedicarbonitrile in the cavity of **32**.[31] According to the modeling, the host adopts the geometry of a molecular box surrounded by four aromatic walls. The guest seems to prefer the two orientations A and B (Figure 2.16) which optimizes attractive π–π-stacking interactions between parallel aromatic surfaces and edge-to-face interactions between perpendicular aromatic rings of the two binding partners. It is difficult to determine experimentally if orientations A and B are preferred in solution. At fast exchange, the ^{1}H NMR spectra reflect a time-averaged complex geometry C with the guest located in the plane passing through the central carbon atoms of the two diphenylmethane units (Figure 2.16).

A preference for orientation C actually seems to exist for steric reasons in the inclusion complexes of the larger arenes, *e.g.* pyrene or fluoranthene, and host **33** (Figure 2.11). This inclusion geometry is experimentally characterized by rather moderate upfield shifts ($\Delta\delta \approx 1$–1.5 ppm at saturation binding) for

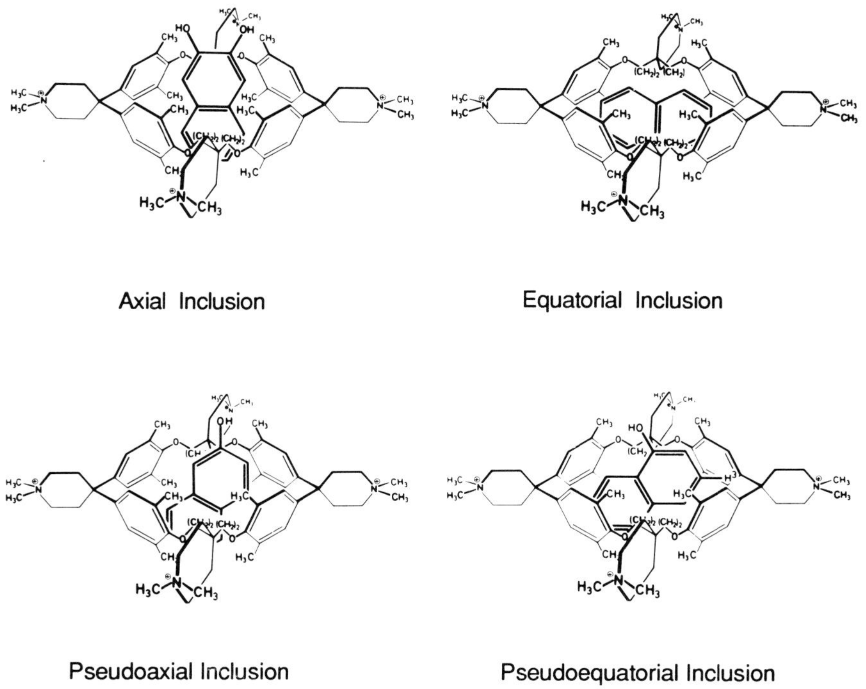

Axial Inclusion

Equatorial Inclusion

Pseudoaxial Inclusion

Pseudoequatorial Inclusion

Figure 2.14 *Different possible positions of naphthalene and substituted naphthalene derivatives in the cavity of **33**. The position of the spiro piperidinium rings fixed in the aliphatic chains of **33** is arbitrarily chosen. The inclusion of the following guests is shown: 2,3-naphthalenediol (axial), naphthalene (equatorial), 2,6-naphthalenediol (pseudoaxial), and 1,5-naphthalenediol (pseudoequatorial)*
(Reproduced with permission from reference 31. Copyright 1988 VCH.)

the guest protons that are located in the cavity and point towards the central sp^3-carbon atoms of the diphenylmethane units.[63] In complexes of benzene or naphthalene derivatives with bis(diphenylmethane) hosts like **32**, **33**, or **40**, however, the guest protons located in the cavity show much larger upfield shifts at saturation binding, usually between $\Delta\delta \approx 2$ and ≈ 3 ppm. This could be taken as experimental evidence for a preference of inclusion geometries A and B (Figure 2.16) in which the guest protons in the cavity are pointing directly into the shielding regions of the diphenylmethane benzene rings. However, a dynamic equilibrium between three energetically similar complex conformations A–C cannot be excluded by ^{1}H NMR spectroscopy.

Figure 2.17 shows the X-ray crystal structures of naphthalene complexed to two differently sized tetraprotonated tetraaza[n.1.n.1] paracyclophanes.[85] In the cavity of the smaller host **19**, naphthalene takes a pseudoaxial location

[85] K. Mori, K. Odashima, A. Itai, Y. Iitaka, and K. Koga, *Heterocycles*, 1984, **21**, 388.

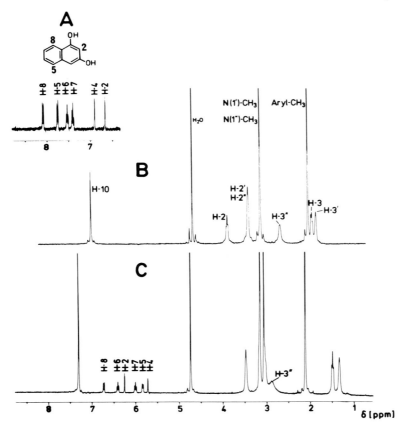

Figure 2.15 *360 MHz ^{1}H NMR spectra of (A) a 0.004 M solution of 1,3-naphthalenediol, (B) a 0.004 M solution of 33, and (C) a 0.004 solution of host and guest (T = 303K). See formula 33 for the numbering of the cyclophane protons in spectrum (B)*

whereas, in the cavity of the [7.1.7.1]paracyclophane **44**, the pseudoequatorial-type orientation is preferred. Naphthalene adopts a different orientation to fill the larger cavity.

Compounds **20**, **45**, and **46** are other tetraazaparacyclophane derivatives for which the inclusion complexation of neutral naphthalene derivatives was observed in aqueous solution.[35a,36] A variety of macrocyclic lactams, *e.g.* **47**, have also been shown by ^{1}H NMR to form inclusion complexes with naphthalenediols in aqueous solution;[83,86] quantitative data on the stability of these complexes have not been reported.

[86] A. Wallon, U. Werner, W. M. Müller, M. Nieger, and F. Vögtle, *Chem. Ber.*, 1990, **123**, 859.

Table 2.3 *Association constants (K_a) and free energies of complexation ($-\Delta G°$) for the 1:1 complexes formed between cyclophanes and neutral naphthalene derivatives in aqueous solutions*

Host	Guest	K_a [L^{-1} mol]	$-\Delta G°$ [kcal mol^{-1}]
33[a]	1,5-Dimethylnaphthalene	3.3×10^4	6.0
	2,6-Dimethylnaphthalene	2.6×10^4	5.9
	2,7-Naphthalenediol	1.9×10^4	5.7
	2,6-Naphthalenediol	1.0×10^4	5.4
	1,3-Naphthalenediol	9.8×10^3	5.4
	1,5-Bis(dimethylamino)naphthalene[b]	9.7×10^3	5.4
	1-(Dimethylamino)naphthalene	9.3×10^3	5.3
19[c,d]	2,7-Naphthalenediol	2.8×10^3	4.7
45[c,d]	2,7-Naphthalenediol	2.6×10^2	3.3
46[c,d]	2,7-Naphthalenediol	4.3×10^3	5.0
20[e]	1-Naphthol	1.4×10^3	4.3
	1-Methylnaphthalene	1.4×10^3	4.3
23[c,f]	1,3-Naphthalenediol	3.3×10^2	3.4

[a] water, $T = 293$–295 K, ref. 31 and 63; [b] 0.015 M K_2CO_3; [c] KCl-HCl, pD 1.9 $T = 298$ K; [d] Ref. 81; [e] D_2O/CD_3OD (80:20, v/v), $T = 298$ K; ref. 36; [f] ref. 38.

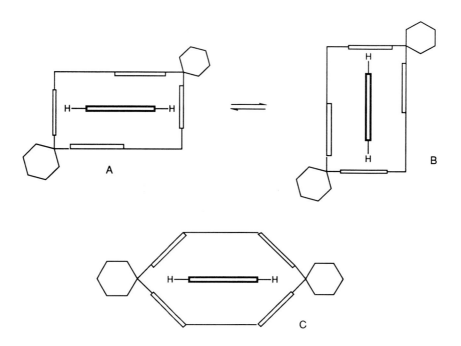

Figure 2.16 *Two preferred orientations (A and B) of naphthalene and benzene substrates in the cavity of bis(diphenylmethane) hosts. The schematic drawing C shows the time-averaged complex geometry reflected by the ^1H NMR spectra at fast exchange between A and B*

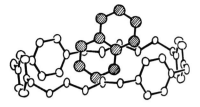

(a) Pseudoaxial Inclusion

(b) Pseudoequatorial Inclusion

Figure 2.17 *Crystal structures of the complexes formed between naphthalene and cyclophanes **19** (a) and **44** (b)*
(Reproduced with permission from reference 85. Copyright 1984 Heterocycles.)

2.7 Complexes of Benzene Derivatives

2.7.1 Complexation of Flat Benzene Derivatives

The complexation of benzene derivatives in water has attracted special interest since many biologically active molecules possess phenyl rings, *e.g.* the amino acids phenylalanine and tyrosine, the neurotransmitters dopamine and adrenaline, and a great variety of pharmaceuticals. Benzene derivatives possess much smaller apolar surfaces than the larger arenes, and therefore strong complexation of neutral benzene derivatives is more difficult to accomplish. The major driving forces for apolar complexation, the desolvation of the complementary host–guest surfaces, and the formation of attractive van der Waals contacts between the binding partners (Chapter 7), are reduced compared to the binding of larger arenes.

The most comprehensive analysis of benzene complexation in water has been reported for the two octamethoxycyclophanes **39** and **40**.[51] The *X*-ray crystal structure of a dihydrate of **39** (Figure 2.18) shows a deep, organized

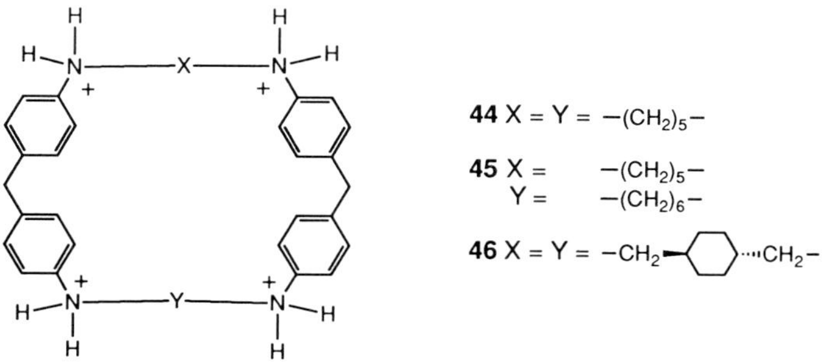

44 X = Y = —(CH$_2$)$_5$—

45 X = —(CH$_2$)$_5$—
    Y = —(CH$_2$)$_6$—

46 X = Y = —CH$_2$—⟨ ⟩···CH$_2$—

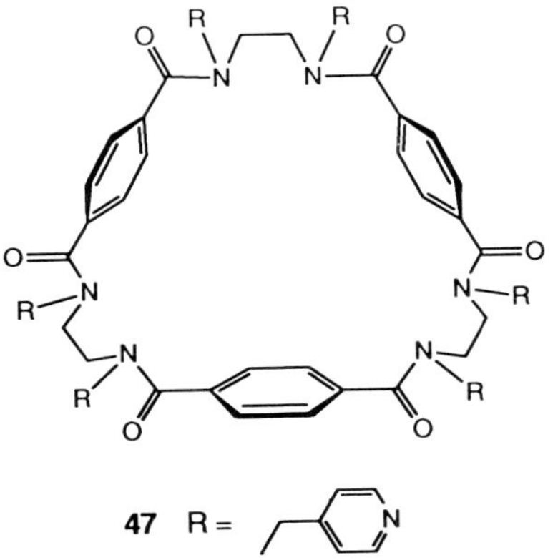

47 R =

cavity in which two highly disordered water molecules (not shown) are located. The eight methoxy groups, oriented in the planes of the benzene rings to which they are attached, provide considerable depth to the binding site. In cyclophane **27**, in which hydrogen atoms replace the eight methoxy groups, the distance between the two *meta*-hydrogens is ≈ 4.3 Å, while in **39** and **40**, the height of the cavity wall is almost doubled with a distance of ≈ 7 Å between the methyl carbon atoms of the two *meta*-methoxy groups.

Computer-assisted modeling plays an increasing role in the search for a microscopic-level understanding of molecular recognition processes.[87,88] A detailed MM2 force field computational study suggested that the *X*-ray geometry of cyclophane **39** with the two cavity-bound waters (Figure 2.18) reflects the structure of a complexed rather than of a free host.[89] The calculated low energy gas phase geometry of free **39** is shown in Figure 2.19. To obtain this structure, starting from the *X*-ray coordinates, the angles of

[87] W. L. Jorgensen, *Acc. Chem. Res.*, 1989, **22**, 184.
[88] R. J. Loncharich, E. Seward, S. B. Ferguson, F. K. Brown, F. Diederich, and K. N. Houk, *J. Org. Chem.*, 1988, **53**, 3479.
[89] D. B. Smithrud, E. M. Sanford, I. Chao, S. B. Ferguson, D. R. Carcanague, J. D. Evanseck, K. N. Houk, and F. Diederich, *Pure Appl. Chem.*, 1990, **62**, 2227.

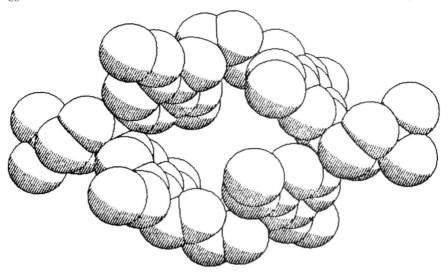

Figure 2.18 *X-ray crystal structure of* **39**. *Two highly disordered water molecules located in the cyclophane cavity are not shown (Reproduced with permission from reference 51. Copyright 1988 ACS.)*

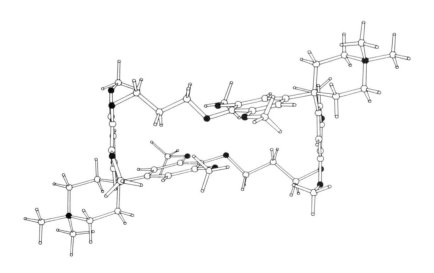

Figure 2.19 *Low-energy gas phase geometry of* **39**

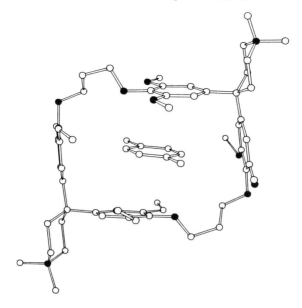

Figure 2.20 *Calculated low-energy geometry of the 39·p-xylene complex in aqueous solution*

both dioxaalkane bridges were simultaneously changed using the dihedral angle drive in the MM2 program. In the free host, the O–$(CH_2)_3$–O bridges adopt torsional angle sequences which provide a less open cavity than in the water solvate. However, the shapes of the two diphenylmethane units in the free and complexed host (Figure 2.20) are very similar and, as already discussed in Chapter 2.3, these spacers provide the necessary degree of preorganization to the cyclophane binding site.

The cyclophanes **39** and **40** form stable 1:1 complexes with neutral *para*-disubstituted benzene derivatives in aqueous solution (Table 2.4).[51] All complexation processes are enthalpically driven. The origin of this enthalpic driving force for tight complexation in water is analyzed in detail in Chapter 7.

Figure 2.20 shows a low energy conformation of the *p*-xylene complex of **39** in the aqueous phase. To obtain this structure, low energy gas-phase conformations of the complex located in a Monte-Carlo search of conformational space were subjected to a surface-area-based solvation treatment.[90] The AMBER force field[91] and the BATCHMIN Monte-Carlo program[92] in the MACROMODEL molecular modeling package[39] were applied to the gas phase conformational search. The calculated geometry of the *para*-xylene complex

[90] T. Ooi, M. Oobatake, G. Némethy, and H. A. Scheraga, *Proc. Natl. Acad. Sci. USA*, 1987, **84**, 3086.
[91] S. J. Weiner, P. A. Kollman, D. A. Case, U. C. Singh, C. Ghio, G. Alagona, S. Profeta, Jr., and P. Weiner, *J. Am. Chem. Soc.*, 1984, **106**, 765.
[92] G. Chang, W. C. Guida, and W. C. Still, *J. Am. Chem. Soc.*, 1989, **111**, 4379.

Table 2.4 *Association constants K_a and enthalpic ($\Delta H°$) and entropic ($T\Delta S°$) contributions to the free energies of complexation ($\Delta G°$) at 293.4 K for complexes of cyclophanes **39** and **40** with 1,4-disubstituted benzene guests in D_2O[a]*

Guest	K_a [L mol^{-1}]	$-\Delta G°$ [kcal mol^{-1}]	$T\Delta S°$ [kcal mol^{-1}]	$\Delta H°$ [kcal mol^{-1}]
		Complexes of **39**		
p-Nitrophenol	2.2×10^3	4.48	-5.6 ± 1.5	-10.1 ± 1.5
p-Nitrotoluene	2.1×10^3	4.47	-4.0 ± 2.5	-8.5 ± 2.5
Dimethyl p-benzene dicarboxylate	2.1×10^3	4.45	-3.5 ± 1.0	-8.0 ± 1.0
p-Xylene	1.3×10^3	4.18	-2.2 ± 1.0	-6.4 ± 1.0
p-Dicyanobenzene	1.0×10^3	4.04	-3.3 ± 1.0	-7.3 ± 1.0
p-Dimethoxybenzene	3.7×10^2	3.45	-2.2 ± 1.5	-5.7 ± 1.5
		Complexes of **40**		
Dimethyl p-benzene dicarboxylate	1.2×10^5	6.81	-4.0 ± 1.0	-10.7 ± 1.0
p-Nitrotoluene	3.0×10^4	6.01	-3.6 ± 3.0	-9.6 ± 3.0
p-Tolunitrile	3.0×10^4	6.01	-3.8 ± 2.5	-9.8 ± 2.5
p-Nitrophenol	2.3×10^4	5.86	-5.8 ± 1.5	-11.7 ± 1.5
p-Dimethoxybenzene	1.0×10^4	5.38	-4.8 ± 2.5	-10.2 ± 2.5
p-Xylene	9.3×10^3	5.33	-2.1 ± 1.0	-7.4 ± 1.0
p-Dicyanobenzene	7.8×10^3	5.23	-4.3 ± 1.0	-9.5 ± 1.0
p-Dinitrobenzene	7.7×10^3	5.22	-4.3 ± 1.0	-9.5 ± 1.0
p-Cresol	3.2×10^3	4.71	-4.4 ± 1.5	-9.1 ± 1.5
p-Diaminobenzene	3.6×10^2	3.43	-3.7 ± 1.5	-7.1 ± 1.5

[a] Uncertainty in K_a-values: ± 10%.

resembles the X-ray structure of the benzene complex shown in Figure 2.7. Both complexes are stabilized by attractive π–π-stacking interactions between parallel aromatic surfaces and by edge-to-face interactions between perpendicularly aligned aromatic rings of host and guest. Specific complexation-induced shifts of both host and guest ^{1}H NMR resonances indicate that similar complex geometries are also favored in aqueous solution (see Chapter 2.6.2).

Table 2.4 shows that cyclophane **39** with the C_3-bridges forms less stable complexes than **40** having C_4-bridges although CPK molecular model examinations had led to the opposite prediction. According to the models, the C_3-bridged host should form tighter complexes allowing more favorable van der Waals contacts. Similar results were obtained in studies with cyclophanes **25**–**27**.[31,93] CPK model examinations suggested that stable complexes should be formed between benzene derivatives and host **25** with its $(CH_2)_2$-bridges and between axially enclosed naphthalene guests and host **26** with its $(CH_2)_3$-bridges. In both cases, no significant complexation was observed. Inclusion

[93] F. Diederich, K. Dick, and D. Griebel, *Chem. Ber.*, 1985, **118**, 3588.

only occurred with larger hosts; benzene derivatives form stable complexes with **26** (Table 2.6), and naphthalene derivatives take axial positions in the cavity of **27** with $(CH_2)_4$-bridges. Following these observations as well as those by Collet and coworkers[94] about discrepancies between CPK molecular model predictions and experimentally observed binding, it can be concluded that a cyclophane designed for inclusion of a neutral organic molecule should have a larger cavity than suggested by CPK model examinations. Several explanations are plausible: (i) The CPK-models do not reproduce space occupancy and free cavity space accurately; atoms in commercial models are dimensioned to approximately 80% of their actual van der Waals radii. (ii) In a very tight complex, the conformational degrees of freedom of host and guest can be severely reduced leading to a very unfavorable entropy term in the binding event. (iii) In a less tight complex, the cyclophane has the flexibility to adopt a conformation with optimum complementarity to the guest or, in other words, the guest can induce a favorable binding conformation in the host (induced fit). (iv) Unfavorable electrostatic interactions between host and guest atoms with partial charges of same sign are better avoided in a more flexible system.[95]

The above conclusion is opposite to the guidelines established from the studies of Cram *et al.* for the design of efficient cation binders.[33,96] With the spherands, they have designed the most efficient systems for the strong, very tight complexation of inorganic cations. CPK molecular models of the spherand complexes accurately predict the tight encapsulation of these ions.

Benzene derivatives have a similar orientation in the complexes of **39** and **26** (Figures 2.16 and 2.20). However, the conformations of the dioxaalkane bridges in the two cyclophanes differ considerably. Methyl (in **32**) or methoxy groups (in **39**) attached to the aromatic rings *ortho* to the bridges not only deepen significantly the cavity binding site; they also influence the torsional angles in the dioxalkane chains and possibly enforce a more preorganized binding site. Without these *ortho* substituents, the $-O-CH_2$ groups of the bridges are aligned in the planes of the aromatic rings to which they are attached (Figures 2.6 and 2.7). With methyl or methoxy groups, for steric reasons, the CH_2 group connected to the aryl ether oxygen either turns inwards into the cavity or outwards (Figures 2.18–2.20). Models of anisole and 2,6-dimethylanisole (Figure 2.21) illustrate this effect.

A detailed conformational analysis has also been undertaken by Wilcox *et al.* for cyclophane receptors shaped by Tröger's base derivatives.[38] When the chiral dibenzodiazocine structural unit[97] is included in a macrocyclic structure like **23**, its two aromatic rings, in contrast to diphenylmethane units, are not able to rotate in relation to the cavity defined by the macrocycle. Nevertheless, X-ray crystal structures and molecular modeling show that dibenzodiazocine

---

[94] J. Canceill, L. Lacombe, and A. Collet, *J. Am. Chem. Soc.*, 1986, **108**, 4230.
[95] C. A. Hunter and J. K. M. Sanders, *J. Am. Chem. Soc.*, 1990, **112**, 5525.
[96] D. J. Cram, T. Kaneda, R. C. Helgeson, S. B. Brown, C. B. Knobler, E. Maverick, and K. N. Trueblood, *J. Am. Chem. Soc.*, 1985, **107**, 3645.
[97] T. H. Webb and C. S. Wilcox, *J. Org. Chem.*, 1990, **55**, 363.

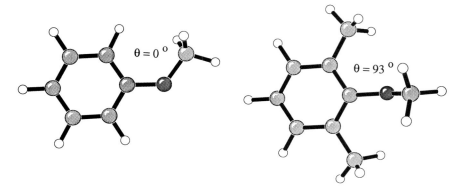

Figure 2.21 *Optimized geometries of anisole and 2,6-dimethylanisole illustrating the differences in the dihedral angles about the aryl ether linkages in hosts* **26** *and* **32**

units possess a certain degree of conformational flexibility.[98] The dihedral angle between the aromatic rings can be changed considerably through substitution as shown in Table 2.5.

Table 2.5 *Dihedral angle between the phenyl rings in Tröger's Base as a function of substituents* [a]/98

R_2	R_3	R_4	R_5	Θ deg
H	H	CH_3	H	92.9 (2) [b]
				97.4 (2) [b]
				102.15 (5) [c]
H	H	CH_2CH_2Br	H	92.73 (10)
CH_3	H	CH_3	H	104.01 (6)
H	CH_2OH	CH_3	H	89.71 (7)
H	CH_3	CH_3	H	88.6 (1)

[a] With one exception, all data were obtained for crystalline racemates; [b] Two types of molecules in unit cell; [c] (+) Enantiomer.

Table 2.6 *Association constants (K_a) and free energies of complexation ($-\Delta G°$) for the 1:1 complexes formed between cyclophanes **23**, **26**, **48** and neutral benzene derivatives in aqueous solutions*

Host	Guest	$K_a [L^{-1} mol]$	$-\Delta G°$ [kcal mol^{-1}]
26[a]	p-Dicyanobenzene	1.5×10^3	4.3
	p-Dinitrobenzene	1.3×10^3	4.2
	p-Nitrotoluene	6.2×10^2	3.7
	Ethyl anthranilate	2.0×10^2	3.1
	p-Dimethoxybenzene	80	2.5
	p-Xylene	< 10	< 1.3
23[b]	4-Cyanophenol	166	2.9
	4-Nitrophenylacetate	143	2.9
	2,4,6-Trimethylphenol	100	2.7
	4-Methoxyphenol	59	2.4
	4-Methylphenol	53	2.3
48[b]	2,4,6-Trimethylphenol	71	2.5
	4-Cyanophenol	66	2.5
	4-Methylphenol	44	2.2

[a] Ref. 31; D_2O, $T = 293$ K; [b] Ref. 38; KCl–HCl in D_2O, pD 1.9, $T = 298$ K.

The receptors **23** and **48** have been investigated for the binding of benzene derivatives in a KCl–DCl buffer at pD 1.9 giving the association constants shown in Table 2.6.[38] A computer modeling study using MACROMODEL[39] revealed the following conformational characteristics for **23**. In the lowest energy conformation, the macrocycle adopts a rectangular box shape (Figure 2.22). Each aliphatic chain adopts one *gauche* torsional angle in addition to having all *anti* angles. An interesting representation of the chain conformations, as shown in Figure 2.22, utilizes schematics introduced by Dale *et al.* for the description of conformers.[99,100] The analysis of ^{1}H NMR complexation-induced shifts indicates that benzene guests take an orientation in the cavities of **23** and **48** similar to that in bis(diphenylmethane) hosts (Figure 2.16).

The comparison of the binding properties of **23** and **48** shows an interesting phenomenon concerning the structural requirements for diphenylmethane units to act as efficient spacers. Host **23** is a consistently better receptor (Table 2.6). The two cyclophanes differ only by the presence of geminal methyl groups at the central diphenylmethane carbon atom in **48**. The diphenylmethane unit in **23** prefers the gable conformation (C_{2v}) over the C_2 or propeller geometry (Scheme 2.3).[101] In contrast, force-field analysis suggests that the 2,2-diphenylpropane unit in **48** prefers the C_2 or propeller geometry by ≈ 3 kcal mol^{-1}. The binding results obtained from hosts **23** and **48** suggest that the

---

[98] I. Sucholeiki, V. Lynch, L. Phan, and C. S. Wilcox, *J. Org. Chem.*, 1988, **53**, 98.
[99] J. Dale, *Top. Stereochem.*, 1976, **9**, 199.
[100] B. B. Masek, B. D. Santarsiero, and D. A. Dougherty, *J. Am. Chem. Soc.*, 1987, **109**, 4373.
[101] J. C. Barnes, J. D. Paton, J. R. Damewood, Jr., and K. Mislow, *J. Org. Chem.*, 1981, **46**, 4975.

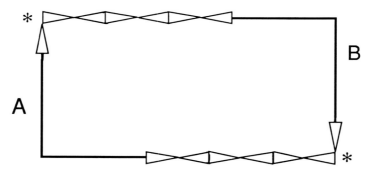

Figure 2.22 *Description of the torsional angles in the bridges of host* **23** *derived from Tröger's base.*[38] *The asterix indicates a gauche dihedral angle. A and B are the two diaryl spacer units*

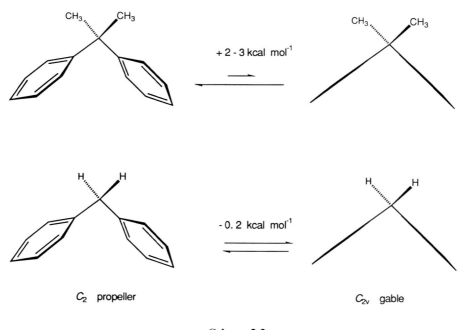

Scheme 2.3

gable conformation is the more favorable for complexation. The stronger binding by **23** compared to **48** can be explained by a better host preorganization. For host–guest complexation, the diphenylpropane unit of **48** needs to rearrange from a propeller to a gable conformation. Obviously, the energy for this rearrangement is paid for by the host–guest binding energy.

The complexation of benzene and flat neutral derivatives in aqueous

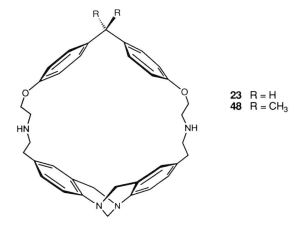

23 R = H
48 R = CH$_3$

solution has also been reported for other cyclophanes.[35b,41,102-104] In aqueous solutions below the CAC of 0.024 M, host **49** forms a 1:1 inclusion complex with *p*-nitrophenol ($K_a = 20$ L mol^{-1}) as well as a 1:1 complex with Cu(II) at the amine binding sites ($K_a = 5 \times 10^2$ L mol^{-1}).[104]

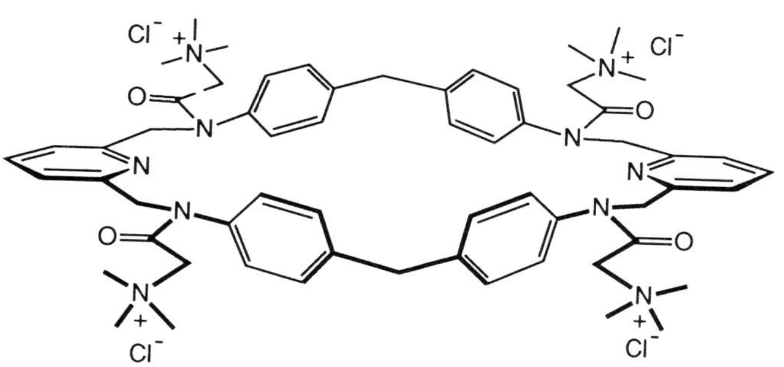

49

2.7.2 Complexes of [*m.n*]Paracyclophanes

Cyclophanes shaped by two diphenylmethane units possess ideal binding sites for the incorporation of flat aromatic rings. In **39**, the distance between the two oxygen atoms of one diphenylmethane unit, which defines the cavity width, is 8.41 Å.[51] At ≈ 10.6–11.2 Å, depending on the host conformers (MMX force

---

[102] F. Vögtle and W. M. Müller, *Angew. Chem.*, 1984, **96**, 711; *Angew. Chem. Int. Ed. Engl.*, 1984, **23**, 712.
[103] H.-J. Schneider, T. Blatter, S. Simova, and I. Theis, *J. Chem. Soc., Chem. Commun.*, 1989, 580.
[104] R. Fornasier, F. Reniero, P. Scrimin, and U. Tonellato, *J. Incl. Phenom.*, 1988, **6**, 175.

field),[105] the O···O distance at the naphthylphenylmethane units[106,107] in **50** is much wider, and this macrocycle provides space for larger, more spherical guests.[108] In D_2O/methanol-d_4 (1:1, v/v) at 293 K, cyclophane **50** forms inclusion complexes of defined geometry with [*m.n*]paracyclophanes. Methanol was added in order to reduce the aggregation potential of this large host. Table 2.7 shows the stabilities of the paracyclophane complexes. The cavity inclusion of the [*m.n*]paracyclophanes **51–57** is supported by strong upfield shifts of the aromatic guest resonances and characteristic up- and downfield shifts of the host resonances in the ^{1}H NMR spectra of solutions of their complexes. For example, the upfield shifts of the aromatic guest resonances at saturation binding of **51** are calculated as 0.86 and 0.63 ppm (1-H, 5-H respectively) and 1.82 and 2.03 ppm (2-H, 4-H respectively). In the spectrum of a solution of **50** (1 mM) and **51** (4 mM), the signals of the bridges in the host are shifted upfield (+0.04 to 0.15 ppm), and those of the phenyl protons are shifted weakly downfield (−0.04 ppm). For the naphthyl resonances of the host, characteristic downfield complexation shifts are observed [−0.14 (1-H), −0.21 (3-H), −0.18 (4-H), −0.13 (5-H), −0.14 (7-H), −0.15 (8-H)]. These shifts indicate a preference for the inclusion geometry shown in Scheme 2.4 in which the four benzene rings of host and guest **53** are aligned in a fourfold π–π stacking array.

Apolar interactions and desolvation, rather than ion pairing, provide the major driving force for efficient [*m.n*]paracyclophane complexation by **50**. This is demonstrated by the favorable binding of the neutral guest **52**. The stability of the complexes formed in the solvent mixture D_2O/CD_3OD (1:1, v/v)

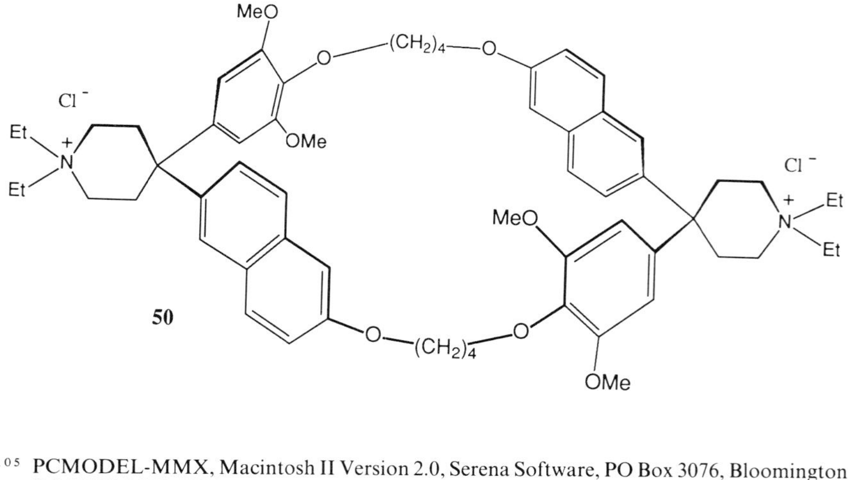

---

[105] PCMODEL-MMX, Macintosh II Version 2.0, Serena Software, PO Box 3076, Bloomington, IN 47402-3076, USA.
[106] H. Kawakami, O. Yoshino, K. Odashima, and K. Koga, *Chem. Pharm. Bull.*, 1985, **33**, 5610.
[107] K. Odashima, H. Kawakami, A. Miwa, I. Sasaki, and K. Koga, *Chem. Pharm. Bull.*, 1989, **37**, 257.
[108] D. R. Carcanague and F. Diederich, *Angew. Chem.*, 1990, **102**, 836; *Angew. Chem. Int. Ed. Engl.*, 1990, **29**, 769.

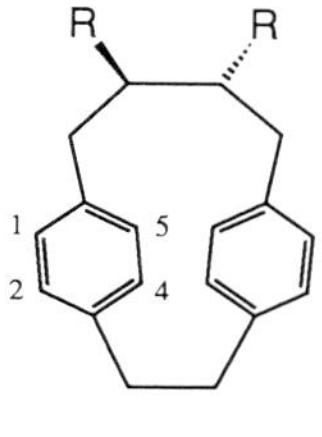

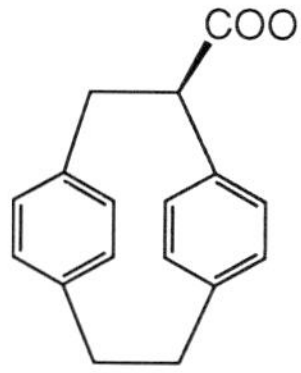

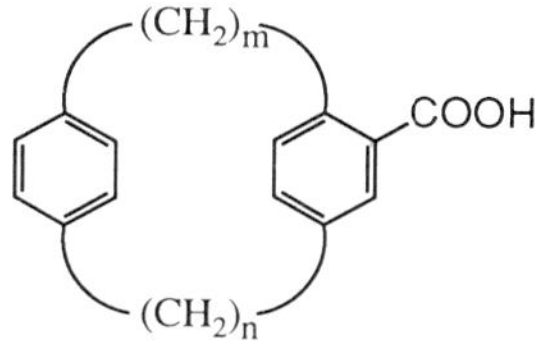

| 51 | R = COOH |
| 52 | R = OH |

53

54	m = n = 2
55	m = n = 3
56	m = 3, n = 4
57	m = n = 4

Table 2.7 *Association constants K_a and free energies of complexation $-\Delta G°$ determined by ^1H NMR for complexes formed between the [m.n]paracyclophanes **51–57** and host **50** in D_2O/CD_3OD (1:1, v/v) containing 0.01 M Na_2CO_3, T= 293 K*

[m.n]Paracyclophane	K_a [L mol^{-1}]	$-\Delta G°$ [kcal mol^{-1}]
51	840	3.92
52[a]	460	3.57
53	530	3.64
54	200	3.07
55	375	3.43
56	440	3.54
57	425	3.54

[a] No Na_2CO_3 added.

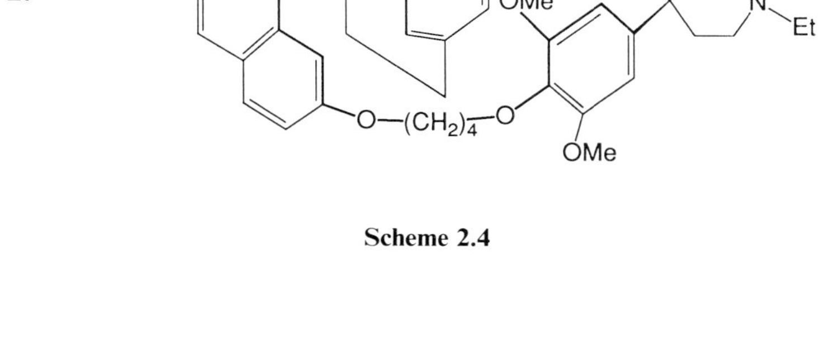

Scheme 2.4

is quite high; association constants for complexation in pure water should be about two orders of magnitude higher than those shown in Table 2.7 for the binary mixture (Chapter 3.6).

In the series of paracyclophanes bearing carboxylates to increase water-solubility, the number and attachment site of these groups are crucial for strong binding. The complexes of **51** and **53** having one or two carboxyl residues attached to the paracyclophane bridges are the most stable. Inclusion also occurs with guests having a single carboxyl residue attached to one of the aromatic rings. In contrast, a [4.4]paracyclophanedicarboxylic acid bearing one carboxyl function at each benzene ring no longer forms a complex with **50**. The two carboxylates of this guest would be deeply positioned into the apolar cavity upon inclusion. Apparently, the costs for desolvation of two carboxyl residues are too large for such a process to take place.

The incorporation of [*m.n*]paracyclophanes into the cavity of **50** shows that the dimensions of its binding site are sufficient enough to incorporate two π-systems in a sandwich-type orientation. Therefore, cyclophane **50** can be used to study the supramolecular catalysis of reactions between two included guest molecules, *e.g.* the Diels–Alder reaction (Chapter 8).[109,110]

2.8 Complexes of Heteroaromatic Substrates

The complexation of neutral aromatic heterocycles in water has not been studied in great detail despite the abundance of these substrates in biological systems. Dougherty *et al.* measured the binding affinities of cyclophane **36** and other related macrorings for indole, quinoline, and isoquinoline derivatives in deuterated borate buffers at pD $\approx$ 9.[42] The association constants of the 1:1 complexes are shown in Table 2.8. The quinoline and isoquinoline derivatives form complexes of similar stability to naphthalene and its functionalized derivatives (Table 2.3). In these complexes, the strongly solvated nitrogen atoms presumably are oriented into the aqueous solution and, therefore, inclusion of the heterocycles is not accompanied by unfavorable desolvation processes. Whitlock and Miller prepared the naphthalenophane **58** which forms an intercalation complex with pyridine in water.[111] Diederich and Sanford investigated the binding between **40** and adenosine.[112] ^{1}H NMR titrations indicated that the structured 1:1 complex **59** was formed (Scheme 2.5) in which, expectedly, the purine base is incorporated deeply into the cavity. Only the moiety around C-1' of the ribose is included within the cavity, while the rest of the sugar ring with its polar functionalities is oriented into the aqueous phase. Scheme 2.5 shows the upfield complexation shifts in the ^{1}H NMR spectrum at saturation binding that support this proposed inclusion structure.

---

[109] D. Rideout and R. Breslow, *J. Am. Chem. Soc.*, 1980, **102**, 7816.
[110] N. K. Sangwan and H.-J. Schneider, *J. Chem. Soc., Perkin Trans. 2*, 1989, 1223.
[111] S. P. Miller and H. W. Whitlock, Jr. *J. Am. Chem. Soc.*, 1984, **106**, 1492.
[112] E. M. Sanford and F. Diederich, unpublished results.

Table 2.8 *Association constants (K_a) and free energies of complexation ($-\Delta G°$) for the 1:1 complexes formed between cyclophanes **36**, **40**, **58** and aromatic heterocycles in aqueous solutions*

Host	Guest	$K_a [L^{-1} mol]$	$-\Delta G°$ [kcal mol^{-1}]
36[a]	Indole	1.3×10^3	4.2
	N-Methylindole	2.2×10^3	4.5
	Quinoline	1.0×10^4	5.4
	2-Methylquinoline	1.2×10^4	5.5
	4-Methylquinoline	3.9×10^4	6.2
	Isoquinoline	4.7×10^4	6.3
	1-Methylisoquinoline	5.5×10^4	6.4
58[b]	Pyridine	10	1.3
40[c]	Adenosine	1.6×10^2	3.0

[a] Ref. 42; deuterated borate buffer, pD ≈ 9, $T = 295$ K; [b] Ref. 111; D_2O, $T = 296$ K; [c] D_2O, $T = 293$ K.

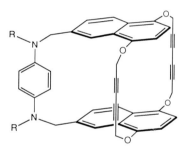

58 R = COCH$_2$CH$_2$COO$^-$ K$^+$

Scheme 2.5

2.9 Complexes of Aliphatic Substrates

Only a few reports describe the inclusion complexation of neutral acyclic molecules. The formation of structured complexes by linear alkanes and other highly flexible aliphatic compounds is challenging. Their incorporation into a tight cavity binding site will be accompanied by a much larger loss in internal rotational entropy compared to the complexation of a rigid aromatic ring.

Collet *et al.* observed the incorporation of dichloromethane and chloroform into the spherical cavity of cryptophane **38** in D_2O at 300 K.[48] The driving force for stoichiometric CH_2Cl_2 inclusion was measured as $\Delta G° = -5.1 \pm 0.5$ kcal mol^{-1}, and for $CHCl_3$ inclusion as $\Delta G° = -5.3 \pm 0.5$ kcal mol^{-1}. These complexes are approximately 1.4–2.0 kcal mol^{-1} more stable than those formed by the same guests and a similar cryptophane (compound **13** in Chapter 3.4) in organic solvents. This reflects the stronger driving forces for neutral molecular complexation in aqueous solutions compared to organic solvents which will be discussed further in Chapter 7.[113]

Busch *et al.* investigated the inclusion of 1-butanol into the cavity of the Co(II) complex **60**.[114-116] In these studies, they measured the effect that paramagnetic metal ion and dioxygen coordination to the host have on the longitudinal relaxation time T_1 of the guest protons. This technique permits an estimate on the distances between the paramagnetic centers and the individual protons of the guest molecules and, therefore, provides insight into the spatial relationship between host and guest. Alternatively, the complexation in the voids of Ni(II) complexes which are structurally related to **60** was studied by ^{13}C NMR titrations.[117] Both techniques provided evidence for the

60

---

[113] S. B. Ferguson, E. M. Seward, E. M. Sanford, M. Hester, M. Uyeki, and F. Diederich, *Pure Appl. Chem.*, 1989, **61**, 1523.
[114] T. J. Meade, W.-L. Kwik, N. Herron, N. W. Alcock, and D. H. Busch, *J. Am. Chem. Soc.*, 1986, **108**, 1954.
[115] T. J. Meade, K. J. Takeuchi, and D. H. Busch, *J. Am. Chem. Soc.*, 1987, **109**, 725.
[116] N. W. Alcock, W.-K. Lin, C. Cairns, G. A. Pike, and D. H. Busch, *J. Am. Chem. Soc.*, 1989, **111**, 6630.
[117] K. J. Takeuchi and D. H. Busch, *J. Am. Chem. Soc.*, 1983, **105**, 6812.

interaction between the cavities of the transition metal complexes with 1-butanol in D_2O. In addition, the formation of interesting ternary complexes, models for cytochrome P-450 enzymes, was observed when the organic guest is incorporated and when dioxygen binds reversibly to the cobalt ion (as $Co(III)O_2^-$).[115]

Although the inclusion complexation of cyclohexane, decaline, and adamantane derivatives, was demonstrated by ^1H NMR for several cyclophanes, the amount of quantitative binding data, especially for neutral cycloalkane derivatives, is still remarkably small. Hosts **20**[103] and **33**[118] bind flat cycloalkanes, but the more spherical adamantane derivatives cannot be fully incorporated into the cavities of bis(diphenylmethane) hosts and presumably only form nesting complexes (Scheme 2.2). A full inclusion of these guests is possible in the wider cavities of **36**[42] and **50**.[108] Table 2.9 shows thermodynamic data for cycloalkane complexes in which apolar surface association represents the major driving force for their formation. Other complexes, in which electrostatic interactions between charged functional groups of the host and guest play a major role, are discussed in Chapter 4.

Table 2.9 *Association constants (K_a) and free energies of complexation ($-\Delta G°$) for the 1:1 complexes formed between cyclophanes and cycloalkanes in aqueous solutions*

Host	Guest	$K_a [L^{-1} mol]$	$-\Delta G°$ [kcal mol^{-1}]
20	*trans*-1,4-Methylcyclohexanol[a]	4	0.8
	trans-Decaline[b]	≈15	≈1.6
33[c]	Cyclohexylacetate	7.4×10^2	3.8
	trans-1,4-Cyclohexanedimethanol	5.0×10^2	3.6
	1-Adamantanol	1.6×10^2	2.9
50[d]	1-Adamantaneacetic acid	1.2×10^2	2.8
	Camphor	1.5×10^2	2.9
63[e]	1-Adamantaneacetic acid	6.0×10^2	3.8

[a] Ref. 103; D_2O/CD_3OD (80:20, v/v), $T=298$ K; [b] Ref. 103; D_2O, $T=298$ K;
[c] Ref. 118; D_2O, $T=293$ K; [d] Ref. 108; D_2O/CD_3OD (50:50, v/v), $T=293$ K;
[e] Ref. 107; D_2O, $T=300$ K.

In an early study of cycloalkane complexation, Vögtle *et al.* obtained an interesting example of high complexation selectivity among isomeric host structures.[119,120] With its two triphenylmethane methyl groups turned outwards, host **61** possesses a spacious cavity and forms inclusion complexes with adamantane and neutral derivatives in acidic aqueous solution. In isomer **62** however, one of the two triphenylmethane methyl groups is turned into the

---

[118] F. Diederich and K. Dick, *Chem. Ber.*, 1985, **118**, 3817.
[119] J. Franke and F. Vögtle, *Angew. Chem.*, 1985, **97**, 224; *Angew. Chem. Int. Ed. Engl.*, 1985, **24**, 219.
[120] H. Schrage, J. Franke, F. Vögtle, and E. Steckhan, *Angew. Chem.*, 1986, **98**, 335; *Angew. Chem. Int. Ed. Engl.*, 1986, **25**, 336.

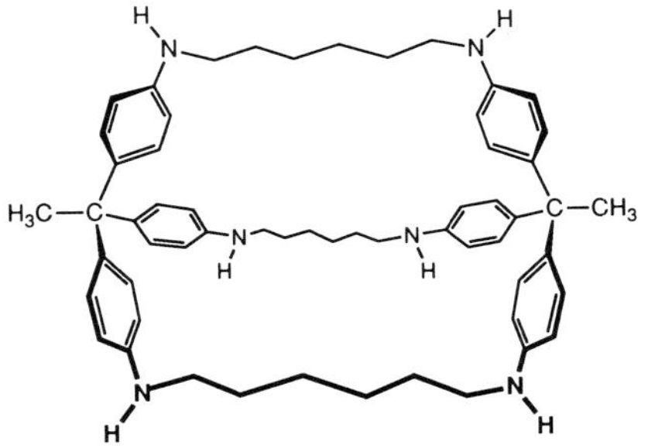

61 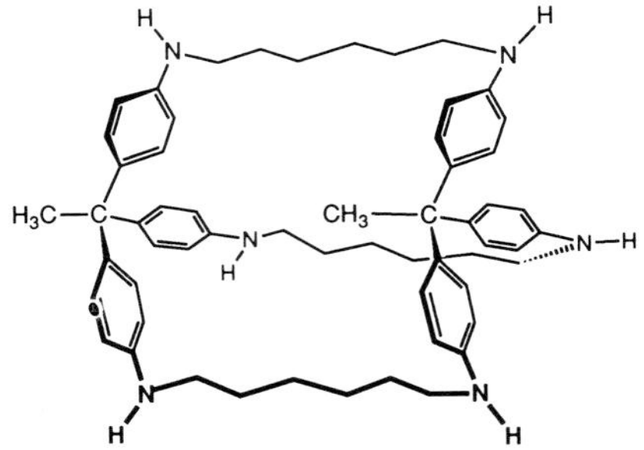 **62**

cavity, which drastically reduces the space available for a guest. Hence, **62** does not complex molecules as large as adamantane.

The bis(diphenylmethane) derivative **20** undergoes complexation with aromatic steroids, *e.g.* estradiol.[121] These complexes resemble those formed by tetralin derivatives, and the flat aromatic steroid ring A is preferentially incorporated into the binding site. On the other hand, the cyclophanes **50** and **63**,[106,107] shaped by larger naphthylphenylmethane units, form stable inclusion complexes with aliphatic steroids. Table 2.10 shows the results of ^{1}H NMR binding titrations executed with **50** and a variety of steroids in D$_2$O/CD$_3$OD (1:1, v/v) at 293 K.[108] Monitoring the isolated signals of the steroid methyl groups was especially easy in these titrations which revealed that **50**

Table 2.10 *Association constants K_a and free energies of formation $-\Delta G°$ for the steroid complexes formed by cyclophane* **50** *and upfield complexation shifts $\Delta\delta$ calculated at saturation binding for the ^{1}H NMR resonances of the steroid methyl groups; D$_2$O/CD$_3$OD (1:1, v/v), T = 293 K*

Steroid	$K_a [L\ mol^{-1}]$	$-\Delta G°[kcal\ mol^{-1}]$	$\Delta\delta$		
			$CH_3\ (19)$	$CH_3\ (18)$	$CH_3\ (21)$
64a[a]	145	2.92	0.73	0.56	0.25
64b[a]	250	3.21	0.76	0.66	0.39
64c[a]	810	3.91	0.47	1.40	0.89
64d[a]	1750	4.35	0.30	1.49	1.23
64e[a]	7075	5.18	0.55	1.49	0.90
65a	1095	4.08	1.44	0.22	
65b	1510	4.26	1.48	0.30	
65c	3545	4.76	1.48	0.43	

[a] Solutions contain 0.01 M Na$_2$CO$_3$.

[121] S. Kumar and H.-J. Schneider, *J. Chem. Soc., Perkin Trans. 2*, 1989, 245.

forms stable complexes with bile acids, corticoids, as well as androgenic steroids. According to CPK model examinations, the steroids are included axially in the cavity of **50** (Figure 2.23), and free axial rotation is maintained. In support of an axial inclusion, the signals of the steroid methyl groups show characteristic upfield shifts in the complexes. This geometry allows the polar, strongly solvated groups at C-3 in ring A and at ring D of the steroids to stay in the aqueous solution.

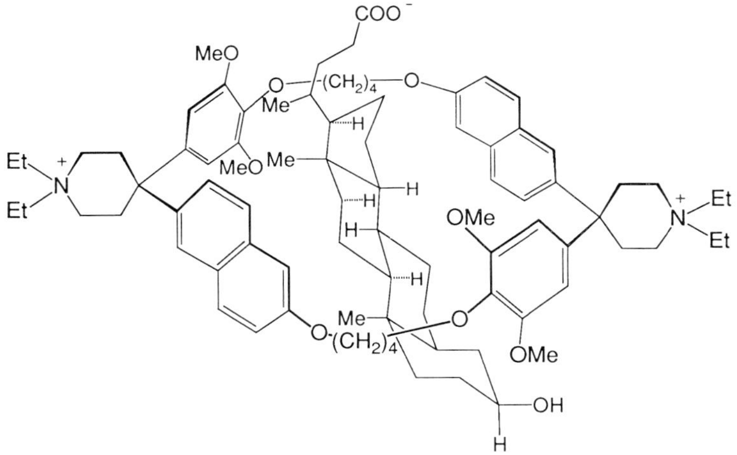

Figure 2.23 *Schematic drawing of the axial inclusion of lithocholic acid (**64e**)*

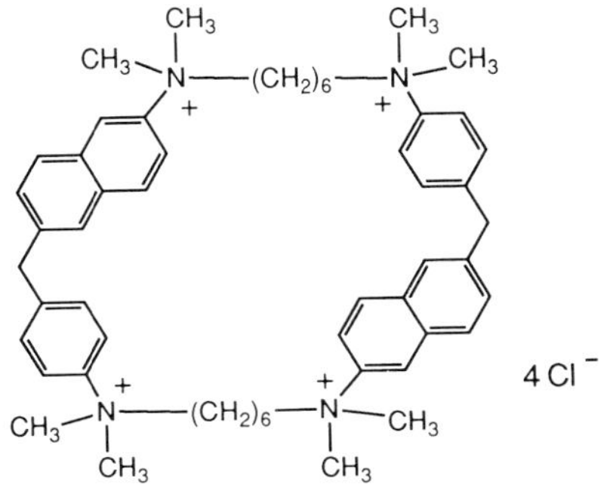

63

High binding selectivity is observed in the series of the structurally related bile acid derivatives **64a–e**. The complex formed by lithocholic acid ($K_a = 7075$ L mol^{-1}) is ≈ 2 kcal mol^{-1} more stable than the complex of deoxycholic acid ($K_a = 250$ L mol^{-1}) which has an additional hydroxy group at C-12α. In the complex of lithocholic acid (**64e**), the apolar surfaces of rings B, C, and D

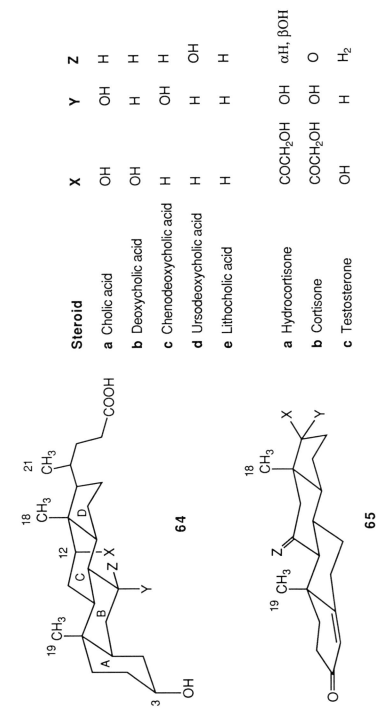

Steroid	X	Y	Z
a Cholic acid	OH	OH	H
b Deoxycholic acid	OH	H	H
c Chenodeoxycholic acid	H	OH	H
d Ursodeoxycholic acid	H	H	OH
e Lithocholic acid	H	H	H
a Hydrocortisone	COCH$_2$OH	OH	αH, βOH
b Cortisone	COCH$_2$OH	OH	O
c Testosterone	OH	H	H$_2$

64

65

interact favorably with the cavity walls of complementary polarity. In contrast, complexes of the other cholic acid derivatives **64a–d** are destabilized by unfavorable interactions between the strongly solvated hydroxy groups at their rings B and C and the apolar cavity walls. Similarly, in the series **65a–c**, testosterone (**65c**) forms a more stable complex than the two corticoids **65a,b** due to polar functional groups on rings B and C.

The presence of polar, highly solvated groups at the central rings B and C of the steroid skeleton not only affects the stability of the complexes but also their geometry. Inclusion occurs preferentially in a way which minimizes the energetically unfavorable desolvation of these functional groups in the complex. The characteristic upfield shifts of the steroid methyl resonances indicate that **64a** and **64b** are preferentially encapsulated with ring B (Table 2.10). The $CH_3(19)$ resonance shows the largest upfield complexation shift indicative of a deep incorporation of this methyl group into the shielding cavity. In contrast, derivatives **64c–e** are preferentially included with rings C and D, whereas the corticoids **65a,b** and testerone (**65c**) are all preferentially incorporated with their rings A and B. Note that the upfield complexation shifts of the $CH_3(19)$ and $CH_3(18)$ resonances are completely reversed in the complexes of **64d** and **65b** (Table 2.10).

A major objective in steroid recognition studies is the selective complexation and sequestration of cholesterol from aqueous solution. At sufficient binding strength, the dissolution of atherosclerotique plaque should even be possible. The data in Tables 2.9 and 2.10 demonstrates high selectivity of **50** for apolar steroids. Polar steroids as well as other large alicyclic compounds such as adamantane derivatives form considerably weaker complexes. This indicates that the selective complexation of cholesterol, the least polar steroid, should be feasible. So far, complexes of **50** with cholesterol are insoluble and, upon formation, precipitate out of solution. Structural modifications of **50** leading to enhanced water solubility and further increased binding strength and selectivity should provide the desirable cholesterol sequestrating agents.

CHAPTER 3
Apolar Complexation in Organic Solvents

3.1 Introduction

Efficient complexation of apolar solutes is not limited to aqueous solutions. This has been demonstrated by the increasing number of recent studies that are the subject of Chapter 3. Prior to 1980, only one article described an apolar inclusion event in organic solvents. In 1975, Siegel and Breslow reported that arenes like toluene and anisole or ferrocene are bound by cyclodextrins in dimethyl sulfoxide (Me_2SO) and in dimethylformamide (DMF).[1] In addition to describing the thermodynamics and the kinetics of complexation between various classes of guests and cyclophanes in organic solvents, Chapter 3 analyzes how the strength of apolar binding depends on solvent polarity. Also, host–guest interactions in binary aqueous solvent mixtures are analyzed. An increased understanding of how changes in the composition of binary solvent mixtures affect apolar binding should be quite useful for a variety of reasons. Biotic and abiotic complexation studies are often executed in aqueous solvent mixtures rather than in pure aqueous solutions for the following reasons. (i) The addition of organic solvent to the aqueous solution can enhance the solubility of substrate, receptor, or complex. (ii) In case of very strong association in water, binding strength might be too high and/or complexation–decomplexation rates too slow for an evaluation of the thermodynamic quantities in ^1H NMR or other spectral binding titrations. The addition of organic solvents lowers the apolar binding strength, fastens the exchange kinetics, and meaningful NMR titrations in adequate concentration ranges become possible (Chapter 2.6.1). (iii) Medium-induced conformational changes are well documented in peptide chemistry. The addition of 2,2,2-trifluoroethanol to aqueous solutions is known to specifically stabilize peptide α-helices.[2–4] Studies of cyclophane-substrate interactions in binary solvent

[1] B. Siegel and R. Breslow, *J. Am. Chem. Soc.*, 1975, **97**, 6869.
[2] A. M. Gronenborn, G. Bovermann, and G. M. Clore, *FEBS Lett.*, 1987, **215**, 88.
[3] Y. Theriault, Y. Boulanger, and J. K. Saunders, *Biopolymers*, 1988, **27**, 1897.
[4] M. Mutter and R. Hersperger, *Angew. Chem.*, 1990, **102**, 115; *Angew. Chem. Int. Ed. Engl.*, 1990, **29**, 185.

mixtures could help in clarifying the nature of such solvent-induced conformational changes.

The observation that stable cyclophane complexes of apolar solutes can form in solvents other than water may open new perspectives for supramolecular catalysis. In nonaqueous solvents, the rates of catalytic processes in supramolecular complexes and the selectivity with regard to reaction, substrate, and stereochemistry should be different than in water. Acid–base catalysis and the solvation of ground states and transition states should be considerably altered in these solvents as compared to water. It is well documented that enzymatic properties such as reaction selectivity and stereoselectivity in organic solvents differ significantly from those known to occur in aqueous solution.[5-7]

3.2 Complexation of Polycyclic Arenes

3.2.1 Structural Factors Determining the Complexation Strength

With the cyclophanes **1–3**, comprehensive studies of polycyclic arene binding in organic solvents were undertaken.[8-10] The geometry of the complexes formed by pyrene and other polycyclic arenes with hosts **1** and **2** are shown in Figures 2.11 and 2.12. Scheme 3.1 depicts the complex (**4**) formed between macrobicycle **3** and pyrene. A comparative study revealed a large difference in the binding ability of the two structurally very similar macrobicyclic receptors **2** and **3**. Compound **3** with its two tertiary cryptand nitrogen atoms is a much better binder in organic solvents than **2** having two amide groups in the periphery of the cavity. The association constants of the perylene complexes formed in methanol ($T = 303$ K) are listed in Table 3.1. The free energy of formation $\Delta G°$ of the **3**·perylene complex is -7.0 kcal mol^{-1}, whereas the complexation of perylene by **2** is less favorable by 3.2 kcal mol^{-1} and thus similar to the binding of this guest by the macromonocyclic compound **1**. This large difference in binding ability between the two macrobicycles **2** and **3**, observed in all solvents and with various arenes as guests, cannot be explained based on geometrical factors. Rather, the difference must be explained by a complexation-induced reduction in the solvation of the two amide groups in the periphery of the cavity of **2**. Generally, if the energetically favorable solvation of a polar functional group of one or both of the two binding partners is reduced upon complexation, and if no new binding interaction in the complex compensates for this loss in solvation energy, a considerably lower complexation strength will be observed. Another example for this

---

[5] A. L. Margolin, D.-F. Tai, and A. M. Klibanov, *J. Am. Chem. Soc.*, 1987, **109**, 7885.
[6] T. Sakurai, A. L. Margolin, A. J. Russell, and A. M. Klibanov, *J. Am. Chem. Soc.*, 1988, **110**, 7236.
[7] L. T. Kanerva and A. M. Klibanov, *J. Am. Chem. Soc.*, 1989, **111**, 6864.
[8] F. Diederich, K. Dick, and D. Griebel, *J. Am. Chem. Soc.*, 1986, **108**, 2273.
[9] D. Smithrud and F. Diederich, *J. Am. Chem. Soc.*, 1990, **112**, 339.
[10] F. Diederich, *Angew. Chem.*, 1988, **100**, 372; *Angew. Chem. Int. Ed. Engl.*, 1988, **27**, 362.

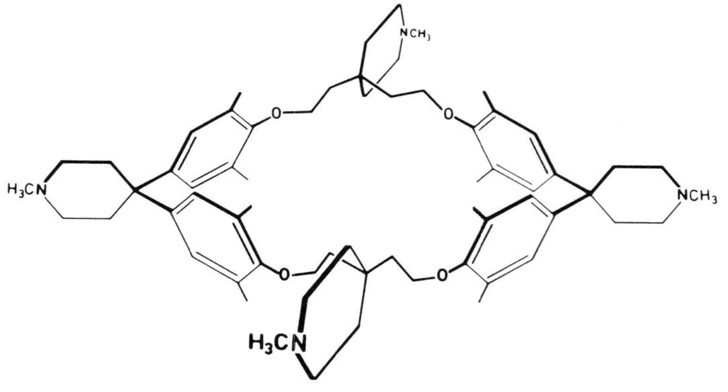

1

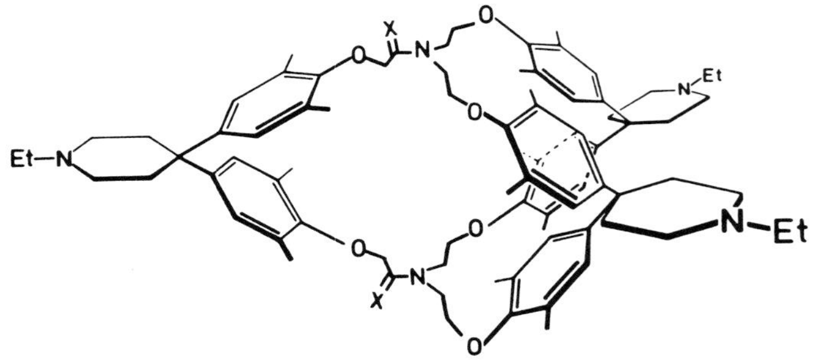

2   X = O
3   X = 2H

general principle was obtained in the complexation of various bile acids by an apolar cyclophane receptor (Chapter 2.9).[11]

Similar to aqueous solutions, the geometrical complementarity between host and guest is, as expected, also crucial in determining the strength of complexation in organic media. The stability sequence shown in Table 3.2 for complexes of **3** in methanol, also observed in all other solvents, exemplifies the importance of complementarity.

3.2.2 The Strength of Molecular Complexation of Arenes in Water and in Organic Solvents is Predictable by Linear Free Energy Relationships

To compare apolar binding strength in aqueous and organic solvents of all polarity (Table 3.3), the pyrene complex **4** was ideal for the following reasons:[9]

---

[11] D. R. Carcanague and F. Diederich, *Angew. Chem.*, 1990, **102**, 836; *Angew. Chem. Int. Ed. Engl.*, 1990, **29**, 769.

Apolar Complexation in Organic Solvents

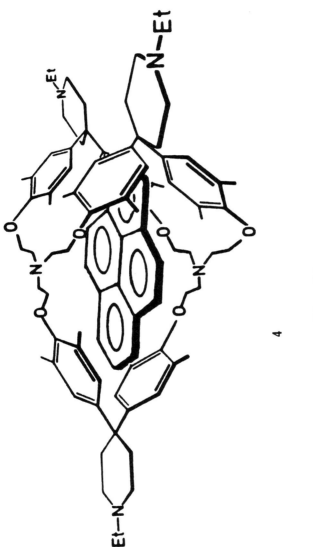

4

Scheme 3.1

Table 3.1 *Association constants (K_a) and free energies of complexation ($-\Delta G°$) for the 1:1 complexes formed between cyclophanes **1–3** and perylene in methanol ($T = 303$ K)*

Host	K_a [L^{-1} mol]	$-\Delta G°$ [kcal mol^{-1}]
1	8.4×10^2	4.0
2	5.6×10^2	3.8
3	1.1×10^5	7.0

Table 3.2 *Association constants (K_a) and free energies of complexation ($-\Delta G°$) for the 1:1 complexes formed between cyclophane **3** and arenes in methanol ($T = 303$ K)*

Guest	K_a [L^{-1} mol]	$-\Delta G°$ [kcal mol^{-1}]
Perylene	1.1×10^5	7.0
Fluoranthene	7.2×10^4	6.7
Pyrene	4.4×10^4	6.4
Naphthalene	1.2×10^2	2.9
Durene	27	2.0

(i) Host **3** and complex **4** are soluble in solvents covering the entire polarity range from water to apolar solvents. In some studies, dimethyl sulfoxide (1 or 10% v/v) was added as a cosolvent to increase the solubility of free pyrene.

(ii) According to extensive ^{1}H NMR studies, the pyrene complex **4** adopts in all solvents the geometry shown in Scheme 3.1.[8] For steric reasons, pyrene can only be incorporated in the cyclophane plane passing through the three spiro carbon atoms.

(iii) Even in apolar solvents, the stability of the pyrene complex is sufficient for a meaningful evaluation of complexation strength.

(iv) Because all solvent molecules in Table 3.3 are small enough to easily enter and exit the large, highly preorganized host cavity without causing major conformational strain, the host cavity is completely solvated when pyrene is not bound. This complete solvation of the host cavity is an important criterion for a meaningful study of solvent polarity effects on apolar complexation. A comparison would not be meaningful if differences in binding strength would result from the fact that one solvent molecule solvates the binding site whereas, another larger solvent molecule does not fit and therefore leaves a nonsolvated, empty cavity.

Table 3.3 shows the stability constants, K_a (L mol^{-1}), and free energies of formation, $-\Delta G°$ (kcal mol^{-1}) of complex **4** in eighteen solvents at $T = 303$ K as well as the empirical solvent polarity parameter $E_T(30)$ (kcal mol^{-1})[12]

---

[12] C. Reichardt, 'Solvents and Solvent Effects in Organic Chemistry', 2nd ed., VCH, Weinheim, 1988.

Apolar Complexation in Organic Solvents

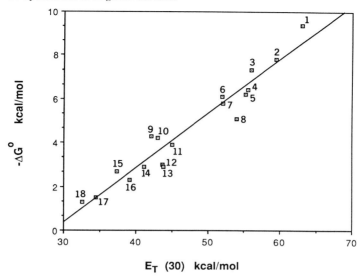

Figure 3.1 *Dependence of the free energy of formation $-\Delta G°$ (kcal mol^{-1}) of complex **4** (T = 303 K) on the solvent polarity as expressed by $E_T(30)$ values (kcal mol^{-1}). The numbers in the graph refer to the entries shown in Table 3.3*
(Reproduced with permission from reference 9. Copyright 1990 ACS.)

measured for these solvents. Figure 3.1 shows the linear free energy relationship between the free energies of formation of **4** and the solvent polarity parameters $E_T(30)$.

The following conclusions are drawn from the data in Figure 3.1 and Table 3.3:[9] Linear Gibbs free energy correlations using $E_T(30)$ values have been successfully applied for predicting solvent effects and spectral absorptions.[12] The $E_T(30)$ values reflect all the nonspecific intermolecular forces between solvent and solute molecules occurring in these processes. Macroscopic solvent properties such as dielectric constant, dipole moment, refractive index, cohesion, and polarizability have been found to be related by E_T parameters.[13,14] Figure 3.1 shows that this empirical solvent polarity parameter is also very useful for predicting the strength of apolar host–guest complexation in solvents of all polarities. The correlation coefficient for the linear free energy relationship shown is $R = 0.934$. This correlation allows predictions for the binding free energy of complex **4** in additional solvents according to the equation $-\Delta G° = 0.25\ E_T - 7.1$ (kcal mol^{-1}). Linear correlations of similar significance are also obtained when the stability of the perylene and fluoranthene complexes in various solvents is plotted against $E_T(30)$ values.

The impact of solvation effects on complexation strength is impressive.

---

[13] M. Chastrette and J. Carretto, *Tetrahedron*, 1982, **38**, 1615.
[14] V. Bekárek and J. Jurina, *Coll. Czech. Chem. Commun.*, 1982, **47**, 1060.

Table 3.3 Association constants K_a ($L\ mol^{-1}$) and free energies of formation $-\Delta G°$ ($kcal\ mol^{-1}$) of complex **4** in eighteen solvents of different polarity as expressed by E_T (30) values ($kcal\ mol^{-1}$), $T = 303K$

Run	Solvent	K_a ($L\ mol^{-1}$)	$-\Delta G°$ ($kcal\ mol^{-1}$)	$E_T(30)$ ($kcal\ mol^{-1}$)
1	Water/1% Me$_2$SO[a]	6.0×10^6	9.4	63.0
2	2,2,2-Trifluoroethanol/1% Me$_2$SO	4.2×10^5	7.8	59.4
3	Ethylene glycol/10% Me$_2$SO	1.8×10^5	7.3	55.9
4	Methanol	4.4×10^4	6.4	55.5
5	Formamide/10% Me$_2$SO	3.0×10^4	6.2	55.2
6	Ethanol	2.5×10^4	6.1	51.9
7	N-Methylacetamide/10% Me$_2$SO	1.5×10^4	5.8	52.1
8	N-Methylformamide/10% Me$_2$SO	4.8×10^3	5.1	54.0
9	Acetone	1.2×10^3	4.3	42.2
10	N,N-Dimethylacetamide/10% Me$_2$SO-d_6	1.1×10^3	4.2	43.0
11	Dimethylsulfoxide-d_6[b]	6.9×10^2	3.9	45.0
12	N,N-Dimethylformamide-d_7/10% Me$_2$SO-d_6	1.6×10^2	3.0	43.7
13	N,N-Dimethylformamide-d_7	1.5×10^2	2.9	43.8
14	Dichloromethane-d_2	1.2×10^2	2.9	41.4
15	Tetrahydrofuran-d_8	8.4×10^1	2.7	37.4
16	Chloroform-d_1	4.3×10^1	2.3	39.1
17	Benzene-d_6	1.2×10^1	1.5	34.5
18	Carbon disulfide	9×10^0	1.3	32.6

[a] The aqueous solution contains [Na$_2$CO$_3$] = 10^{-3} mol L^{-1} to prevent protonation of the pentamine host; [b] H/D solvent isotope effects are below the error in K_a.

Upon changing from the most polar solvent, water, to the least polar solvent considered in this study, carbon disulfide, complexation free energies decrease from $-\Delta G° = 9.4$ kcal mol^{-1} to $-\Delta G° = 1.3$ kcal mol^{-1}. The attractive host–guest interactions that stabilize complex **4** with its very tight geometric complementarity are London dispersion interactions and local dipole–induced dipole interactions. Since the geometry of complex **4** is very similar in all 18 solvents, a large difference in attractive host–guest interactions in the various environments cannot be at the origin of the observed changes in binding strength. Hence, the large difference in free binding energy of $\Delta(\Delta G°) = 8.1$ kcal mol^{-1} observed upon changing from water to carbon disulfide is predominantly due to solvation effects. These effects are subject to detailed analysis in Chapter 7.

The linear free energy relationship also holds for water. Binding strength decreases regularly from water to polar protic solvents to dipolar aprotic and to apolar solvents. Water does not promote apolar complexation beyond the level expected on the basis of its physical properties such as dielectric constant, polarizability, or dipole moment expressed in the empirical solvent polarity parameter.

Of great interest is the finding that some organic solvents approach water in their potential for promoting apolar complexation. Very strong pyrene complexation is observed in 2,2,2-trifluoroethanol ($-\Delta G° = 7.8$ kcal mol^{-1}; run 2) and in ethylene glycol ($-\Delta G° = 7.3$ kcal mol^{-1}; run 3). The complex stability in these solvents is higher than in methanol (run 4). The amide solvents demonstrate diverse properties. While the complexation strength in formamide (run 5) is comparable to those in methanol and ethanol (run 6), binding in the N-alkylated amides (runs 7,8,10,12) becomes increasingly unfavorable.

The association constants in Table 3.3 below $K_a \approx 5 \times 10^3$ L mol^{-1} were determined by ^{1}H NMR titrations.[8,9] All higher association constants were obtained in optical titrations. When host 3 was added to solutions of perylene, pyrene, or fluoranthene, changes in the optical spectra of these guests occurred. Upon addition of 3, the p-bands[15] of these guests show significant bathochromic shifts in addition to considerable reductions in their molar extinction coefficients. These complexation-induced spectral changes are utilized to evaluate the association strength in binding titrations. As an example, the p-bands of free pyrene appear in pure methanol at $\lambda_{max} = 334$, 318, and 304 nm. In a methanolic solution of complex 4, they are shifted to $\lambda_{max} = 341$, 325, and 311 nm. Interestingly, the shape and position of the α-bands of pyrene at longer wavelength (between $\lambda_{max} = 351$ and 372 nm in pure methanol) are almost unchanged in the complex. Figure 3.2 shows the electronic absorption and emission titrations used to determine the stability of complex 4 in ethanol. The K_a values obtained from both titrations (4.4×10^4 and 3.4×10^4 L mol^{-1}, respectively, at 303 K) are in excellent agreement. In the absorption titration, several isosbestic points indicate the exclusive formation of a 1:1 complex.

The strong binding of pyrene was utilized to inhibit the formation of an intramolecular excimer by 1,3-bis(1-pyrenyl)propane (5).[16] After excitation, compound 5 forms an intramolecular excimer (Chapter 1.3.2) in very dilute methanol solutions ([5] = 2.4×10^{-6} mol L^{-1}, $\lambda_{exc} = 340$ nm, $T = 292.5$ K). The excimer fluorescence at these conditions is shown in spectrum 1 in Figure 3.3. Upon addition of 3, the excimer emission decreases, and the monomer emission increases. With [3] $\approx 5 \times 10^{-4}$ mol L^{-1}, the excimer emission has completely disappeared, and only the monomeric fluorescence of pyrene is visible. After encapsulation of one pyrene moiety of 5 in the cavity of 3, the geometric proximity and orientation of the two pyrene moieties, required for the formation of an intramolecular excimer, can no longer be established.

3.3 Electron Donor–Acceptor Interactions Stabilize Inclusion Complexes of Aromatic Guests

The formation of intermolecular charge-transfer complexes between electron-rich (donor) and electron-deficient (acceptor) π-systems has been the subject

---
[15] E. Clar, 'Polycyclic Aromatic Hydrocarbons', Academic Press, London, 1964, Vols. 1 and 2.
[16] K. Zachariasse and W. Kühnle, *Z. Phys. Chem. NF*, 1976, **101**, 267.

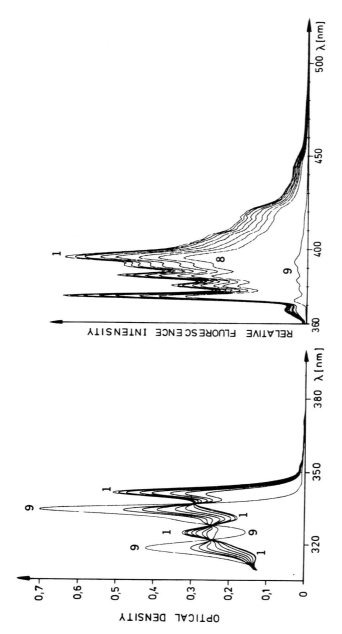

Figure 3.2 Electronic absorption (A) and emission (B) titrations to determine the association constant of complex **4** in ethanol (T = 303 K). The absorption titration was evaluated at $\lambda = 341$ nm, $d = 5$ cm; the emission titration was evaluated at $\lambda_{exc} = 341$ nm and $\lambda_{em} = 395$ nm. In both titrations, the total guest concentration $[G_0]$ is 2.99×10^{-6} mol L^{-1}, and the total host concentration $[H_0]$ is the following from spectrum 1 to spectrum 9: 3.00, 2.41, 1.80, 1.20, 0.90, 0.60, 0.45, 0.30, and 0×10^{-4} mol L^{-1} (Reproduced with permission from reference 8. Copyright 1986 ACS.)

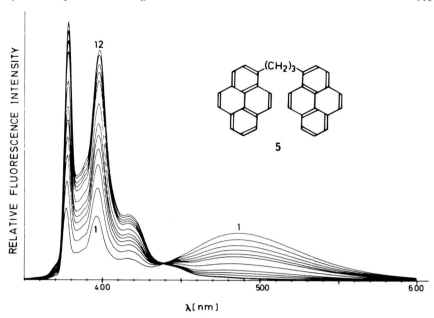

Figure 3.3 *Electronic emission spectra in methanol of 1,3-bis(1-pyrenyl)propane (5) in the absence (spectrum 1) and in the presence of host 3 (spectra 2–12). $[G_0]$ is 2.4×10^{-6} mol L^{-1}; $T = 292.5$ K; $\lambda_{exc} = 341$ nm. From spectrum 2 to spectrum 12, the concentration of 3 is the following: 1.47, 2.94, 4.41, 5.88, 7.35, 12.0, 24.0, 36.0, 60.0, 206.0, and 515.0×10^{-6} mol L^{-1} (Reproduced with permission from reference 8. Copyright 1986 ACS.)*

of intensive investigations since the 1950s.[17-19] At the same time, the importance of electron donor–acceptor (EDA) interactions for selective intermolecular association in biological systems was also recognized.[18,20] These forces are of key importance in chromatographic optical resolutions on chiral stationary phases, especially on Pirkle phases.[21] In this monograph, studies on the nature of EDA interactions,[22] their dependency on donor and acceptor strength as well as on distance and orientation of the interacting chromophores[23] were discussed in Chapters 1.3.3 and 1.5.1.

To explore the contributions of EDA interactions to the stability of inclusion complexes formed by cyclophanes, the binding between host 6 and a series of 2,6-disubstituted naphthalene derivatives, *e.g.* 7a–i, was studied in

[17] G. Briegleb, 'Elektronen-Donator–Acceptor-Komplexe', Springer, Berlin, 1961.
[18] R. Foster, 'Organic Charge-Transfer-Complexes,', Academic Press, London, 1969.
[19] M. W. Hanna and J. L. Lippert, in 'Molecular Complexes', ed. R. Foster, Elek Science, London, 1973, Vol.1, p. 1.
[20] B. Pullman and A. Pullman, 'Quantum Biochemistry', Wiley Interscience, New York, 1963.
[21] W. H. Pirkle and T. C. Pochapsky, *Chem. Rev.*, 1989, **89**, 347.
[22] K. Morokuma, *Acc. Chem. Res.*, 1977, **10**, 294.
[23] H. A. Staab, C. P. Herz, C. Krieger, and M. Rentea, *Chem. Ber.*, 1983, **116**, 3813.

methanol and Me_2SO (Table 3.4).[24] 2,6-Disubstituted naphthalene derivatives rather than benzene derivatives were chosen for this study since their substituents are located more outside of the cavity of **6**. Thus, contributions to observed binding differences from steric and polarization interactions of the various guest substituents with the host and from complexation-induced changes in the solvation of the polar groups (Chapter 3.2.1) are minimized.

In the cavity of **6**, the naphthalene derivatives adopt a pseudoaxial position, which allows for efficient $\pi-\pi$ host–guest interactions (Figures 2.14 and 2.16). ^{1}H NMR binding titrations revealed that the electron-deficient guests **7g–i** with two acceptor substituents form more stable complexes than the electron-rich guests **7a–c**, while the complexes of guests having one donor and one acceptor substituent (**7d–f**) demonstrate an intermediate stability. The differences in complex stability are as large as ≈ 1.5 kcal mol^{-1} and result from a major contribution of intermolecular EDA interactions. With its four trialkyl-substituted anisole units, compound **6** can be considered as a donor host. As expected for EDA interactions, the most stable complexes are formed with the acceptor guests **7g–i**. The extra stabilization results from electrostatic and polarization interactions; a charge-transfer band is not observed in the electronic absorption spectra of these complexes.

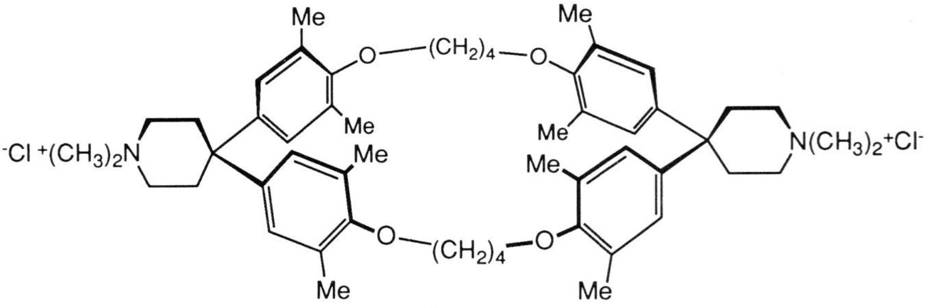

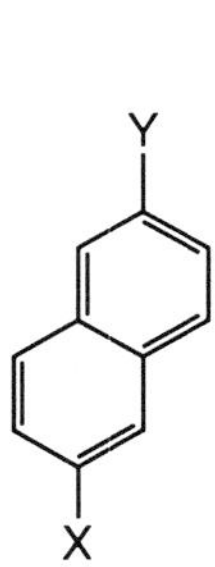

[24] S. B. Ferguson and F. Diederich, *Angew. Chem.*, 1986, **98**, 1127; *Angew. Chem. Int. Ed. Engl.*, 1986, **25**, 1127.

Table 3.4 Association constants (K_a) and free energies of complexation ($-\Delta G°$) for the 1:1 complexes formed between cyclophane **6** and the naphthalene derivatives **7a–i** in methanol-d_4 ($T = 303$ K)

Guest	X	Y	K_a [L^{-1} mol]	$-\Delta G°$ [kcal mol^{-1}]
(a) Donor–donor guests:				
7a	OH	OH	24	1.91
7b	NH$_2$	NH$_2$	33	2.11
7c	OCH$_3$	OCH$_3$	47	2.32
(b) Donor–acceptor guests:				
7d	NH$_2$	NO$_2$	102	2.78
7e	OCH$_3$	NO$_2$	109	2.82
7f	OCH$_3$	CN	117	2.87
(c) Acceptor–acceptor guests:				
7g	COOCH$_3$	COOCH$_3$	188	3.15
7h	NO$_2$	NO$_2$	213	3.23
7i	CN	CN	277	3.39

The conclusion that the differences in complexation observed in methanol-d_4 are a result of EDA interactions and not from special solvation effects of the polar protic solvent is clearly demonstrated by studies in Me$_2$SO-d_6. While overall binding is considerably weaker in this solvent than in methanol, the same trends in complexation strength as shown in Table 3.4 are observed. Acceptor–acceptor guests form the strongest complexes with **6** while donor–acceptor and donor–donor guests form progressively weaker complexes.

In aqueous solution, EDA interactions also represent an important force contributing to the stability of inclusion complexes. As an example, the data shown in Table 2.6 for cyclophane complexes of benzene derivatives can readily be explained by contributions of EDA interactions to the binding strength. However in water, it appears that EDA effects can well be masked by the energetics resulting from specific changes in solvation of the binding partners in the complexation process. For example, the differences in binding strength shown in Table 2.4 for cyclophane complexes of benzene derivatives cannot be explained by EDA interactions.

The difference in stability of the complexes formed by cyclophane **8** with quinoline and indole derivatives in aqueous borate buffer (Table 2.8) was attributed by Dougherty *et al.* to contributions from EDA interactions.[25] In its rhomboid conformation shown in Figure 3.4, host **8** is ideally suited to tightly bind naphthalene-sized guests and to promote π–π-stacking interactions between the binding partners. It was found that the electron-deficient quinolines and isoquinolines bind to **8** with much higher affinities [$\Delta(\Delta G°) > 1$ kcal mol^{-1}] than electron-rich indoles that are quite similar in size and shape (Table 2.8). The enhanced binding of the quinoline and isoquinoline guests is best explained by EDA interactions between the electron-rich dialkylanisole

[25] M. A. Petti, T. J. Shepodd, R. E. Barrans, Jr., and D. A. Dougherty, *J. Am. Chem. Soc.*, 1988, **110**, 6825.

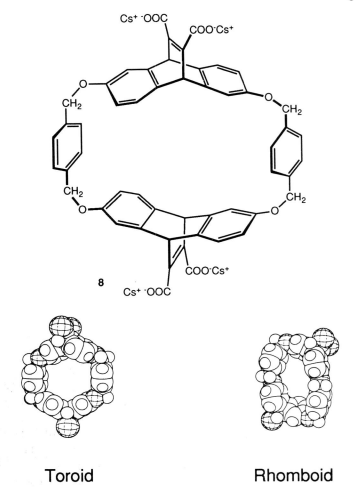

Figure 3.4 *Space-filling models of cyclophane 8 in the toroid and rhomboid conformations, generated by use of the program BIOGRAF (Biodesign, Inc., Pasadena, CA). Oxygens in the macrocycle are hatched and the carboxylates are truncated to hatched spheres for clarity (Reproduced with permission from reference 25. Copyright 1988 ACS.)*

rings of the ethenoanthracene units of **8** and the electron-deficient heterocycles in the inclusion complexes.

The electron-rich macrocycle **9** and its electron-deficient counterpart **10** have been prepared by Inazu *et al.* to form charge-transfer inclusion complexes.[26] However, inclusion complexation could not be observed, presumably as a result of the binding site being too narrow in these

---

[26] T. Shinmyozu, T. Sakai, E. Uno, and T. Inazu, *J. Org. Chem.*, 1985, **50**, 1959.

Apolar Complexation in Organic Solvents 119

[3.3.3.3]paracyclophanes (see discussion in Chapter 2.2). In contrast, cyclophane **11**, in which two paraquat units are bridged by two *p*-phenylene bridges forms intensively colored charge-transfer inclusion complexes with neutral electron-rich benzene derivatives in acetonitrile solutions as well as in the solid state.[27,28] Figure 3.5 shows the *X*-ray crystal structure of the intense-red complex formed by *p*-dimethoxybenzene. The inclusion geometry is ideal for promoting strong π–π-interactions between the two electron-deficient paraquat units and the electron-rich guest. Table 3.5 shows the maxima of the charge-transfer bands and the stabilities of the complexes between **11** and electron-rich benzene derivatives in acetonitrile. The inclusion geometry of these complexes in solution is strongly supported by characteristic complexation shifts in the ^{1}H NMR spectra. The resonances of the aromatic guest protons move sharply upfield upon complexation since they are pointing into

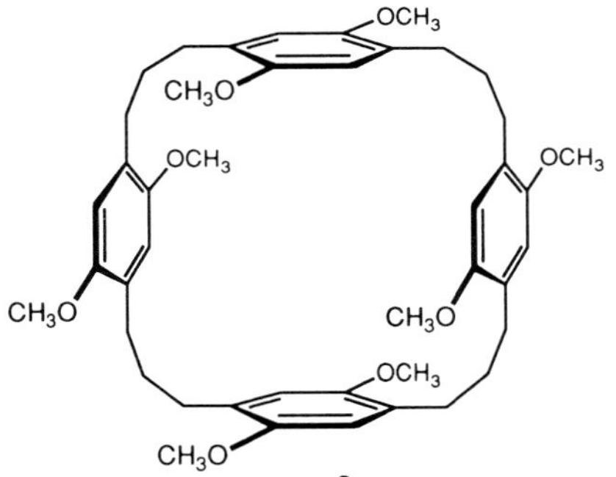

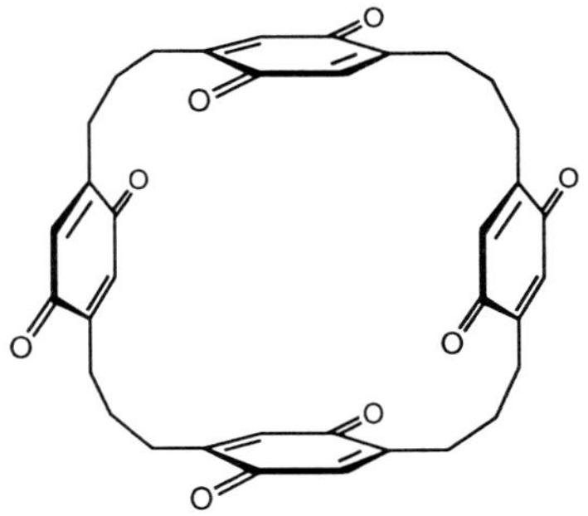

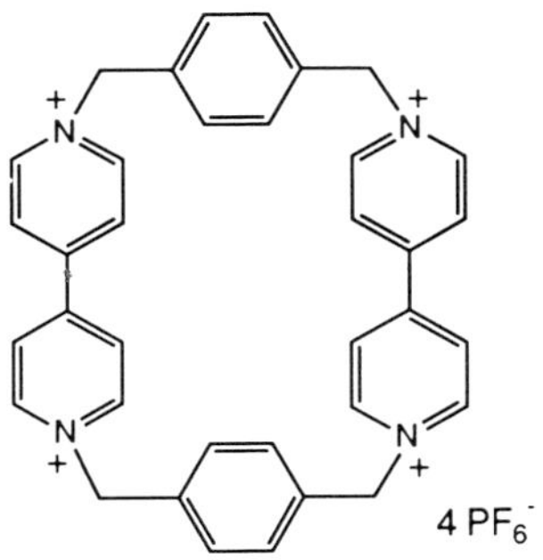

[27] B. Odell, M. V. Reddington, A. M. Z. Slawin, N. Spencer, J. F. Stoddart, and D. J. Williams, *Angew. Chem.*, 1988, **100**, 1605; *Angew. Chem. Int. Ed. Engl.*, 1988, **27**, 1547.
[28] P. R. Ashton, B. Odell, M. V. Reddington, A. M. Z. Slawin, J. F. Stoddart, and D. J. Williams, *Angew. Chem.*, 1988, **100**, 1608; *Angew. Chem. Int. Ed. Engl.*, 1988, **27**, 1550.

Table 3.5 *Charge-transfer absorptions, association constants (K_a), and free energies of complexation ($-\Delta G°$) for the 1:1 complexes formed between cyclophane **11** and dimethoxybenzenes in acetonitrile ($T = 298$ K)*

Guest	λ_{max}	ϵ_{max}	K_a [L^{-1} mol]	$-\Delta G°$ [kcal mol^{-1}]
p-dimethoxybenzene	478	631	17.0 ± 0.5	1.68
m-dimethoxybenzene	484	392	8.0 ± 0.3	1.23
o-dimethoxybenzene	462	1025	8.1 ± 0.2	1.24

the shielding regions of the p-phenylene bridges of the host. The central paraquat protons (*meta* to the nitrogens) of **11** also show upfield shifts, whereas the p-phenylene protons, deshielded by the perpendicularly oriented guest, move downfield.

3.4. The Cryptophanes: Shape-selective Inclusion Complexation of Methane Derivatives

With the cryptophanes, *e.g.* **12–14**, Collet *et al.* prepared a remarkable series of cyclophanes with spherical cavities shaped by two cyclotrivetratrylenyl units.[29] The *X*-ray crystal structures could be obtained for many cryptophane complexes and, hence, the host–guest interaction geometries are very well defined (Figure 2.8). Cryptophanes show excellent shape-selectivity in inclusion complexation. Compounds **12** and **13** have a marked preference for CH_2Cl_2 over smaller or bulkier guests, whereas **14** preferentially encapsulates a molecule of chloroform.[30] 1,1,2,2-Tetrachloroethane was chosen as the solvent in the binding studies because it is too large to fit into the cavity and, therefore, does not compete with the guests for the binding site. The complexation–decomplexation rates are slow on the ^{1}H NMR time scale at room temperature. Resonances of free and complexed guest appear side-by-side in spectra recorded in the presence of excess guest. Inclusion complexation is clearly indicated by the very large upfield shifts (> 4 ppm) measured for the proton resonances of complexed guests.

The cryptophane binding studies provide additional support for the statement made in Chapter 2.7.1 that it is wise to design more spacious cavity binding sites for apolar substrates than suggested by CPK model examinations. The smaller cryptophane **12** had been designed for $CHCl_3$, whereas experimentally, this host shows a marked preference for the smaller guest CH_2Cl_2. Also, **14** was designed for CCl_4 but the $CHCl_3$ complex is approximately two orders of magnitude more stable than the one formed by CCl_4.[30]

[29] A. Collet, *Tetrahedron*, 1987, **43**, 5725.
[30] (*a*) J. Canceill, M. Cesario, A. Collet, J. Guilhem, C. Riche, and C. Pascard, *J. Chem. Soc., Chem. Commun.*, 1986, 339; (*b*) J. Canceill, L. Lacombe, and A. Collet, *J. Am. Chem. Soc.*, 1986, **108**, 4230.

Apolar Complexation in Organic Solvents 121

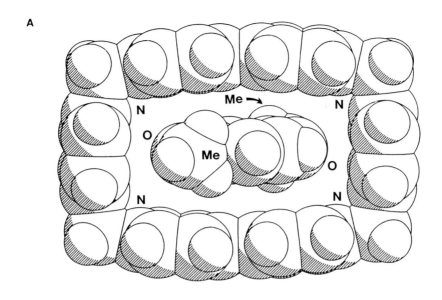

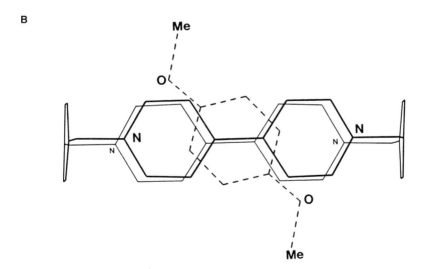

Figure 3.5 *Structure of the complex between cyclophane 11 and p-dimethoxybenzene in the crystal. A) Space-filling view into the binding site. B) Side view of the complex*
(Reproduced with permission from reference 28. Copyright 1988 VCH.)

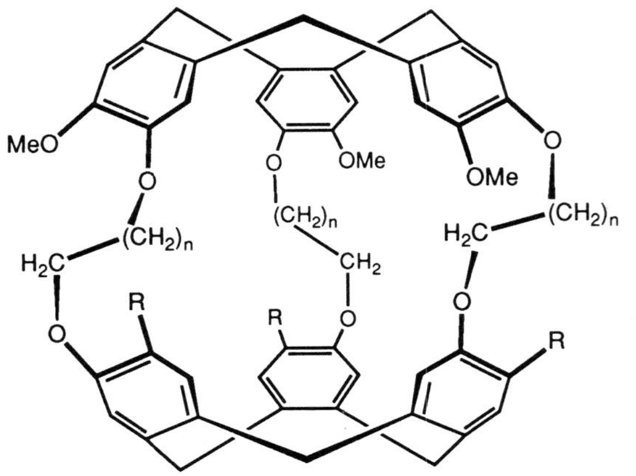

12 R = H,   n = 1
13 R = OMe, n = 1
14 R = OMe, n = 2

Table 3.6 Guest van der Waals volume, complexation-induced upfield shift $\Delta\delta$ of the guest ^1H NMR resonances, barriers for inclusion $\Delta G_i^{\neq}$, and thermodynamic quantities $\Delta G°$, $\Delta H°$, and $T\Delta S°$ obtained for complexes of cryptophane **14** by ^1H NMR in $CDCl_2CDCl_2$ at 200 MHz, T = 300 K

Guest	V_{vdW} Å^3	$\Delta\delta$	$\Delta G_i^{\neq}$ kcal mol^{-1} 300 K		$\Delta G°$ kcal mol^{-1} 300K	$\Delta H°$ kcal mol^{-1}	$T\Delta S°$ kcal mol^{-1}
				330K			
CH_3I	54.5	3.70	13.6		−2.4		
CH_2Cl_2	57.6	4.19	13.3		−2.8	+1.0	+1.8
CH_2Br_2	65.5	4.18			−3.0		
CH_3COCH_3	70.0	3.44			−1.3		
$CHCl_3$	72.2	4.44	13.3	14.4	−3.7	−6.0	−2.1
$CHCl_2Br$	76.1	4.42		14.9	−3.4	−5.2	−1.8
$CH(CH_3)_3$	79.4	4.25 2.95	13.9		−2.8	−3.8	−0.9
$CHClBr_2$	80.1	4.41		14.8	−2.9	−1.5	+1.2
$CHBr_3$	84.0	4.35		15.1	−2.3	−1.4	+1.2
CCl_4	86.8				−1.2		
$C(CH_3)Cl_3$	89.2	3.55			−0.2		
$C(CH_3)_2Cl_2$	91.6	3.45			+0.1		
$C(CH_3)_3Cl$	93.9	3.18			+0.8		

An impressive study of shape selectivity in inclusion complexation was undertaken with cryptophane **14**.[31] Figure 3.6 shows the X-ray crystal structure of the chloroform complex of this host. Table 3.6 includes the thermodynamic and kinetic data for a series of 12 complexes formed by substituted methanes in 1,1,2,2-tetrachloroethane-d_2. The free energies of

[31] J. Canceill, M. Cesario, A. Collet, J. Guilhem, L. Lacombe, B. Lozach, and C. Pascard, *Angew. Chem.*, 1989, **101**, 1249; *Angew. Chem. Int. Ed. Engl.*, 1989, **28**, 1246.

Apolar Complexation in Organic Solvents 123

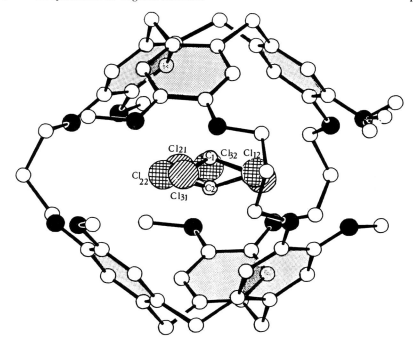

Figure 3.6 *Side view of the 14·CHCl$_3$ complex in the crystal. The two almost equivalent and equiprobable orientations of the encaged chloroform molecule are shown*
(Reproduced with permission from reference 31. Copyright 1989 VCH.)

formation of these complexes are plotted in Figure 3.7 as a function of the van der Waals volume of the guests. The following conclusions are drawn from these results:

(i) Figure 3.7 shows a smooth and steep curve relating to size recognition. As an example, **14** discriminates between CHCl$_3$ and CHCl$_2$Br by 0.3 kcal mol^{-1}, a significant energy difference considering that the difference in volume of these two molecules is only 5%.

(ii) With its spherical cavity, host **14** recognizes tetrahedral (sp^3) *versus* flat (sp^2) molecules. Acetone, with almost the same size as chloroform, is much more weakly bound.

(iii) The barriers of inclusion $\Delta G_i^{\neq}$ for all guests are similar. Therefore, in the series of guests studied, selectivity is not controlled by the cross section of the entrance to the cavity.

(iv) The thermodynamic driving force for complexation depends on the guest size. The strongest complexation is enthalpically driven, and this enthalpy is partially compensated by a negative entropy term. Tight complexes are stabilized by numerous favorable van der Waals contacts, but with tightness the two binding partners lose translational and

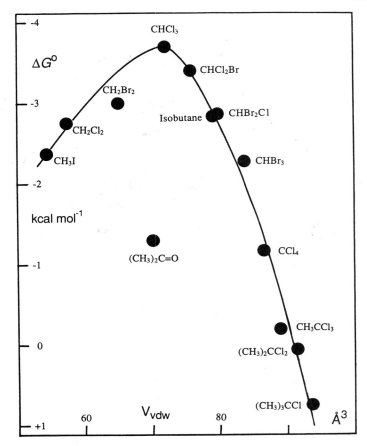

Figure 3.7 *Free energy of formation $\Delta G°$ at 300 K of a series of complexes of **14** as a function of the size of the guests, expressed by their van der Waals volume V_{vdw}*
(Reproduced with permission from reference 31. Copyright 1989 VCH.)

rotational entropy. By contrast, the complexation of the weaker binding guests, with less complementarity to the cavity, is not enthalpically driven. A good explanation for this interesting observation is yet to be found.

Upon electrochemical oxidation, **14** easily forms a radical cation and functions as a donor in unique three-dimensional charge-transfer salts.[32] By constant-current electrolysis of a $CH_2Cl_2/CHCl_3$ (4:1, v/v) mixture containing **14** (10^{-3} mol L^{-1}) and $NBu_4^+ PF_6^-$ as supporting electrolyte (10^{-1} mol

---

[32] A. Renault, D. Talham, J. Canceill, P. Batail, A. Collet, and J. Lajzerowicz, *Angew. Chem.*, 1989, **101**, 1251; *Angew. Chem. Int. Ed. Engl.*, 1989, **28**, 1249.

L^{-1}), blue-green single crystals of the charge-transfer salt [(**14**$^{\cdot+}$)(PF$_6$$^-$)(CHCl$_3$)] were grown at the platinum wire electrode. The X-ray crystal structure showed that chloroform is included in the cavity of **14** and that the complex prefers to be closely packed in a three-dimensional, salt-like array. ESR spectra of single crystals demonstrated the existence of quasi-free π-electrons delocalized over the entire molecule. The paramagnetic signal was still present when the crystal was dissolved in dichloromethane which shows that **14**$^{\cdot+}$ is a stable species.

3.5 From Cavitands to Hemicarcerands to Carcerands: Increasing the Barriers for Escape of Encapsulated Organic Molecules

3.5.1 Cavitands for Small Linear Guests

In 1985, Cram *et al.* reported on the synthesis of **15**, a host containing a cylindrical well with a diameter limiting inclusion to small linear guests.[33] Hosts similar to **15** possessing enforced cavities had previously been named as 'cavitands'.[34,35] The main structural unit in **15** as well as in the other cyclophanes described in the following sections is the conformationally flexible octol compound **16** reported by Högberg.[36] ^{1}H NMR titrations in chloroform demonstrated the enthalpically driven formation of weak 1:1 inclusion complexes between **15** and small linear guests, *e.g.* carbon disulfide, methylacetylene, and oxygen. The thermodynamic data for carbon disulfide (CS$_2$) binding are measured as $K_a = 2.6$ L mol^{-1} ($\Delta G° = -0.4$ kcal mol^{-1}, $T = 212$ K, $\Delta H° = -3.5$ kcal mol^{-1}, $T\Delta S° = -3.1$ kcal mol^{-1}). The inclusion of oxygen was apparent from the broadening of the ^{1}H NMR resonances specifically of those host protons that converge into the cavity. Figure 3.8 shows the X-ray crystal structure of the **15**·carbon disulfide complex. Complexation kinetics are fast on the ^{1}H NMR time scale, as would be expected for a readily accessible bowl-shaped binding site such as in **15**.

With **17**–**19**, cavitands containing two binding cavities, one shaped like a box and the other shaped like a bowl, were prepared. Interesting solid-state complexes were formed with these systems.[37] For example, it was shown by X-ray crystallography, that **17** crystallized from CH$_3$COCH$_3$/CH$_2$Cl$_2$ with CH$_3$COCH$_3$ in the bowl and CH$_2$Cl$_2$ in the box. Cavitand **18** crystallized from CH$_3$C$_6$H$_5$ to give a complex with a CH$_3$C$_6$H$_5$ molecule in each of the two cavities. Each toluene has its methyl group in the cavity since the binding areas are not large enough to allow phenyl ring penetration. The binding of CD$_3$CN

---

[33] D. J. Cram, K. D. Stewart, I. Goldberg, and K. N. Trueblood, *J. Am. Chem. Soc.*, 1985, **107**, 2574.
[34] D. J. Cram, *Science (Washington, DC)*, 1983, **219**, 1177.
[35] J. R. Moran, S. Karbach, and D. J. Cram, *J. Am. Chem. Soc.*, 1982, **104**, 5826.
[36] A. G. S. Högberg. *J. Org. Chem.*, 1980, **45**, 4498.
[37] J. A. Tucker, C. B. Knobler, K. N. Trueblood, and D. J. Cram, *J. Am. Chem. Soc.*, 1989, **111**, 3688.

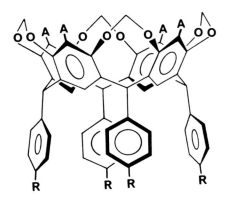

	A	B
17	H	H
18	Br	CH$_3$CH$_2$
19	BuOCH$_2$	CH$_3$CH$_2$

by each cavity of cavitand **19** was estimated by ^{1}H NMR titrations. The association constant at 300 K was calculated as $K_a \approx 89$ L mol^{-1} for the bowl inclusion complex and as $K_a \approx 45$ L mol^{-1} for the box inclusion complex. Both association processes are enthalpy driven. For the bowl cavity binding, the authors measured by van't Hoff analysis $\Delta H° = -6.2 \pm 1.2$ kcal mol^{-1} and $\Delta S° = -11.4 \pm 4$ cal mol^{-1} K^{-1} and for the box cavity binding $\Delta H° = -6.4 \pm 1.5$ kcal mol^{-1} and $\Delta S° = -13.6 \pm 5$ cal mol^{-1} K^{-1}.

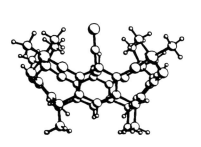 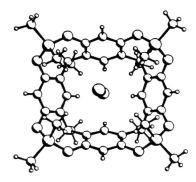

Figure 3.8 *Side and top views of the X-ray crystal structure of the 15·CS$_2$ complex (Reproduced with permission from reference 33. Copyright 1985 ACS.)*

3.5.2 Strong Dimer Formation between Kite-type Molecules in Organic Solvents

Extended structured lipophilic surfaces form stable dimers in organic solvents such as chloroform. This fascinating observation was made by Cram *et al.* with selected derivatives of the mobile system **20** (kite)⇌**20** (vase).[38,39] Above −38 °C in CDCl$_3$, only the vase form **20** was detected by ^{1}H NMR. Below −38 °C, only the kite conformation is detected. No **20** (kite) ·**20** (kite) association was observed at any temperature.

This situation changes dramatically, if extra alkyl groups are introduced giving **22** and **23**. These groups sterically inhibit entirely *vase* formation. Compound **22** exists only as dimer in CDCl$_3$. The crystal structure of this dimer is shown in Figure 3.9. As many as 70 intermolecular atom-to-atom van der Waals contacts are measured; 40 more are within contact distance plus 0.1–0.2 Å. Many of these close contacts occur between the quinazoline units of the two dimer components. ^{1}H NMR dimerization shifts and NOE difference spectroscopy suggest that the structure of the **22·22** dimer in solution strongly resembles the solid state structure shown in Figure 3.9. In sharp contrast, **23** is only detectable, as a kite monomer in CDCl$_3$.

Why does **22** dimerize when such association is not observed for **20** (kite) and **23**? All three possess roughly planar rectangular faces of 15 × 20 Å in the *kite* conformation. The face in **22** contains two protruding up-methyl groups and two methyl-sized indentations lined by a sloping aryl face, an out methyl, and two oxygens. Scheme 3.2 shows the dimerization of two monomers of **22** rotated by 90° to each other. Superposition as shown produces a face-to-face dimer in which four methyl groups occupy four methyl-specific indentations, producing a unique fit. This locking of four CH$_3$ groups into four cavities of

---

[38] J. A. Bryant, C. B. Knobler, and D. J. Cram, *J. Am. Chem. Soc.*, 1990, **112**, 1254.
[39] J. A. Bryant, J. L. Ericson, and D. J. Cram, *J. Am. Chem. Soc.*, 1990, **112**, 1255.

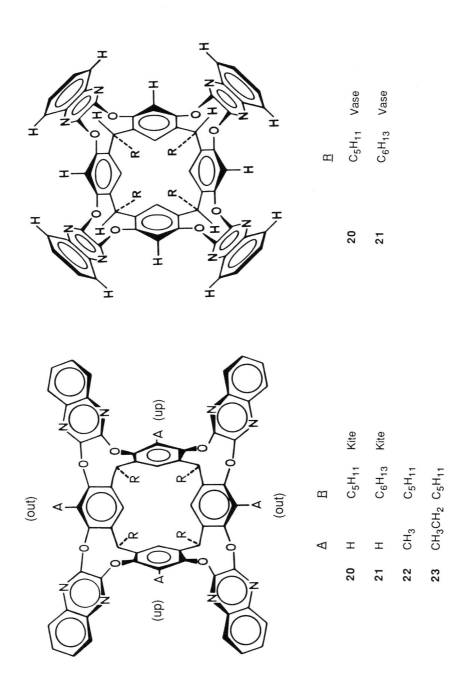

	A	R	
20	H	C_5H_{11}	Kite
21	H	C_6H_{13}	Kite
22	CH_3	C_5H_{11}	
23	CH_3CH_2	C_5H_{11}	

	R	
20	C_5H_{11}	Vase
21	C_6H_{13}	Vase

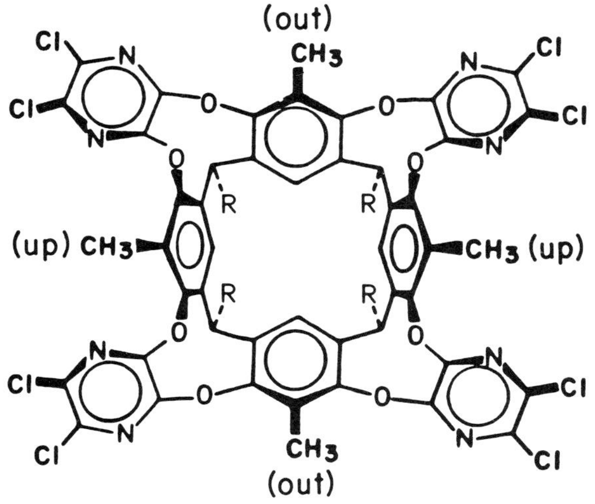

Apolar Complexation in Organic Solvents

24   R = C$_5$H$_{11}$

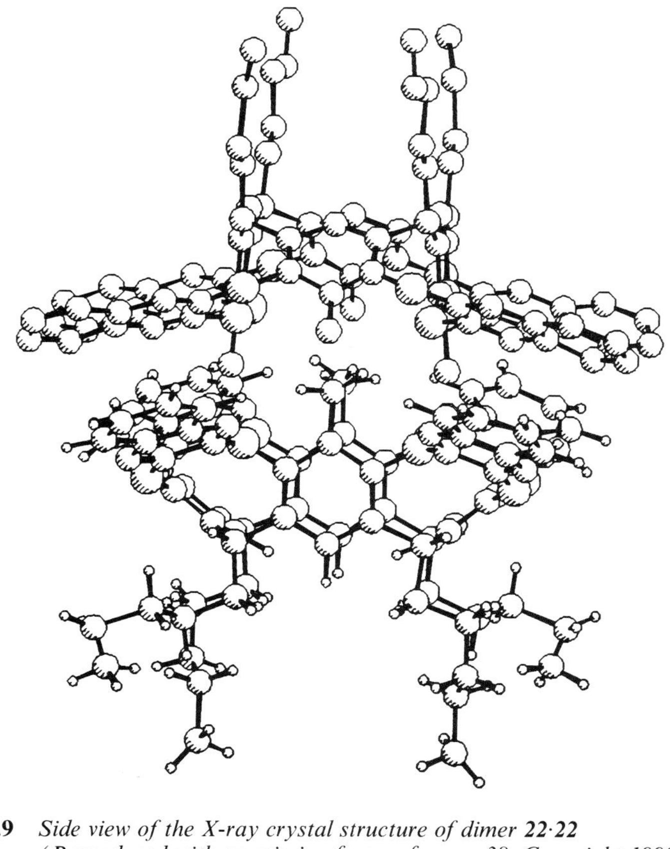

Figure 3.9 *Side view of the X-ray crystal structure of dimer **22·22**
(Reproduced with permission from reference 38. Copyright 1990 ACS.)*

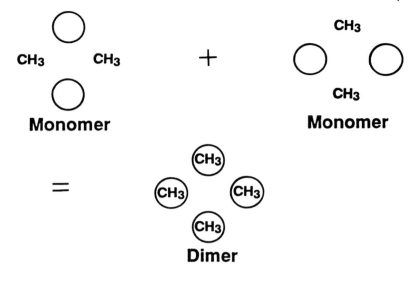

Scheme 3.2

Table 3.7 *Association constants (K_a) and free energies of complexation ($-\Delta G°$) for the dimers formed by the kite structures 22 and 24*

Dimer	Solvent	Temp [K]	$K_a [L^{-1} mol]$	$-\Delta G°$ [kcal mol^{-1}]
22·22	CDCl$_3$	285	8.7×10^4	6.6 ± 0.2
	CDCl$_3$	255	6.5×10^5	6.9 ± 0.2
22·24	CDCl$_3$	255	2.6×10^5	6.3 ± 0.2
24·24	CDCl$_3$	255	3.1×10^3	4.1 ± 0.2
	CDCl$_3$/(CD$_3$)$_2$CO (3:1 v/v)	255	2.3×10^4	5.1
	CDCl$_3$/CD$_3$NO$_2$ (3:1 v/v)	255	3.4×10^4	5.3
	CDCl$_3$/CD$_3$OD (3:1 v/v)	255	$> 6.3 \times 10^4$	> 5.6
	CD$_3$C$_6$H$_5$	227	2.3×10^3	3.5
	CDCl$_3$	227	8.8×10^3	4.1
	CD$_2$Cl$_2$	227	9.5×10^3	4.3

the dimer would be absent in a dimer of **20** (kite) with aryl-hydrogens instead of aryl-methyls and, therefore, **20** (kite) does not form a stable dimer. The dimerization of **23** is sterically inhibited by the inability of the ethyls to enter the methyl-sized indentations in the complementary surface. Similar to **22**, the pyrazine derivative **24** with its aryl-methyl groups also undergoes dimerization. Mixed dimers form between **22** and **24**, since both surfaces have outreaching methyl groups and complementary indentations. No mixed dimers form between the ethyl derivative **23** and **22** or **24**.

The driving force for the formation of the **22·22** and **24·24** dimers is very large and highly solvent dependent.[39] Association constants measured in various solvents are shown in Table 3.7. The association strength of the dimers in various solvents follows the general correlation between apolar binding strength and solvent polarity shown in Figure 3.1. The quinazoline kite **22** with its larger surface forms a more stable dimer than the smaller pyrazine derivative **24**. The thermodynamic quantities for the formation of dimer **24·24** in $CDCl_3$ were determined as $\Delta H° = -3.8 \pm 0.5$ kcal mol^{-1} and $\Delta S° = 1.1 \pm 3$ cal mol^{-1} K^{-1}. The small favorable entropic component of the dimerization process is remarkable. It shows (i) that the two associating surfaces are very rigid, and that the amount of vibrational and rotational entropy lost during dimerization is small, and (ii) that considerable entropy is gained upon desolvation of two large surfaces.

Another remarkable feature of the dimerization process is the slow dissociation rate.[39] At 285 K, the barrier for the dissociation of **24·24** was calculated as $\Delta G^{\ddagger} = 14.0 \pm 0.3$ kcal mol^{-1} and the first-order dissociation rate constant as $k_{-1} = 106$ s^{-1}. This gives a second-order association rate constant of 1.4×10^5 L mol^{-1} s^{-1}, much slower than diffusion controlled. The authors explain the relatively slow dissociation kinetics by the absence of incremental solvation–desolvation of the two faces involved in dimerization. The locking of four CH_3 groups into complementary indentations prevents easy monomer-to-monomer slippage. The insertion of a solvent molecule requires a clamlike opening of the dimer costing a large portion of the attractive binding forces between the two surfaces.

As mentioned above, compounds **20** and **21** prefer the *vase* conformation in solution at room temperature. Figure 3.10 shows the X-ray crystal structure of a solvate of **21** (vase) with three molecules of acetone.[40] Stoichiometric complexes between **21** (vase) and benzene derivatives seem to form in acetone as a solvent. Desorption chemical ionization mass spectrometry provided evidence for the association between negative ions of cavitands related to **21** and benzene derivatives in the gas phase.[41]

3.5.3 Carcerands: Closed-surface Hosts that Imprison Guests behind Covalent Bars

With **25**, Cram *et al.* prepared the first closed-surface container with an enforced interior large enough to incarcerate permanently suitably sized molecules [$(CH_3)_2NCHO$, $(CH_2)_4O$, $ClCF_2CF_2Cl$], a gas (Ar), or ions (Cs^+, Cl^-).[42] These species are trapped out of solutions in which **25** is formed by shell closure of two bowl-shaped cavitands derived from octol **16**. Closed-surface hosts like **25** have been named as carcerands. An escape of incarcerated guests from the interior requires the cleavage of covalent bonds of the host

---

[40] E. Dalcanale, P. Soncini, G. Bacchilega, and F. Ugozzoli, *J. Chem. Soc., Chem. Commun.*, 1989, 500.
[41] M. Vincenti, E. Dalcanale, P. Soncini, and G. Guglielmetti, *J. Am. Chem. Soc.*, 1990, **112**, 445.
[42] (*a*) D. J. Cram, S. Karbach, Y. H. Kim, L. Baczynskyj, and G. W. Kalleymeyn, *J. Am. Chem. Soc.*, 1985, **107**, 2575; (*b*) D. J. Cram, S. Karbach, Y. H. Kim, L. Baczynskyj, K. Marti, R. M. Sampson, and G. W. Kalleymeyn, *J. Am. Chem. Soc.*, 1988, **110**, 2554.

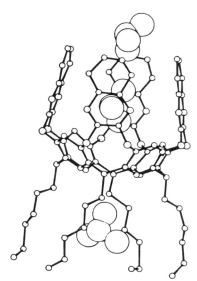

Figure 3.10 *X-ray crystal structure of **21**(vase)·3 (CH$_3$)$_2$CO. The acetone molecules are depicted in their van der Waals radii form (Reproduced with permission from reference 40. Copyright 1989 Royal Society of Chemistry, London.)*

shell. Due to severe solubility problems, evidence for the structure of **25** and for guest incarceration was obtained by elemental analysis and fast atom bombardment (FAB) mass spectrometry only. Extreme insolubility also prevented the separation of the complexes with varying contents of guest.

By introducing phenethyl-groups, the solubility of the carcerands was

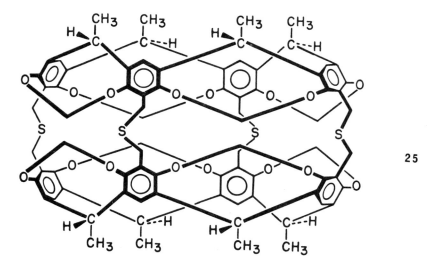

25

Apolar Complexation in Organic Solvents 133

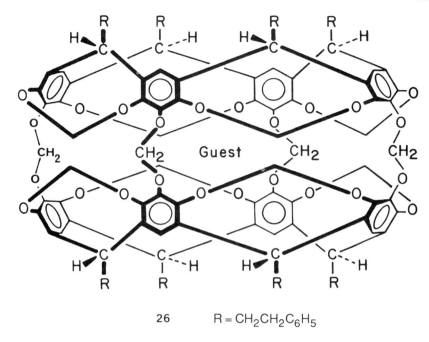

26   R = CH$_2$CH$_2$C$_6$H$_5$

increased.[43] From the shell closure reactions shown in Scheme 3.3, complexes of **26**, the first closed-surface container with good solubility in organic solvents, were isolated. Depending on the solvent used for the shell-closure, a molecule of (CH$_3$)$_2$SO, (CH$_3$)$_2$NCOCH$_3$, or (CH$_3$)$_2$NCHO was incarcerated. The good solubility of these complexes ('carceplexes') allowed their purification by chromatography on silica gel (CHCl$_3$/hexane) and crystallization (CHCl$_3$/CH$_3$CN). An isolable amount of empty carcerand was not formed in any of the reactions. Apparently, an encapsulated solvent molecule is required for a favorable solvation of the transition states of the final S_N2-type shell closure reactions leading to **26**. In support of this explanation, attempts to prepare **26** failed in (CH$_2$)$_5$NCHO, a solvent too large for incarceration. The reaction in (CH$_2$)$_5$NCHO/(CH$_3$)$_2$NCHO (99.5:0.5, v/v) led to **26**·(CH$_3$)$_2$NCHO as the only formed carceplex (10% yield). A run in (CH$_3$)$_2$NCOCH$_3$/(CH$_3$)$_2$NCHO (1:1, v/v) gave 10% of **26**·(CH$_3$)$_2$NCOCH$_3$ and **26**·(CH$_3$)$_2$NCHO in the ratio of 5.3 to 1. These two complexes were chromatographically separated which demonstrates a differential communication of the encarcerated guests with their host's environment. In the ^{1}H NMR spectra of the carceplex solutions in CDCl$_3$, the resonances of the entrapped guests appear upfield shifted by 1–4 ppm from their normal positions. Also, the ^{1}H NMR integration supported the 1:1 stoichiometry of the formed carceplexes. When a solution of **26**·(CH$_3$)$_2$SO in (CH$_3$)$_2$NCHO

---

[43] J. C. Sherman and D. J. Cram, *J. Am. Chem. Soc.*, 1989, **111**, 4527.

134 Chapter 3

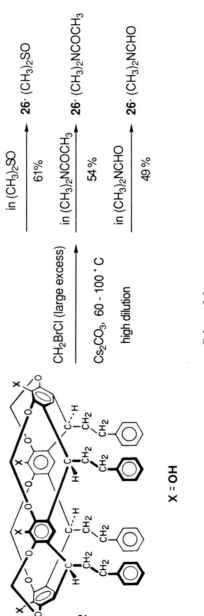

Scheme 3.3

Apolar Complexation in Organic Solvents

was heated to reflux for 12 h, no guest exchange occurred showing that guests cannot enter or depart the carceplexes without breaking covalent bonds of the closed-shell host. The observation that the rates of rotation about the amide bonds in the encarcerated amides differ from the rates measured for gas phase and solutions led Cram *et al.* to conclude that the carceplexes represent a new state of matter and that the interiors of carcerands are best viewed as new phases.[43]

3.5.4 Hemicarcerands: Guest Exchange with High Structural Recognition and Activation Free Energies

Compounds **25** and **26** are carcerands which entrap guests during shell closure reactions and imprison them permanently. With **27**, a hemicarcerand was reported with a shell hole large enough to permit entrance and egress of suitable guests.[44] Shell closures similar to those shown in Scheme 3.3 in $(CH_3)_2SO$ gave **27**·$(CH_3)_2SO$ (51% yield), in $(CH_3)_2NCOCH_3$ gave **27**·$(CH_3)_2NCOCH_3$ (42% yield), and in $(CH_3)_2NCHO$ gave **27**·$(CH_3)_2NCHO$ (20% yield). Figure 3.11 shows the crystal structure of **27**·$(CH_3)_2NCHO$ as a solvate with 2 CH_3CN and 2 $CHCl_3$.

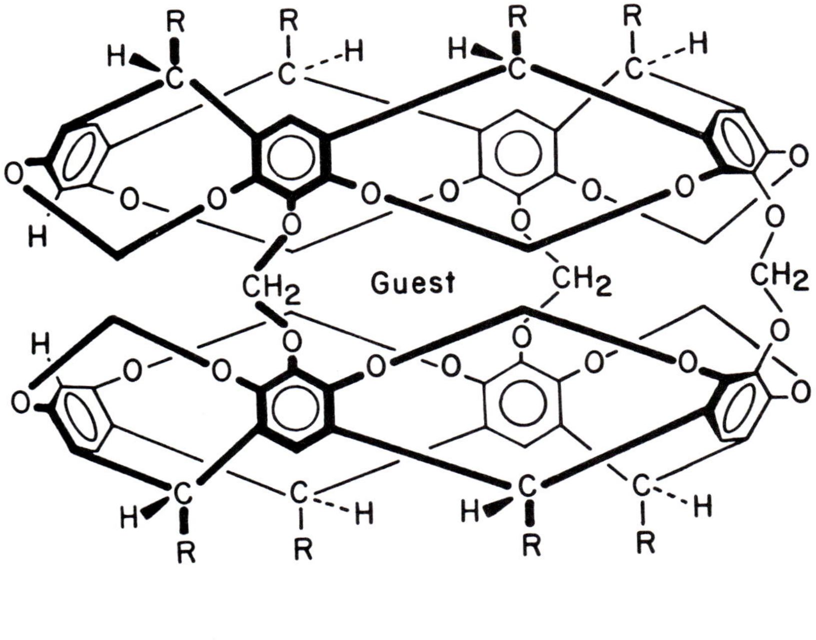

27 $R = CH_2CH_2C_6H_5$

---

[44] M. Tanner, C. B. Knobler, and D. J. Cram, *J. Am. Chem. Soc.*, 1990, **112**, 1659.

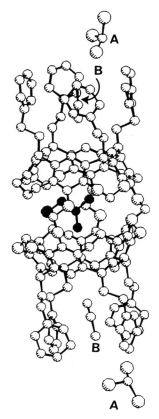

Figure 3.11 *X-ray crystal structure of the hemicarceplex* **27**·*(CH₃)₂NCHO. The guest is darkened. Solvent molecules are marked as A (CDCl₃) and B (CD₃CN)*
(Reproduced with permission from reference 44. Copyright 1990 ACS.)

Upon heating the hemicarcerand complexes ('hemicarceplexes') in solvents too large to become guests, the guests entrapped during synthesis are expelled, and free hemicarcerand **27** is obtained. The activation energies for these decomplexation processes are very high: **27**·(CH₃)₂SO required 214 °C for 48 h in 1,2,4-trichlorobenzene; **27**·(CH₃)₂NCOCH₃ needed 165 °C for 24 h in mesitylene, and **27**·(CH₃)₂NCHO 165 °C for 12 h. Subsequently, the empty hemicarcerand could be filled with new guests. Treating free **27** in CH₃CN, CS₂, pyridine, or CH₂Br₂ led to the formation of 1:1 complexes of these solvent molecules. Xenon was included when **27** was treated with a 0.14 molar solution of xenon in CDCl₃.

Hemicarcerands are true carcerands at room temperature where no guest exchange occurs. As a significant advantage over the carcerands, they allow the rational encapsulation and release of guests at high temperature. The carcerand concept opens fascinating research perspectives. The formation of carceplexes and hemicarceplexes provides matrix isolation at room temperature. The structural, photochemical, and photophysical properties as well as the chemical reactivity of individual molecules, isolated in unique and well defined environments, can now be studied at room temperature.

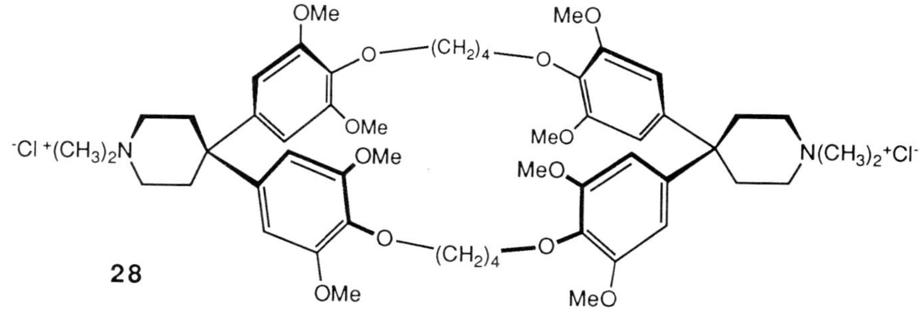

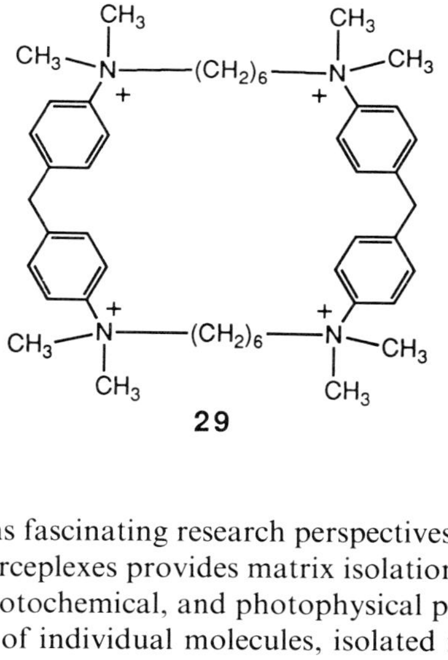

3.6 Apolar Complexation Strength in Binary Solvent Mixtures

As discussed in the introduction to Chapter 3, there is considerable interest in understanding and predicting, how apolar binding phenomena changes with the composition of binary solvent mixtures. The groups of *Schneider*[45] and *Diederich*[9,46] explored linear free energy relationships (LFER) between solvent polarity parameters and the composition of binary solvent mixtures. Both groups found that the complexation of neutral aromatic guests in binary

---

[45] H. J. Schneider, R. Kramer, S. Simova, and U. Schneider, *J. Am. Chem. Soc.*, 1988, **110**, 6442.
[46] E. M. Sanford and F. Diederich, unpublished results.

Table 3.8 *Linear correlations coefficients for the LFERs between free energies of complexation $\Delta G°$ in D_2O–CD_3OD mixtures and $E_T(30)$*[45,46]

Complex	Linear correlation coefficient R
28·p-benzodinitrile	0.995
28·p-dimethoxybenzene	0.992
29·1-(dimethylamino)naphthalene-5-sulfonamide	0.997
29·naphthalene	0.995
29·p-nitrophenyl acetate	0.992

solvent systems by host **28**, with a pronounced apolar binding site, and by **29**, with a cavity aligned with quaternary nitrogens, is accurately predictable in LFERs with $E_T(30)$ as a solvent polarity parameter. Table 3.8 shows the linear correlation coefficients obtained for the LFERs between the association constants (log K_a) for a variety of complexes in CD_3OD–D_2O mixtures and the $E_T(30)$ parameters of these solutions. Figure 3.12 shows the individual correlations of the binding free energy of the **28**·p-benzodinitrile complex and $E_T(30)$ with the solvent composition and the LFER between the two quantities. A similarly good LFER was obtained for the binding of p-benzodinitrile in D_2O–DMSO mixtures.

Studies with host **29** seem to suggest that $E_T(30)$ is not as suitable for predicting the complexation of charged guests in binary solvent mixtures.[45] In complexes formed between **29** and ionic guests, ion pairing as well as attractive pole–dipole interactions between the charged center of the guest and the aromatic π-systems of the host represent additional binding interactions (Chapter 4). These interactions apparently are not as well reflected by $E_T(30)$ as are apolar (hydrophobic) interactions resulting in lower linear correlation coefficients for LFERs with this solvent parameter. Schneider *et al.* showed that the binding of both neutral and anionic guests to host **29** is best correlated with the solvophobicity parameter S_p, defined as $S_p = 1.00$ in water and $S_p = 0.0$ in *n*-hexane.[45,47] Clearly, more of these useful correlations are desirable, especially for those solvent mixtures most often applied in studies of biological systems, *e.g.* water–2,2,2-trifluoroethanol, water–formamide, and the binary mixtures with alkylated formamide derivatives.[2-4]

[47] M. H. Abraham, *J. Am. Chem. Soc.*, 1982, **104**, 2085.

Figure 3.12 *(A) Free energy of formation of the **28**·p-benzodinitrile complex in D_2O-CD_3OD mixtures, T = 293 K. (B). Solvent polarity parameter E_T(30) as a function of the composition of D_2O–CD_3OD mixtures. (C) LFER between the free energy of formation of the **28**·p-benzodinitrile complex in D_2O–CD_3OD mixtures and $E_T(30)$*

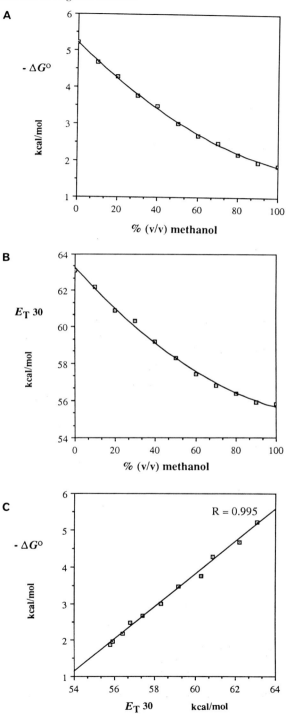

CHAPTER 4

Cyclophane Complexes of Charged Organic Guests

4.1 Introduction

By far the majority of cyclophane receptors has been prepared for ionic guests. Since the appearance of the seminal papers by Charles Pedersen[1,2] in 1967, the complexation of alkali and alkaline earth metal ions, transition metal and rare earth metal ions, and primary ammonium ions has been accomplished by a large number of benzocrown ethers, aryl-fused cryptands, spherands, and macrocyclic Schiff bases. Even though all these compounds should be classified as cyclophanes, metal and ammonium cation binding processes will not be the subject of this monograph. Cyclophane frameworks, which organize a converging array of oxygen, nitrogen, sulfur, and other donor centers for metal or ammonium ion complexation may dramatically alter the thermodynamics and kinetics of bonding.[3] Nevertheless, the key binding force in these cyclophane complexes remains the interaction between the crown-type array of donor centers and the metal or ammonium ion. Therefore, these complexation processes are discussed in the volume on crown ethers in this series.[4] Since the author of this monograph loves the esthetics of supramolecular structure, a handful of examples in Scheme 4.1 should pay tribute to some marvelous cyclophane chemistry not covered in this monograph.[5-9] Also,

---

[1] C. J. Pedersen, *J. Am. Chem. Soc.*, 1967, **89**, 2495.
[2] C. J. Pedersen, *J. Am. Chem. Soc.*, 1967, **89**, 7017.
[3] D. J. Cram, *Angew. Chem.*, 1986, **98**, 1041; *Angew. Chem. Int. Ed. Engl.*, 1986, **25**, 1039.
[4] G. W. Gokel, 'Crown Ethers', Royal Society of Chemistry, London, forthcoming in this series.
[5] R. C. Helgeson, T. L. Tarnowski, J. M. Timko, and D. J. Cram, *J. Am. Chem. Soc.*, 1977, **99**, 6411.
[6] D. J. Cram, T. Kaneda, R. C. Helgeson, and G. M. Lein, *J. Am. Chem. Soc.*, 1979, **101**, 6752.
[7] A. Hamilton, J.-M. Lehn, and J. L. Sessler, *J. Am. Chem. Soc.*, 1986, **108**, 5158.
[8] J.-C. Rodriguez-Ubis, B. Alpha, D. Plancherel, and J.-M. Lehn, *Helv. Chim. Acta*, 1984, **67**, 2264.
[9] C. O. Dietrich-Buchecker, J. P. Sauvage, and J. P. Kintzinger, *Tetrahedron Lett.*, 1983, **24**, 5095.

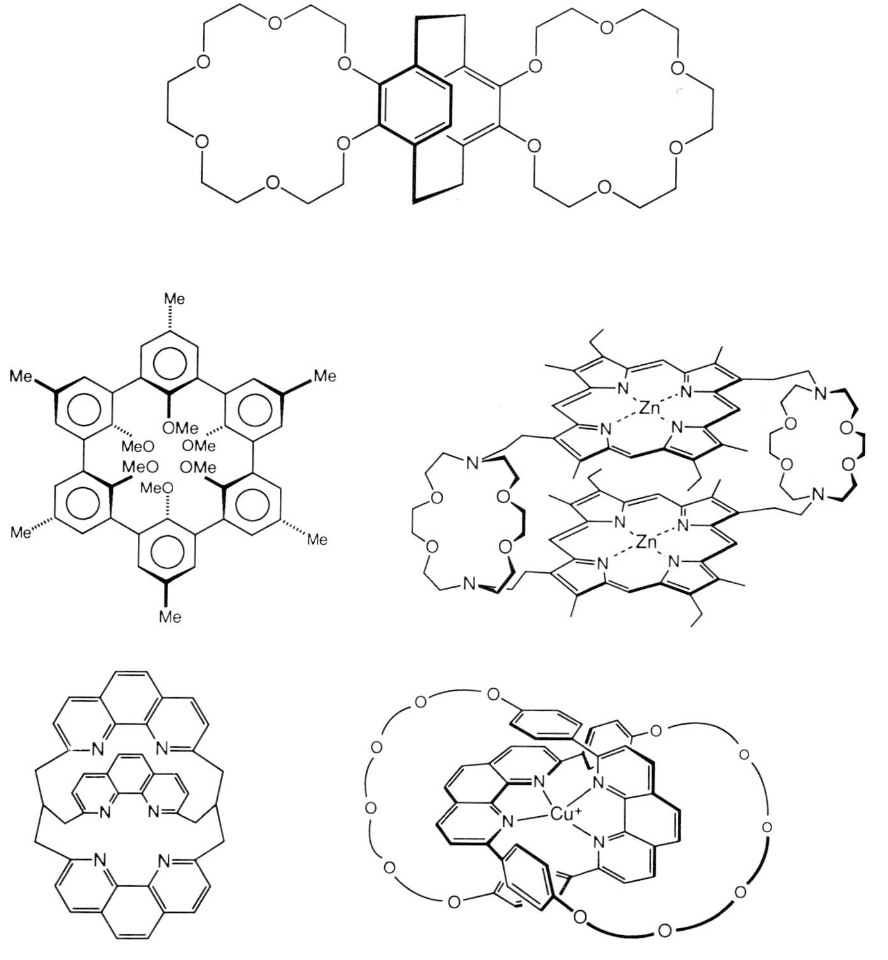

Scheme 4.1

the reader is referred to two recent monographs on crown ethers and analogues.[10,11]

In the complexes of charged organic molecules discussed in this Chapter, the cyclophane frame provides an essential contribution to stable bonding. First, it will be shown how additional Coulombic charge–charge interactions stabilize complexes in which apolar binding in the cyclophane cavity

---

[10] L. F. Lindoy, 'The Chemistry of Macrocyclic Ligand Complexes', Cambridge University Press, Cambridge, 1989.

[11] E. Weber, J. L. Toner, I. Goldberg, F. Vögtle, D. A. Laidler, J. F. Stoddart, R. A. Bartsch, and C. L. Liotta, 'Crown Ethers and Analogs', eds. S. Patai and Z. Rappoport, Wiley, Chichester, 1989.

represents the primary binding mode. In many of the complexes discussed below, the aromatic rings of the cyclophanes not only shape the recognition site but participate in important stabilizing interactions with the charged guests. The nature of these ion–dipole interactions between charged centers and aromatic rings in cyclophane complexes is analyzed. The following section describes binding processes predominantly driven by attractive charge–charge interactions, *i.e.* molecular cation binding by host anions and molecular anion binding by cationic systems. Polytopic receptors with at least one binding site for charged guests are the subject of the following section. A fascinating story tells about the design of cyclophane binders of the charged herbicides diquat and paraquat which, subsequently, led to the development of a new organic material with catenane structure. Chapter 4 finishes with the discussion of systems at the interface between molecular association and polymolecular aggregation.

4.2 High Guest Selectivity in the Inclusion Complexation of Charged Aromatic Compounds

Organic molecules with cationic or anionic centers have been used as guests in many complexation studies with cyclophanes in aqueous solution. Often, the only function of the charges is to solubilize the guest in aqueous environments. For example, *anionic* steroids and paracyclophanes had been chosen as guests mainly for solubility reasons in studies of a cyclophane host with a large apolar binding site (Chapter 2).[12] Good water solubility in addition to useful optical properties has made the fluorescence probes 6-(4-methylphenyl)amino-2-naphthalenesulfonate (TNS) and 8-phenylamino-1-naphthalenesulfonate (ANS) very popular charged guests for inclusion in apolar cyclophane cavities in aqueous solutions.[13-15] The stability constants of ANS and TNS complexes in water are readily determined from fluorescence titrations. In addition, changes in fluorescence intensity and emission maxima of these probes during binding titrations provide valuable information on the micropolarity of the binding sites. Upon transfer from water into a less polar environment, the fluorescence quantum yields of the two probes dramatically increase, and their emission maxima are shifted to higher energy (hypsochromic shift).

The complexes of main interest to this section of Chapter 4 are those for which comparative binding studies revealed a significant stabilization or destabilization as a result of Coulombic host–guest charge–charge interactions. Table 4.1 shows the stability data for complexes of cyclophanes **1** and **2** in aqueous solutions. At pH 1.9, the tetraprotonated tetraazaparacyclophane

---

[12] D. R. Carcanague and F. Diederich, *Angew. Chem.*, 1990, **102**, 836; *Angew. Chem. Int. Ed. Engl.*, 1990, **29**, 769.
[13] F. Diederich, *Angew. Chem.*, 1988, **100**, 372; *Angew. Chem. Int. Ed. Engl.*, 1988, **27**, 362.
[14] U. Werner, W. M. Müller, H.-W. Losensky, T. Merz, and F. Vögtle, *J. Incl. Phenom.*, 1986, **4**, 379.
[15] H.-J. Schneider and J. Pöhlmann, *Bioorg. Chem.*, 1987, **15**, 183.

1,8-ANS

2,6-TNS

1 forms substantially more stable complexes with naphthalenemonosulfonates and disulfonates than with neutral guests, *e.g.* 2,7-naphthalenediol.[16,17] Only weak complex formation was observed for naphthalene derivatives having positively charged centers. The observed binding selectivity is readily explained by the attractive or repulsive Coulombic interactions between the positively charged centers aligning the host cavity and the charged functional groups of the guests bound in the cavity.

A similar result was obtained with cyclophane **2** (Table 4.1).[13,18,19] The complexes with anionic guests are more stable than those of neutral naphthalene derivatives (Table 2.3). The complexes of naphthalenemonosulfonates and disulfonates are stabilized by apolar binding interactions in the cavity along with attractive Coulombic interactions between the anionic residues of the guests and the quaternary nitrogen atoms of the piperidinium rings attached to the aliphatic bridges of the host. This is schematically shown in Figure 4.1 for the **2**·2,6-naphthalenedisulfonate complex.[20] Coulombic interactions severely destabilize the complexes of **2** and cationic naphthalene derivatives. Table 4.1 shows that, for a series of structurally very similar guests, additional charge–charge interactions can lead to differences in free binding energy $\Delta(\Delta G°)$ of more than 6.5 kcal mol^{-1}.

Charge–charge interactions are also observed in the complexes of the anionic cyclophane **4**.[21] The complex of the 1-trimethylammoniumnaphthalene cation and **4** is more stable than the complexes of neutral naphthalene derivatives. These, in return, are more stable than the complex of 2-naphthalenesulfonate, since this guest undergoes repulsive charge–charge interactions with the sulfonate residues of the host.

---

[16] K. Odashima, T. Soga, and K. Koga, *Tetrahedron Lett.*, 1981, **22**, 5311.
[17] K. Odashima and K. Koga, *Heterocycles*, 1981, **15**, 1151.
[18] F. Diederich and K. Dick, *J. Am. Chem. Soc.*, 1984, **106**, 8024.
[19] F. Diederich and K. Dick, *Chem. Ber.*, 1985, **118**, 3817.
[20] F. Diederich and D. Griebel, *J. Am. Chem. Soc.*, 1984, **106**, 8037.
[21] H.-J. Schneider and T. Blatter, *Angew. Chem.*, 1988, **100**, 1211; *Angew. Chem. Int. Ed. Engl.*, 1988, **27**, 1163.

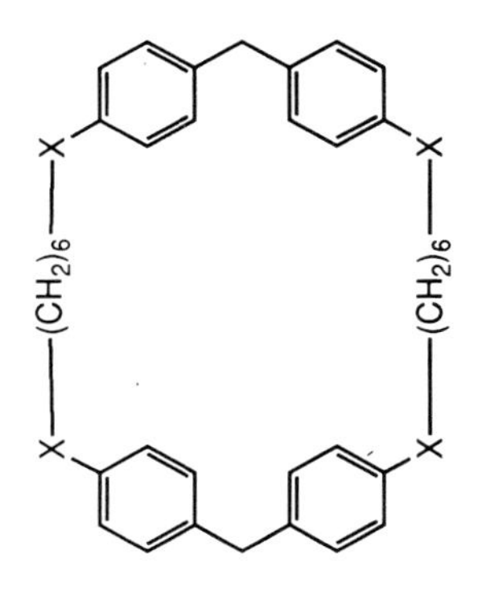

3 X = $^+$NMe$_2$

4 X = NSO$_2$-C$_6$H$_4$-SO$_3^-$ (meta)

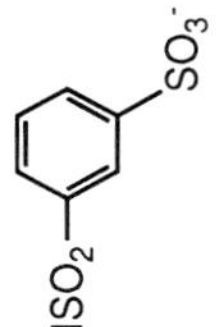

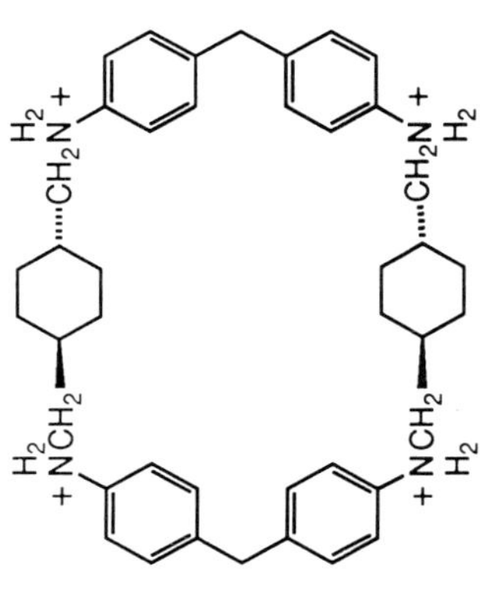

1

2   4 Cl$^-$

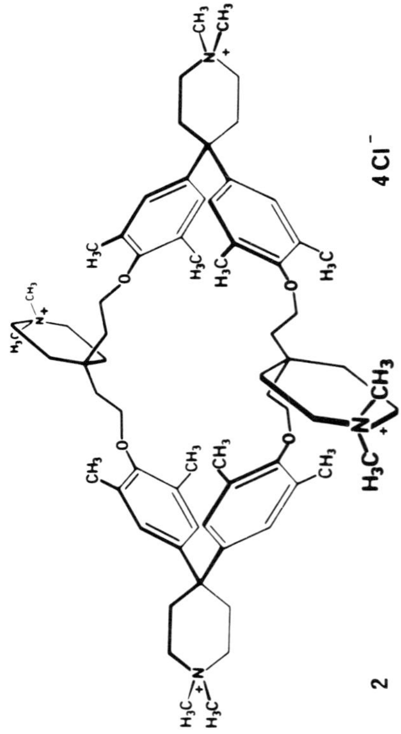

Cyclophane Complexes of Charged Organic Guests

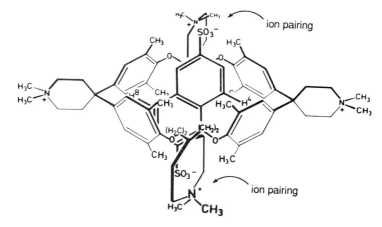

Figure 4.1 *Schematic drawing of the geometry of the 2·2,6-naphthalenedisulfonate complex which, according to the ^{1}H NMR spectrum is highly favored in aqueous solution. Observed upfield complexation shifts of the guest resonances in an aqueous solution with [host] = [guest] = 4×10^{-3} mol L^{-1}: $\Delta\delta$ (in ppm) = 1.36 (1-H), 0.94 (3-H), and 3.09 (4-H)*

Table 4.1 *Association constants (K_a) and free energies of complexation ($-\Delta G°$) for the 1:1 complexes formed between cyclophanes **1** and **2** and naphthalene derivatives in aqueous solutions*

Guest	K_a [L^{-1} mol]	$-\Delta G°$ [kcal mol^{-1}]
*(a) Complexes of **1** (298 K, KCl–HCl buffer, pH 1.9)*		
1,5-Naphthalenedisulfonate	1.4×10^6	8.4
2,6-Naphthalenedisulfonate	3.2×10^5	7.5
1-Naphthalenesulfonate	5.3×10^4	6.4
2-Naphthalenesulfonate	3.0×10^4	6.1
2,7-Naphthalenediol	4.3×10^3	5.0
2,7-Bis(dimethylammonium)naphthalene	very small	
*(b) Complexes of **2** (293 K, water)*		
2,6-TNS	5.0×10^6	9.0
1,8-ANS	3.2×10^6	8.7
2,6-Naphthalenedisulfonate	$>10^6$	>8.0
1,5-Naphthalenedisulfonate	4.4×10^5	7.6
2-Naphthalenesulfonate	4.0×10^5	7.5
1-Naphthalenesulfonate	3.5×10^5	7.4
Naphthalene	1.2×10^4	5.5
1-Trimethylammoniumnaphthalene fluorosulfate	1.7×10^3	4.3
1,5-Bis(dimethylammonium)naphthalene bis(deuterium chloride)[a]	$\approx < 10$	$\approx < 1.3$

[a] in D_2O/DCl/KCl, pD = 1.2.

4.3 Ion–Dipole Effect as a Force for Molecular Recognition

Dougherty et al. found that the anionic cyclophane **5** is a better receptor in aqueous borate buffers (pD 9) for positively charged ammonium and immonium guests than for analogous neutral derivatives.[22a] Table 4.2 shows that the methylquinolinium ion binds stronger by 2.2 kcal mol^{-1} to **5** than quinoline itself. Binding strength is very high; the methylquinolinium complex is stabilized by a free energy of $-\Delta G°_{295} = 7.6$ kcal mol^{-1} ($K_a = 4 \times 10^5$ L mol^{-1}). Experimentally, the ^{1}H NMR complexation shifts demonstrate that onium substrates prefer to be incorporated into the cyclophane cavity with their charged center rather than with their apolar moieties. Table 4.3 shows the upfield shifts calculated for saturation binding of 1-trimethylammoniumadamantane iodide. Clearly, the complexation-induced shifts of the resonances close to the ammonium center are much larger than the shifts encountered by other protons in the molecule.

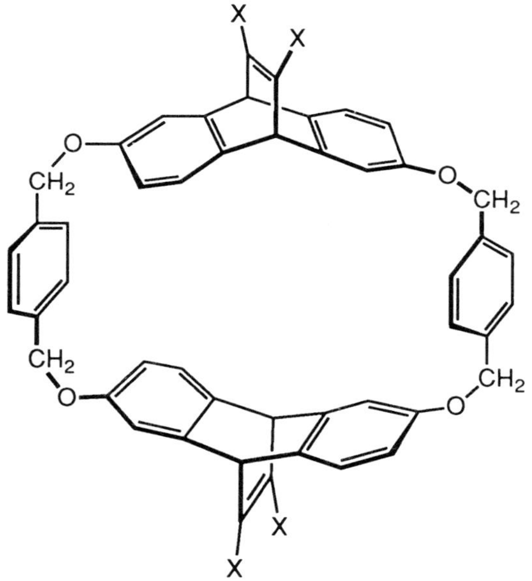

5 X = COO$^-$ Cs$^+$

6 X = COOCH$_3$

These results, surprising at first, cannot be explained by Coulombic attractions in the complexes since the rigid structure of **5** prevents the water-solubility providing carboxylates from establishing close contacts with the guest onium center. The aromatic rings in the bridges between the two ethenoanthracenes have proven to be crucial since onium substrate binding

---

[22] (a) M. A. Petti, T. J. Shepodd, R. E. Barrans, Jr., and D. A. Dougherty, *J. Am. Chem. Soc.*, 1988, **110**, 6825; (b) D. A. Stauffer and D. A. Dougherty, *Tetrahedron Lett.*, 1988, **29**, 6039.

Table 4.2 *Free energies of complexation $-\Delta G°$ (kcal mol^{-1}) at T = 295 K for the 1:1 complexes of cyclophane **5** in aqueous deuterated borate buffer, pD = 9, and of cyclophane **6** in CDCl$_3$*

Guest	$-\Delta G°$ (kcal mol^{-1}) **5** in pD = 9 buffer	$-\Delta G°$ (kcal mol^{-1}) **6** in CDCl$_3$
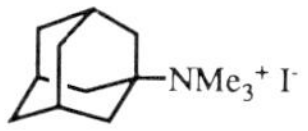 adamantyl-NMe$_3^+$ I$^-$	6.7	2.1
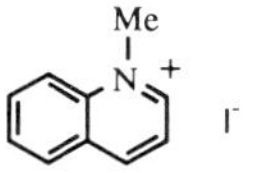 N-methylquinolinium I$^-$	7.6	3.5
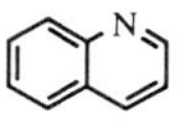 quinoline	5.4	0.0
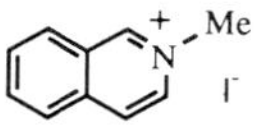 N-methylisoquinolinium I$^-$	7.2	2.5
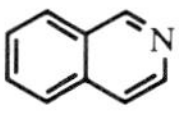 isoquinoline	6.3	0.2

Table 4.3 *Calculated maximum upfield shifts (ppm) upon binding by host **5** in pD 9 buffer and by host **6** in chloroform*

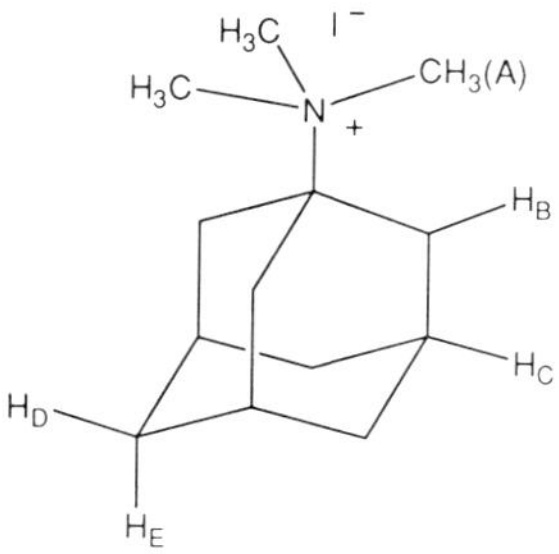

Proton	CDCl$_3$	Aqueous
A	2.99	1.87
B	2.92	2.99
C	1.11	1.18
D	1.07	1.29
E	0.67	0.73

strength is substantially reduced if the *p*-xylylene groups in **5** are replaced by 1,4-cyclohexyldimethylene units. The authors explained the preferred, tight binding of onium ions by **5** and related cyclophanes by an ion–dipole effect, in which the positive charge of the guest is 'solvated' by the electron-rich faces of the aromatic rings of the host. Similar attractive ion–dipole interactions have been advanced as a force for stabilizing the secondary structure of proteins.[23]

The relevance of ion–dipole interactions between onium ions and the aromatic π-systems of cyclophane **5** was demonstrated in an impressive way in studies with host **6** in organic solvents like chloroform.[22b] Here, unlike in water, the desolvation of apolar host and guest surfaces is not a substantial

---

[23] S. K. Burley and G. A. Petsko, *FEBS Lett.*, 1986, **203**, 139.

complexation driving force. Dougherty *et al.* found that onium derivatives form inclusion complexes with **6** in chloroform, whereas substrates without the quaternary ammonium center do not bind at all (Table 4.2). Similar to water, ^{1}H NMR complexation shifts in CDCl$_3$ show that the quaternary ammonium center of 1-trimethylammoniumadamantane iodide is more deeply buried in the host cavity than the rest of the molecule (Table 4.3). This onium ion–aromatic π-system interaction which leads to a complexation strength of 2–3 kcal mol^{-1} in chloroform is a remarkable, previously poorly recognized, driving force for molecular recognition.

In the complexes of water-soluble **5** discussed above, ion–dipole stabilization is effected through the interactions of the charged center of the guest with the aromatic π-systems of the host. A similar stabilization should be operative if a host orients charged centers in close vicinity to the π-systems of encapsulated aromatic guests. This was indeed demonstrated by Schneider *et al.* in studies with the tetraaza[8.1.8.1]paracyclophanes **3** and **4**.[21] The association constant at 298 K for the naphthalene complex of cyclophane **3** ($K_a = 920$ L mol^{-1}) in D$_2$O/CD$_3$OD (80:20, v/v) is five times larger than the association constant for the complex of **4** ($K_a = 180$ L mol^{-1}). If apolar binding would represent the sole inclusion force, host **4** with its apolar cavity

Table 4.4 *Complexation free energies* $-\Delta G°$ *(kcal mol^{-1}) at 298 K in D$_2$O for complexes of **3** and **4**. The values for the naphthalene derivatives are estimates in D$_2$O. All other values are measured in D$_2$O/CD$_3$OD (80:20, v/v)*

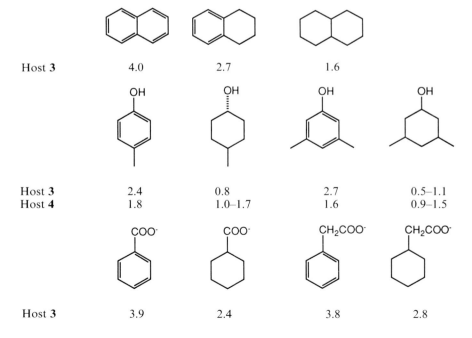

would have formed the more stable complex. Again, ion–π-system host–guest interactions (Figure 4.2) give the best explanation for the larger stability of the 3·naphthalene complex. A similar complex stability is observed if the $-NMe_2^+-$ centers in **3** are replaced by $-NH_2^+-$ groups.

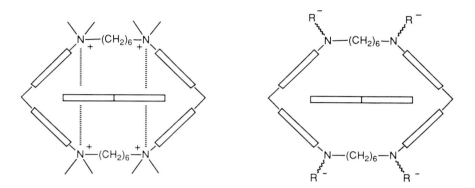

Figure 4.2 *Schematic drawings of the naphthalene complexes of cyclophanes **3** and **4**. In the complex of **3**, the quaternary nitrogens of the host are in close vicinity of the π-system of the guest, and ion–dipole interactions are effective*

In an attempt to further separate the ion–dipole term from the apolar binding term, Schneider *et al.* compared the complexation of similarly sized aromatic and aliphatic substrates by cyclophanes **3** and **4**.[24] Table 4.4 shows that complexes of **3** with aromatic guests generally are considerably more stable than the complexes of aliphatic guests. The transition from an aromatic to a similarly sized aliphatic guest is accompanied by a decrease in complex stability amounting to 1–1.3 kcal mol^{-1} per benzene ring. Complexes of aromatic guests and **3** are specifically stabilized by the $N^+\cdots\pi$-electron attractions. This stabilization is not operative in complexes of host **4** since its charged centers are too remote (Figure 4.2). Also, the negatively charged centers in **4** should be repulsive to the π-cloud of the aromatic guests. Therefore, the difference in stability between the complexes of **4** with aromatic and aliphatic substrates is much smaller (Table 4.4).

These results suggest that the good binding of neutral aromatics by the protonated or quaternized tetraaza[*n*.1.*n*.1]paracyclophane hosts introduced by Koga[25] and by other cyclophane receptors with positively charged centers in the binding site periphery (see Chapter 2) may also result to a large extent from ion–dipole interactions.

[24] H.-J. Schneider, T. Blatter, S. Simova, and I. Theis, *J. Chem. Soc., Chem. Commun.*, 1989, 580.
[25] K. Odashima and K. Koga in 'Cyclophanes, Vol. 2', eds. P. M. Keehn and S. M. Rosenfeld, Academic, New York, 1983, pp. 629.

4.4 Attractive Coulombic Interactions at the Origin of Molecular Cation and Molecular Anion Complexation

4.4.1 Complexes of Onium Ions

Receptors **5** and **6**, discussed in Chapter 4.3, derive their onium ion specificity from ion–dipole interactions between the aromatic π-systems of the host and the charged ammonium group of the encapsulated guest. In contrast, attractive charge–charge interactions predominantly determine the bonding between the anionic hosts **7**,[26] **8**,[27] and **9** (as the tetraphenolate)[28,29] and onium ions in aqueous solutions. Ion–dipole and dispersion interactions are additional, secondary driving forces for the formation of these complexes. Table 4.5 summarizes some of the interesting binding results obtained in aqueous solutions with **7** (at neutral pH) and **9** (at basic pH).

In aqueous solution at pH 7, host **7** forms 1:1 complexes with a variety of quaternary ammonium and diammonium salts.[26] Large complexation-induced upfield shifts of the ^{1}H NMR resonances of the guests are taken as strong indication for cavity inclusion. Of particular interest, from a biological point of view, is the complex formed by the neurotransmitter acetylcholine (**10**). The forces leading to the recognition of this compound at receptor binding sites and enzyme active sites, *e.g.* of acetylcholinesterase, are not well known. Studies with synthetic acetylcholine binders such as **7** could provide new insights into the origin of specific neurotransmitter–neuroreceptor interactions. Methylviologen, MV^{++} (paraquat, **11**), is also strongly bound by **7** ($K_a > 10^5$ L mol^{-1}). This redox-active compound has been intensively studied as an electron carrier[30] in electron transfer and photochemical energy storage processes and is of practical use as a herbicide.[31]

In alkaline solution, cyclophane **9** forms a tetraphenolate which adopts the bowl-type structure **12**, stabilized by a cyclic array of four strong hydrogen bonds. Further deprotonation of **12** is extremely difficult. The remaining four protons are not removed even by an excess of NaOD or by NaOCD$_3$ in a CD$_3$OD solution. In near diffusion-controlled processes, **12** forms a variety of stable 1:1 complexes with onium salts (Table 4.5). The stability of these complexes is remarkable in view of the open bowl-type binding site. Trimethylammonium guests, *e.g.* choline and carnitine, form more stable compounds than the guests with larger alkyl residues at the ammonium

---

[26] M. Dhaenens, L. Lacombe, J.-M. Lehn, and J.-P. Vigneron, *J. Chem. Soc., Chem. Comm.*, 1984, 1097.
[27] F. Vögtle, T. Merz, and H. Wirtz, *Angew. Chem.*, 1985, **97**, 226–227; *Angew. Chem. Int. Ed. Engl.*, 1985, **24**, 221.
[28] H.-J. Schneider, D. Güttes, and U. Schneider, *Angew. Chem.*, 1986, **98**, 635; *Angew. Chem. Int. Ed. Engl.*, 1986, **25**, 647.
[29] H.-J. Schneider, D. Güttes, and U. Schneider, *J. Am. Chem. Soc.*, 1988, **110**, 6449.
[30] I. Willner, N. Lapidot, and A. Riklin, *J. Am. Chem. Soc.*, 1989, **111**, 1883.
[31] H. Colquhoun, F. Stoddart, and D. Williams, *New Scientist*, May 1, 1986, 44.

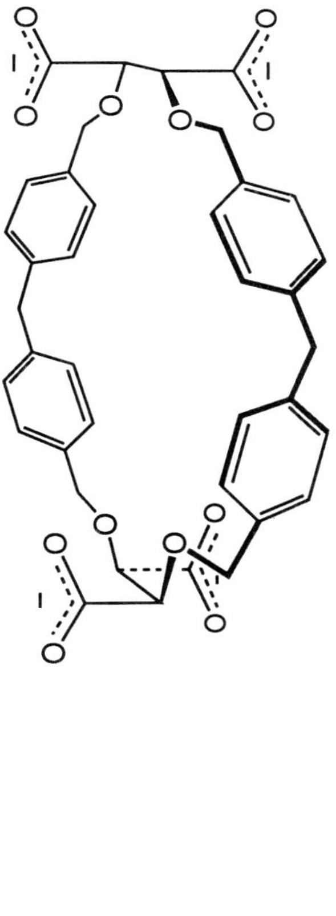

Cyclophane Complexes of Charged Organic Guests

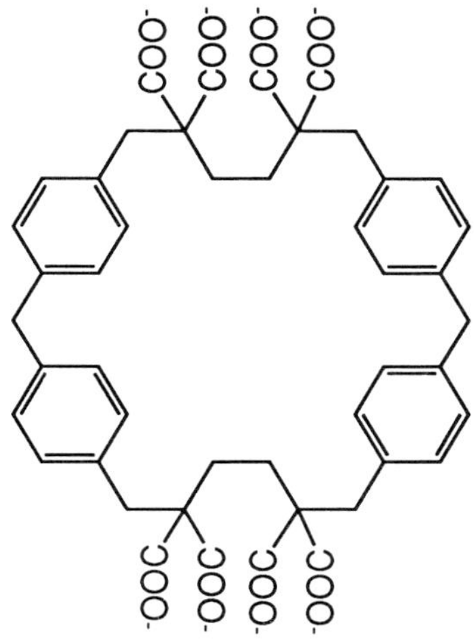

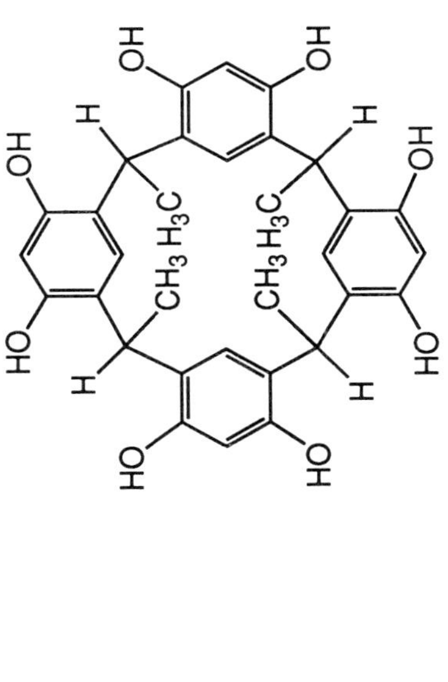

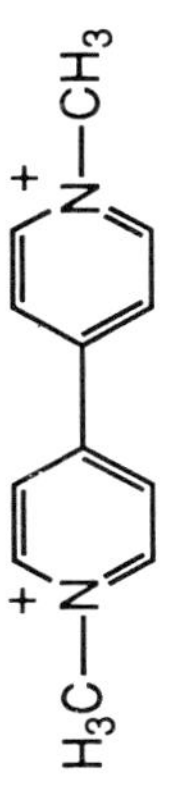

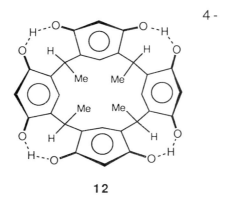

12

Table 4.5 *Association constants (K_a) and free energies of complexation ($-\Delta G°$) for the 1:1 complexes formed between cyclophanes* **7** *and* **9** *and onium ions in aqueous solutions*

Guest	K_a [L^{-1} mol]	$-\Delta G°$ [kcal mol^{-1}]
(a) Complexes of **7** *(296 K, pH 7.0)*		
NMe_4^+	2.5×10^2	3.2
$MeCO_2CH_2CH_2NMe_3^+$ (**10**)	5.0×10^2	3.6
$PhCO_2CH_2CH_2NMe_3^+$	8.0×10^2	3.9
$^+Me_3N(CH_2)_3NMe_3^+$	2.0×10^4	5.8
Methylviologen (**11**)	$\geq 1 \times 10^5$	>6.7
(b) Complexes of **9** *(298K, 0.5 N NaOH)*		
NMe_4^+	2.9×10^4	6.1
NEt_4^+	3.4×10^3	4.8
$N(n\text{-}C_3H_7)_4^+$	30	2.0
$N(n\text{-}C_4H_9)_4^+$	<2	<0.5
$^+Me_3NCH_2CH_2OH$ (choline)	5.0×10^4	6.4
$^+Me_3NCH_2CH(OH)CH_2COO^-$ (carnitine)	6.5×10^3	5.2

nitrogen. Only the charged headgroups of trimethylammonium guests can penetrate the cavity establishing close proximity to the anionic centers of the host (Figure 4.3A). Electrostatic and anisotropic shielding by the host induces ^{1}H NMR upfield complexation shifts of ≈ 2 ppm for the headgroup methyl protons. Guests having larger ammonium headgroups do not fit into the bowl, and the resulting larger separation of the charged centers in the complex leads to weaker binding (Figure 4.3B). Bifunctional ammonium compounds act as ditopic substrates and form 2:1 in addition to 1:1 host–guest complexes (Figure 4.3C). The complexes of **12** are certainly stabilized to a large extent by Coulombic interactions between the charged centers. However, the findings of

Dougherty et al. suggest[22] that a significant ion–dipole effect (Chapter 4.3) should also exist in those complexes in which the trimethylammonium headgroup binds into the aromatic bowl. Apolar binding is of reduced importance; substrates without ammonium center, e.g. tert-butanol (binding free energy $-\Delta G° = 1.1$ kcal mol^{-1}) are only bound very weakly.

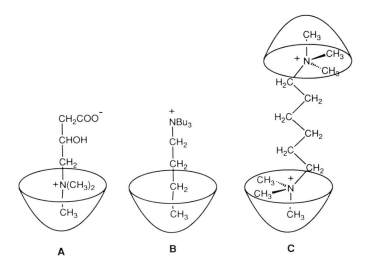

Figure 4.3 *Schematic drawings of the complexes of tetraphenolate 12: (A) Strong binding of the tetramethylammonium headgroup of carnitine; (B) Weak binding of the larger tetra(n-butyl)ammonium ion, and (C) 2:1 host–guest complexation with ditopic guests*

4.4.2 Anion Receptors

Anion recognition is an abundant process in nature since anionic substrates participate in approximately 70% of all enzymatic reactions.[32] It is, therefore, not surprising that the study of anion complexation by synthetic organic receptors has developed into a major area of molecular recognition.[33] By complexing a chloride ion in the cavity of a diprotonated macrobicyclic diamine, Simmons and Park in 1968 reported the first example of anion complexation by a synthetic receptor.[34] Subsequently, a rich variety of molecular shapes for selective anion recognition have been developed.[32,33]

---

[32] F. P. Schmidtchen, *Nachr. Chem. Tech. Lab.*, 1988, **36**, 8.
[33] J.-M. Lehn, *Angew. Chem.*, 1988, **100**, 91; *Angew. Chem. Int. Ed. Engl.*, 1988, **27**, 89.
[34] H. E. Simmons and C. H. Park, *J. Am. Chem. Soc.*, 1968, **90**, 2428.

Only a few synthetic anion binders possess cyclophane structures, and almost all of them were prepared in the group of J.-M. Lehn.[35-38]

Protonation of the six secondary amine groups in compounds **13** and **14** yields macrobicyclic hexa-ammonium ions which form complexes with a variety of anionic substrates in aqueous solution.[35a] The stability constants K_a for the complexes of the monovalent anions Cl^-, NO_3^-, and N_3^- vary between 3×10^2 and 1×10^4 L mol^{-1}. Divalent anions, e.g. SO_4^{2-}, $S_2O_6^{2-}$, and oxalate $C_2O_4^{2-}$ form more stable complexes with K_a values between 1×10^5 and 3×10^6 L mol^{-1}. In the ^{1}H NMR spectra of solutions containing an excess of [(**13**-6H$^+$)(CF$_3$SO$_3^-$)$_6$] and oxalate or nitrate as guests, the signals of both complexed and free ligand are observed at 25 °C, which is indicative of slow anion exchange.

The structures of the complexes formed by **13**-6H$^+$ and **14**-6H$^+$ are not well defined. The anions could form inclusion complexes with full cavity incorporation. Alternatively, they could bind outside the cavity through hydrogen-bonding to the NH$_2^+$-sites on one face of the macrobicycles. Inclusion-type binding should preferentially occur with the larger host **14**-6H$^+$. Smaller anions could fit into the cavity of the smaller receptor **13**-6H$^+$. But larger anions (like SO_4^{2-}, $S_2O_6^{2-}$, tosylate) presumably are bound in the exclusive mode. In the X-crystal structures of the hexanitrate and hexatosylate of **13**-6H$^+$, the anions all are located outside the cavity.

The bis(diazapyrene) derivatives **15** and **16** were designed to provide space for the intercalative inclusion of flat substrates.[36] This complexation mode was nicely confirmed by the X-ray crystal structure of the complex formed between **15** and nitrobenzene. Figure 4.4 shows the orientation of the nitrobenzene molecule sandwiched between the two 1,4,5,8-naphthalenetetracarboxylic acid diimides in **15**. In chloroform, an excess of nitrobenzene quenches the host fluorescence which was taken as an indication for intercalation occurring also in solution.

In the tetraprotonated form, cyclophane **16** is water soluble and acts as an anion binder. Marked upfield shifts of the aromatic ^{1}H NMR resonances of 1,4-benzenedicarboxylate ($\Delta\delta = 0.83$) and 2,6-naphthalenedicarboxylate in the presence of **16**·4H$^+$ are indicative of host–guest complexation. The analysis of the NMR data revealed a complex stoichiometry close to 1:1 and gave approximate association constants of $K_a \approx 10^4$ mol L^{-1} ($T \approx 298$ K). Although the intercalative binding of nitrobenzene by **15** in the X-ray crystal structure suggests a similar binding mode for the aromatic anions by **16**·4H$^+$, external π–π-stacking (Chapter 2.2) has not been excluded as an alternative binding geometry.

---

[35] (a) D. Heyer and J.-M. Lehn, *Tetrahedron Lett.*, 1986, **27**, 5869; (b) J. Jazwinski, J.-M. Lehn, R. Méric, J.-P. Vigneron, M. Cesario, J. Guilhem, and C. Pascard, *Tetrahedron Lett.*, 1987, **28**, 3489.

[36] J. Jazwinski, A. J. Blacker, J.-M. Lehn, M. Cesario, J. Guilhem, and C. Pascard, *Tetrahedron Lett.*, 1987, **28**, 6057.

[37] T. Fujita and J.-M. Lehn, *Tetrahedron Lett.*, 1988, **29**, 1709.

[38] R. A. Pascal, Jr., J. Spergel, and D. van Engen, *Tetrahedron Lett.*, 1986, **27**, 4099.

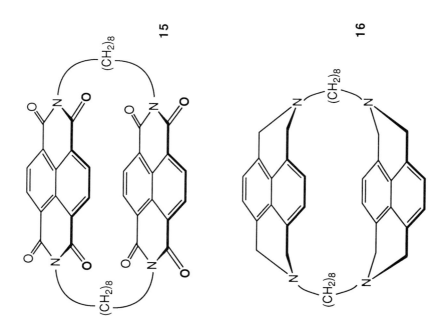

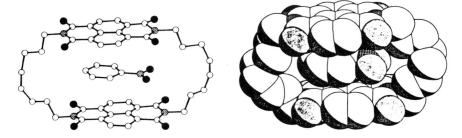

Figure 4.4 *Intercalative binding by **15** as revealed by the X-ray crystal structure of the **15**·nitrobenzene complex (Reproduced with permission from reference 36. Copyright 1987 Pergamon Press.)*

With **17**-3H$^+$ and **18**-6H$^+$, dome-shaped structures with an intramolecular cavity for anion inclusion were prepared.[37] Having three-fold symmetry, these receptors were designed to preferentially recognize trigonal anions of compatible size such as nitrate. ^{1}H NMR binding studies did reveal that the two receptors do form stable and symmetrical complexes with the nitrate anion, in which NO$_3^-$ is presumably included in the ligand cavity.

17 X = O; n = 4

18 X = 2H; n = 3

---

[39] H.-J. Schneider and I. Theis, *Angew. Chem.*, 1989, **101**, 757; *Angew. Chem. Int. Ed. Engl.*, 1989. **28**, 753.

A remarkable result was obtained by Schneider et al. in an analysis of the energetics of salt bridges for thirty five different synthetic host–guest complexes, including those formed by tetracationic **3** and the tetraphenolate **12**.[39] For this analysis, they separated, on the basis of experimental comparisons, the overall binding free energy into a Coulombic term (for the salt bridges) and a hydrophobic term (for dispersion interactions and desolvation driving forces). When they plotted the Coulombic term as a function of the number of salt bridges, they obtained a surprisingly uniform value of 1.2 ± 0.24 kcal mol^{-1} as the measurable electrostatic interaction per ion pair in aqueous solution.[39] This result is surprising considering that the cation–anion combinations had quite different sizes, polarizabilities, and charge densities. Examples are $^+$NH$_2$R$_2$, $^+$NR$_4$ (R = alkyl), aromatic positively charged nitrogen heterocycles, R–CO$_2^-$, RO–PO$_3$H$^-$, RO–PO$_3^{2-}$, and charge-delocalized phenolates. Apparently, it is sufficient for a net ion pairing interaction of ≈ 1.2 kcal mol^{-1} to occur if the two oppositely charged centers are able to align in a strainfree way in close vicinity to each other. This good correlation, which includes many complexes with several charged centers at each binding partner (e.g. in **3** and **12**) also suggests that only the charge–charge interactions between nearest neighbors are the most relevant which is in agreement with the distance dependency expressed in Coulomb's law.

The incremental value of 1.2 ± 0.24 kcal mol^{-1} for salt bridges in water should be quite useful for the design of new synthetic host–guest complexes and for predicting relative binding strengths. In addition, it should have relevance for molecular recognition events in biological systems where salt bridges are abundant. For example, ion pairing occurs in the complexes of charged substrates and coenzymes with enzymes and other receptors. Also, multiple salt bridges stabilize protein dimers and complexes formed between different proteins.

4.5 Cyclophane Subunits as Apolar Binding Sites in Polytopic Receptors

Receptors with multiple binding sites in a well defined spatial orientation are rare. However, for several reasons, they are among the most worthwhile targets for future molecular recognition studies.[33,40] Multiple binding sites lead to increased binding affinity. Breslow et al. showed that cyclodextrin dimers form extraordinarily stable complexes ($K_a > 10^8$ L mol^{-1}) in water with appropriate substrates that interact simultaneously with both cavities.[41] This antibody-type binding affinity is reached since the combination of two apolar binding sites offers a larger surface area for attractive host–guest dispersion interactions and for favorable desolvation processes. Binding at multiple recognition sites should also exhibit very high guest selectivity. It is poorly understood how the binding of a guest at one complexation site

---

[40] J. Rebek, Jr., *Acc. Chem. Res.*, 1984, **17**, 258.
[41] R. Breslow, N. Greenspoon, T. Guo, and R. Zarzycki, *J. Am. Chem. Soc.*, 1989, **111**, 8296.

influences the binding affinity at a second, adjacent site. Studies with di or polytopic hosts should mimic cooperativity and allosteric effects in biological recognition.[42] Synthetic systems should allow to explore how binding information can be transferred selectively from one binding site to a second binding site over large distances in a molecule. Finally, the prospect of catalyzing reactions between two different guests bound to two proximate recognition sites adds further interest to the development of polytopic hosts.[43]

Koga et al. prepared the bis(paracyclophane) 19 with two independent binding sites and examined the complexation of the ditopic guest 21 by ^{1}H NMR in D_2O.[44] The analysis of the complexation-induced upfield shifts of the guest resonances indicated the formation of a stable 1:1 host–guest complex in low concentration ranges (e.g. $H_0 = G_0 = 1 \times 10^{-3}$ mol L^{-1}). In this complex, the two phenylsulfonate groups of the guest occupy the two cyclophane cavities of 19 (Scheme 4.2). In contrast, a mono-paracyclophane host, identical to one of the binding sites in 19, does not lead to efficient 2:1 host–guest complex formation with 21 in the same millimolar concentration range where the ditopic binding by 19 is very efficient. Even in the presence of a large excess (20 mM) of the monotopic host, the formation of a 1:1 complex is largely favored over the formation of a 2:1 host–guest complex. This can be taken as a strong indication for cooperativity in the binding of the two aromatic groups of 21 by the ditopic host 19. If the length of the chain between the two phenylsulfonate residues in the guest is too small, e.g. in 20, double inclusion of these groups by 19 seems to be prevented for reasons of steric and electronic repulsion (Scheme 4.2).

Several approaches towards molecular receptors containing two different recognition sites, one of them being an apolar cyclophane cavity, and the other a crown-type cation binding site, have been described. In compound 22, a tetraza[3.3.3.3]paracyclophane is capped by a diaza-18-crown-6.[45] Whereas binding of alkylammonium ions to the crown binding site was demonstrated by ^{1}H NMR, evidence for apolar guest inclusion into the apolar site was not reported (see Chapter 2.2).

The apolar binding site in the ditopic receptor 23, prepared by Saigo et al.,[46,47] is considerably larger than in 22. The authors studied the interaction between 23 and a series of (ω-phenylalkyl)ammonium picrates (24a–g) in ^{1}H NMR titrations in $CDCl_3/CD_3OD$ (4:1, v/v). All guests form 1:1 complexes with 23. Interestingly, the association constants for the complexes of the (5-phenylpentyl)ammonium (24c, $K_a = 1730$ L mol^{-1}) and (6-phenylhexyl)ammonium picrates (24d, $K_a = 1700$ L mol^{-1}) are approximately 3 times

---

[42] A. Fersht, 'Enzyme Structure and Mechanism', 2nd ed., Freeman, New York, 1985.
[43] T. R. Kelly, C. Zhao, and G. J. Bridger, J. Am. Chem. Soc., 1989, 111, 3744.
[44] C.-F. Lai, K. Odashima, and K. Koga, Tetrahedron Lett., 1985, 26, 5179.
[45] A. D. Hamilton and P. Kazanjian, Tetrahedron Lett., 1985, 26, 5735.
[46] K. Saigo, R.-J. Lin, M. Kubo, A. Youda, and M. Hasegawa, J. Am. Chem. Soc., 1986, 108, 1996.
[47] K. Saigo, N. Kihara, Y. Hashimoto, R.-J. Lin, H. Fujimura, Y. Suzuki, and M. Hasegawa, J. Am. Chem. Soc., 1990, 112, 1144.

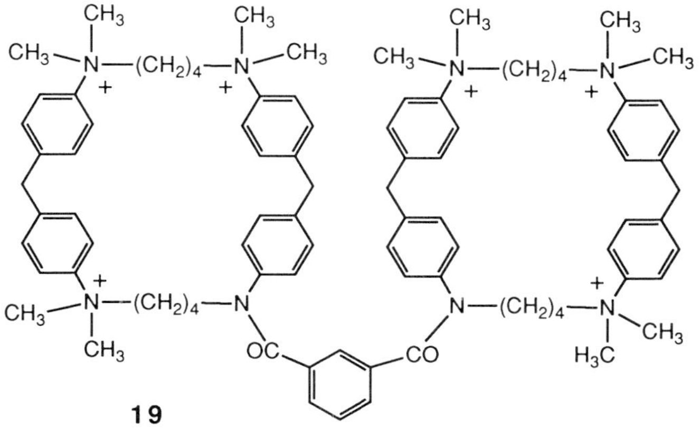

Cyclophane Complexes of Charged Organic Guests   159

19

Na$^+$ $^-$O$_3$S—⟨C$_6$H$_4$⟩—O—(CH$_2$CH$_2$O)n—O—⟨C$_6$H$_4$⟩—SO$_3^-$ $^+$Na

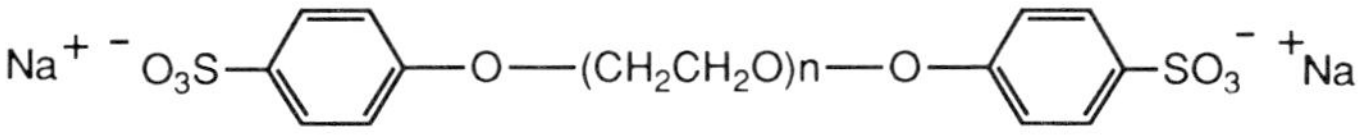

20 n = 1
21 n = 5

larger than those for the residual complexes (K_a's between 260 and 620 L mol^{-1} at 293 K). In all complexes, the anchoring of the primary ammonium ion of the guest to the crown ether of the host represents the major association force. The complexes of **24c** and **24d** apparently encounter additional stabilization through interactions between the cyclophane in **23** and the phenyl groups of these guests (Scheme 4.3). It appears that the alkyl chain between ammonium center and phenyl ring in these guests is long enough to locate the aromatic ring into the cyclophane cavity while maintaining favorable *anti* torsional angles in the chain. In **24a** and **24b**, the alkyl chains are too short for cyclophane–phenyl contacts. The apparent inefficiency of ditopic binding in the complexes of guests **24e–g** could be explained by unfavorable torsional angles of the longer alkyl chains. These chains presumably would need to twist in order to establish the specific distance between phenyl and ammonium group that is required for ditopic binding.

Ditopic binding leads to a more ordered, less flexible complex geometry. Therefore, the binding free energies for the complexes of **24c** and **24d** are characterized by very large negative entropy components. For example, the thermodynamic characteristics for complexation of **24c** at 298 K are $\Delta G° = -4.3$ kcal mol^{-1}, $\Delta H° = -6.7$ kcal mol^{-1}, and $\Delta S° = -7.6$ cal mol^{-1} K^{-1} as compared to $\Delta G° = -3.4$ kcal mol^{-1}, $\Delta H° = -1.9$ kcal mol^{-1}, and $\Delta S° = +5.2$ cal mol^{-1} K^{-1} for the binding of **24a**.

160   Chapter 4

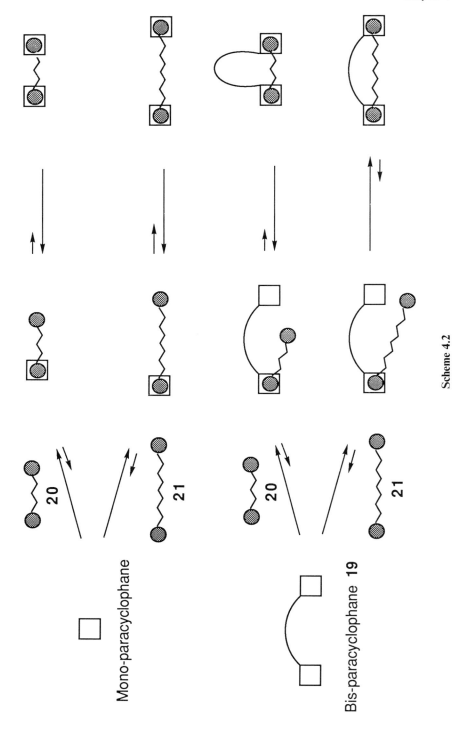

Scheme 4.2

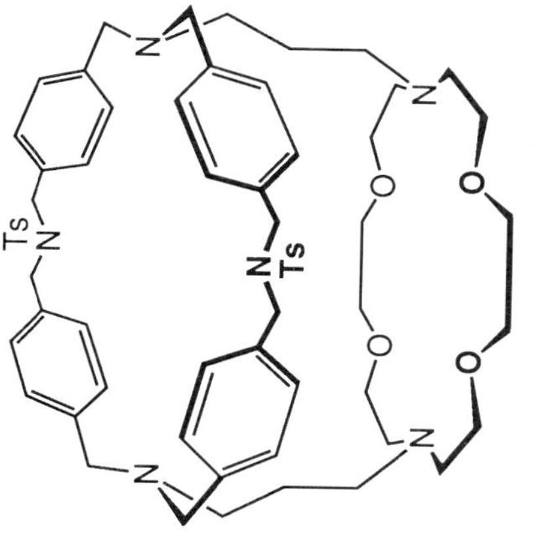

Cyclophane Complexes of Charged Organic Guests

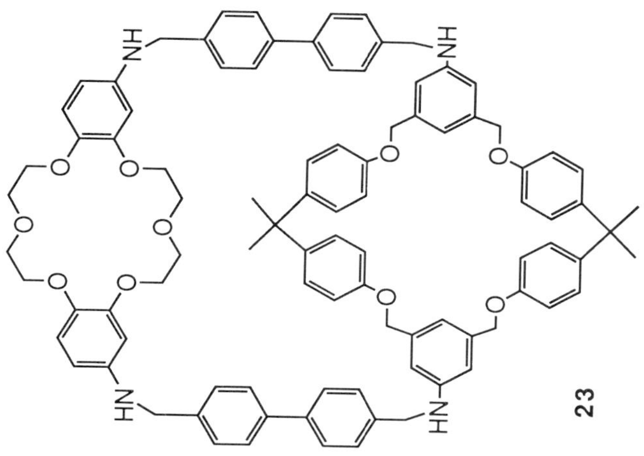

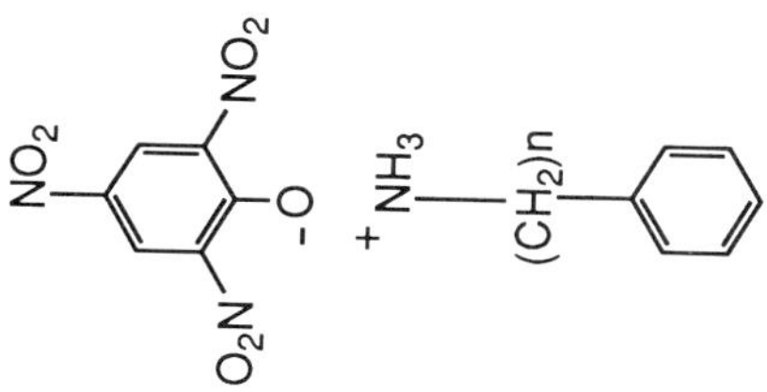

n	
24a	3
b	4
c	5
d	6
e	7
f	8
g	9

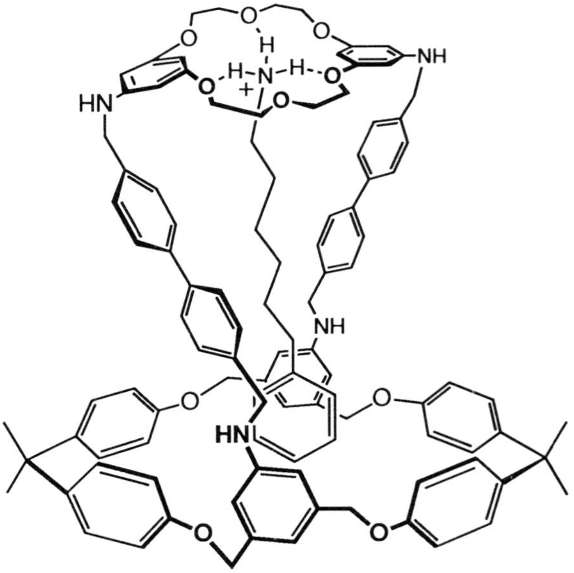

Complex 23·24c

Scheme 4.3

In the ditopic cyclophane **25**,[48,49] the 2,2′,7,7′-tetrahydroxy-1,1′-binaphthyl unit shapes at its minor groove (Scheme 4.4) an efficient cation binding site,[50,51] whereas the much wider major groove shapes an apolar binding cavity suitable for aromatic guests, *e.g.* benzene and naphthalene derivatives. Other cyclophanes that incorporate this versatile chiral spacer are the subject of Chapter 6. The complexation of an alkali metal cation and a neutral aromatic substrate are characterized by very different solvent dependencies. Therefore, studies with **25** were designed to explore whether the two opposite solvation characteristics would allow for the formation of a stable 1:1:1 (host–guest$_1$–guest$_2$) complex. The formation of 1:1 complexes between **25** and potassium metal ions or **25** and 6-methoxy-2-naphthonitrile was investigated by ^{1}H NMR in various water–methanol solvent mixtures. The stability of the complex with 6-methoxy-2-naphthonitrile decreases with increasing methanol content of the binary mixture (Chapter 3.6), whereas the stability of the potassium cation complex increases with increased methanol content. Figure 4.5 shows the opposite solvation characteristics of the two complexation processes. It is obvious that **25** can be switched from an efficient binder of

[48] F. Diederich, M. R. Hester, and M. A. Uyeki, *Angew. Chem.*, 1988, **100**, 1775; *Angew. Chem. Int. Ed. Engl.*, 1988, **27**, 1705.
[49] M. R. Hester, M. A. Uyeki, and F. Diederich, *Isr. J. Chem.*, 1989, **29**, 201.
[50] D. J. Cram and J. M. Cram, *Acc. Chem. Res.*, 1978, **11**, 8.
[51] D. J. Cram and K. N. Trueblood, *Top. Curr. Chem.*, 1981, **98**, 43.

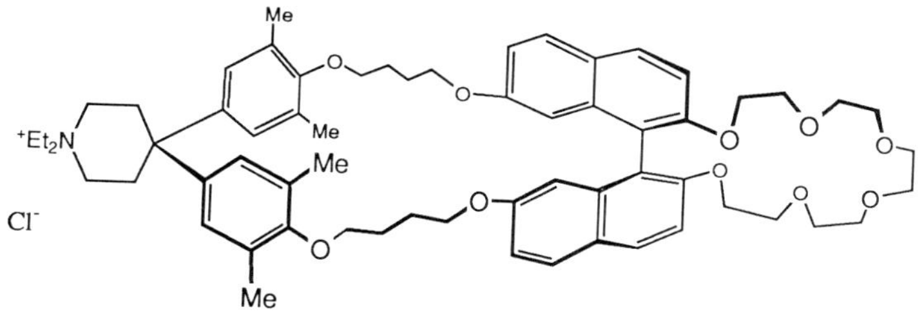

25

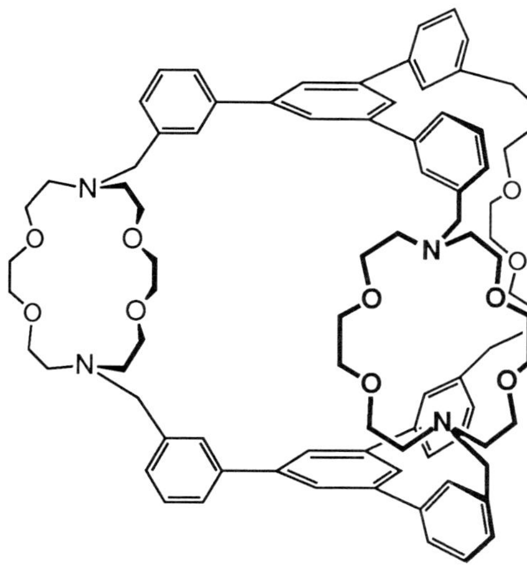

26

neutral aromatics to a good host for potassium cations by decreasing the water content of the water–methanol mixture.

The formation of a 1:1:1 (**25**·K$^+$·6-methoxy-2-naphthonitrile) complex was observed in all four solvent mixtures shown in Figure 4.5. Although the two binding sites in **25** are connected by the conformationally flexible binaphthyl unit,[49,51] their activities are almost independent of each other. Complexation of each guest in the 1:1:1(host–guest$_1$–guest$_2$) complex is approximately as strong as in the corresponding 1:1 complexes.

Compound **26**[52] is another polytopic receptor[53,54] with one cyclophane cavity for the complexation of organic guest molecules and three crown ether

---

[52] F. Vögtle, A. Wallon, W. M. Müller, U. Werner, and M. Nieger, *J. Chem. Soc., Chem. Commun.*, 1990, 158.
[53] G. R. Brown, S. S. Chana, J. F. Stoddart, A. M. Z. Slawin, and D. J. Williams, *J. Chem. Soc., Perkin Trans. 1*, 1989, 211.
[54] P. D. Beer, *Chem. Soc. Rev.*, 1989, **18**, 409.

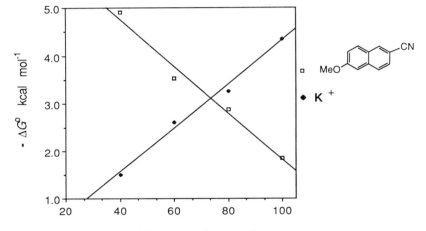

Figure 4.5 *Complexation of potassium cations and 6-methoxy-2-naphthonitrile by cyclophane 25 in various D_2O/methanol-d_4 mixtures (Reproduced with permission from reference 48. Copyright 1988 VCH.)*

binding sites for cation co-complexation. ^{1}H NMR complexation shifts provided evidence for the cavity inclusion of 2,6- and 2,7-naphthalenediols in acidic aqueous solution (D_2O/DCl, pD 1.2). Also, complexes of 1:3 host–guest stoichiometry were formed in ethyl acetate with Rb^+ and guanidinium cations. The co-complexation of neutral organic substrates and these cations, subject to the opposite solvation characteristics shown in Figure 4.5, was not reported.

4.6 From Second-sphere Coordination to Catenane Formation

4.6.1 Second-sphere Coordination

Ligands in the primary coordination sphere of a transition metal complex can interact in an ordered manner with neutral molecules or charged species to give second-sphere or outer-sphere complexes. This concept was first advanced in 1912 by Alfred Werner[55] and has found broad application in supramolecular chemistry. The variety of complexes in which a first-sphere ligand interacts *via* weak non-covalent interactions with a second ligand has been comprehensively reviewed by Stoddart and coworkers.[56] Chapter 4.6 describes, exclusively, fascinating work performed by these researchers.[57]

Crown ethers are very suitable for forming adducts with neutral and cationic transition metal complexes carrying protic ligands (*e.g.* NH_3, H_2O, CH_3CN) in their first coordination sphere. X-Ray analysis showed that the second-sphere complex between *trans*-Pt(PMe$_3$)Cl$_2$(NH$_3$) and dibenzo-18-crown-6 (**27**) is held together predominantly by tripod hydrogen-bonding between the NH_3-ligand of the transition metal and the ether oxygens of the crown.[58]

A variety of attractive interactions stabilize the second coordination sphere complexes formed between [Pt(bipy)(NH$_3$)$_2$]$^{2+}$ and larger dibenzocrown ethers, *e.g.* dibenzo-24-crown-8 (**28**) and dibenzo-30-crown-10 (**29**) in hydrocarbon and halocarbon solvents.[59,60] Figure 4.6 shows the X-ray crystal structure of the complex formed with the larger crown **29**. The two amine ligands at platinum are directed towards the polyether chains forming hydrogen bonds to oxygen atoms. The bipy ligand is π-stacked between the two crown catechols with short contacts between the two aromatic systems measuring 3.48–3.52 Å. Electron donor–acceptor (EDA) interactions (Chapter 3.3) between the π-electron-deficient bipy ligand and the π-electron-rich catechols stabilize the complex. The EDA interaction includes a charge-transfer (CT) component; solutions of the complex in acetonitrile are deep yellow, and the electronic absorption spectrum shows a broad CT band centered at 350 nm. By monitoring changes in the charge-transfer absorption through optical binding titrations, the association constant and the free binding energy for the 1:1 complex of [Pt(bipy)(NH$_3$)$_2$]$^{2+}$ and **29** were determined as $K_a = 1.9 \times 10^5$ L mol^{-1} and $-\Delta G° = 7.2$ kcal mol^{-1} ($T = 298$ K). Substantial upfield shifts of the aromatic ^{1}H NMR resonances of both binding partners indicate that this stable complex adopts in solution a geometry similar to the solid state geometry shown in Figure 4.6.

4.6.2 Complexes of the Herbicides Diquat and Paraquat

The bipyridinium herbicide diquat [DQT]$^{2+}$ possesses structural and electronic similarity to the transition metal complex [Pt(bipy)(NH$_3$)$_2$]$^{2+}$ and,

---

[55] A. Werner, *Ann. Chem.*, 1912, **386**, 1.
[56] H. M. Colquhoun, J. F. Stoddart, and D. J. Williams, *Angew. Chem.*, 1986, **98**, 483; *Angew. Chem. Int. Ed. Engl.*, 1986, **25**, 487.
[57] J. F. Stoddart, *Pure Appl. Chem.*, 1988, **60**, 467.
[58] H. M. Colquhoun, D. F. Lewis, J. F. Stoddart, and D. J. Williams, *J. Chem. Soc., Dalton Trans.*, 1983, 607.
[59] H. M. Colquhoun, J. F. Stoddart, D. J. Williams, J. B. Wolstenholme, R. Zarzycki, *Angew. Chem.*, 1981, **93**, 1093; *Angew. Chem. Int. Ed. Engl.*, 1981, **20**, 1051.
[60] H. M. Colquhoun, S. M. Doughty, J. M. Maud, J. F. Stoddart, D. J. Williams, and J. B. Wolstenholme, *Isr. J. Chem.*, 1985, **25**, 15.

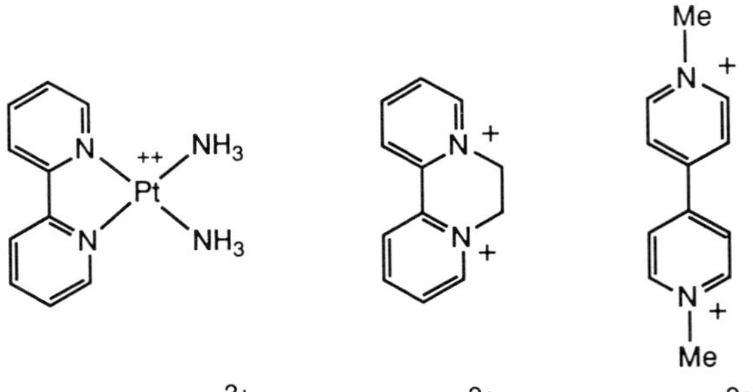

[Pt(bipy)(NH₃)₂]²⁺   [DQT]²⁺   [PQT]²⁺

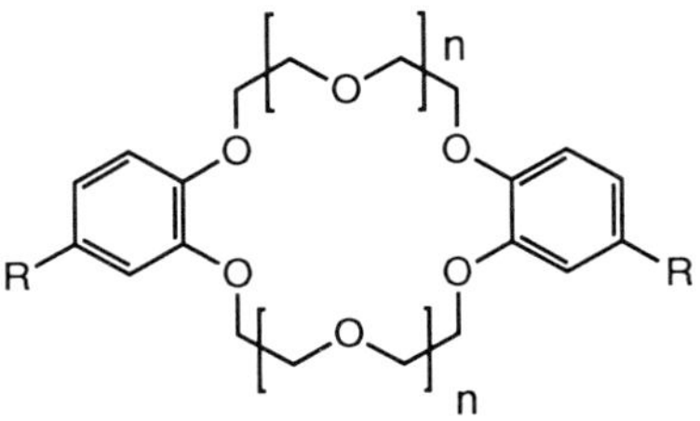

27 n = 1   R = H
28 n = 2   R = H
29 n = 3   R = H
30 n = 3   R-R = CH$_2$—O⟨C$_6$H$_4$⟩O—CH$_2$

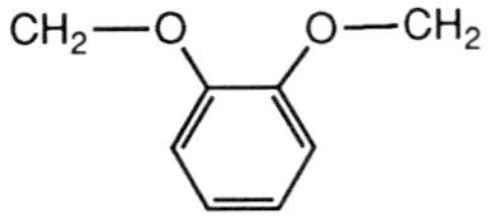

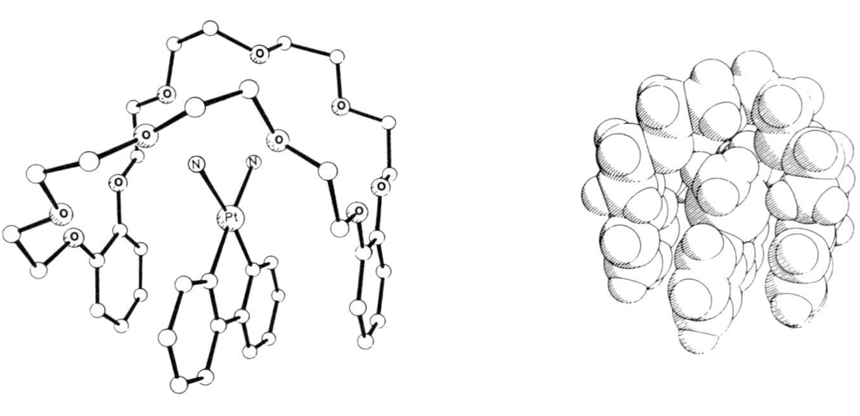

Figure 4.6   *X-ray crystal structure of the complex of dibenzo-30-crown-10 (**29**) and $[Pt(bipy)(NH_3)_2]^{2+}$*
(Reproduced with permission from reference 59. Copyright 1981 VCH.)

therefore, similarities in the binding of these two compounds to dibenzo-crowns could be expected. Not surprisingly, [DQT]$^{2+}$ forms stable 1:1 complexes both in solution (acetonitrile, acetone) and in the solid state with a variety of macromonocyclic and macrobicyclic dibenzocrowns. Figures 4.7 and 4.8 show the X-ray crystal structure of the complexes with **29**[61,62] and the macrobicyclic receptor **30**.[63] Characteristic ^{1}H NMR complexation shifts support similarly ordered structures for the stable complexes that form in acetone solution. These complexes are stabilized by: (i) strong ion–dipole interactions between the onium nitrogens and the polyether chains, (ii) weak hydrogen-bonding between ether oxygens and the hydrogen atoms on carbons α to the charged nitrogens, (iii) dispersion forces, and (iv) EDA interactions from stacking between the π-electron-deficient bipyridyl and the electron-rich catechol units. The EDA interactions include a charge-transfer component, and the complexes are highly colored both in solution and in the solid state. A similar combination of association forces stabilize other diquat and paraquat complexes described in the following paragraphs.

The strongest binding in acetone ($T = 298$ K) occurs between diquat and dibenzo-30-crown-10 (**29**) ($K_a = 1.7 \times 10^4$ L mol^{-1}, $-\Delta G° = 5.8$ kcal mol^{-1}). The complexes of smaller or larger dibenzocrowns are less stable. At $K_a > 9 \times 10^5$ L mol^{-1} ($-\Delta G° > 8.1$ kcal mol^{-1}), the binding constant for the complex of the macrobicyclic ligand **30** in acetone is even larger. The macrobicyclic ligand **30** is more preorganized than **29**. Hence, a smaller amount of the intrinsic host–guest binding energy needs to be invested to rearrange **30** into a productive conformation for the accommodation of diquat. The X-ray structures of free receptor **30** and its diquat complex (Figure 4.8) demonstrate that the host is indeed largely preorganized for the complexation event.[63]

Whereas diquat [DQT]$^{2+}$ forms stable complexes with dibenzocrowns and in particular with **29**, paraquat [PQT]$^{2+}$ (**11**, Chapter 4.4.1), another bipyridinium herbicide, is not efficiently complexed by these ligands. However, stable complexes with both [DQT]$^{2+}$ and [PQT]$^{2+}$ are formed by crown ethers incorporating two hydroquinone residues, e.g. the bisparaphenylene-34-crown-10 derivative **31**.[64,65] Figure 4.9 shows the X-ray crystal structure of the **31**·paraquat complex which also forms in acetone solution ($K_a = 730$ L mol^{-1}, $-\Delta G° = 3.9$ kcal mol^{-1}, 298 K).[65] The planar electron-deficient bipyridinium dication is sandwiched face-to-face between the two π-electron-rich hydroquinone ether rings. Decreasing or increasing the macro-

---

[61] H. M. Colquhoun, E. P. Goodings, J. M. Maud, J. F. Stoddart, D. J. Williams, and J. B. Wolstenholme, *J. Chem. Soc., Chem. Commun.*, 1983, 1140.
[62] F. H. Kohnke, J. F. Stoddart, B. L. Allwood, and D. J. Williams, *Tetrahedron Lett.*, 1985, **26**, 1681.
[63] B. L. Allwood, F. H. Kohnke, J. F. Stoddart, and D. J. Williams, *Angew. Chem.*, 1985, **97**, 584; *Angew. Chem. Int. Ed. Engl.*, 1985, **24**, 581.
[64] B. L. Allwood, N. Spencer, H. Shahriari-Zavareh, J. F. Stoddart, and D. J. Williams, *J. Chem. Soc., Chem. Commun.*, 1987, 1061.
[65] B. L. Allwood, N. Spencer, H. Shahriari-Zavareh, J. F. Stoddart, and D. J. Williams, *J. Chem. Soc., Chem. Commun.*, 1987, 1064.

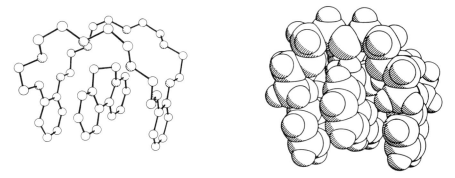

Figure 4.7 *X-ray crystal structure of the complex of dibenzo-30-crown-10 (**29**) and the diquat dication*
(Reproduced with permission from reference 61. Copyright 1983 Royal Society of Chemistry, London.)

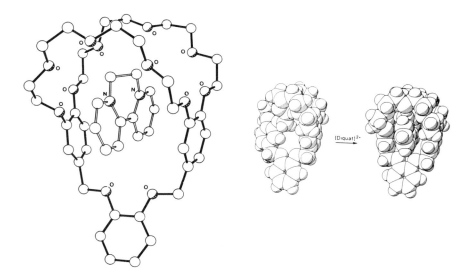

Figure 4.8 *X-ray crystal structure of the complex of the macrobicyclic receptor **30** and the diquat dication*
(Reproduced with permission from reference 63. Copyright 1985 VCH.)

ring size leads to reduced complexation strength. Crystals of the **31**·paraquat complex are red, and the charge-transfer absorption maximum in acetone is found at $\lambda = 436$ nm. Figure 4.10 shows the superposition of the X-ray crystal structures of **31** and its paraquat complex. The preorganization of the free ligand **31** in the solid state is impressive. Interestingly, the X-ray structure in

Figure 4.9 shows that the cyclophane aryl–ether oxygens are placed above and below the carbon atoms α to the nitrogen atoms in $[PQT]^{2+}$. In contrast, the predominant orientation in the corresponding $[DQT]^{2+}$ complex[64] is the co-linearity of two phenolic ether oxygen atoms with the nitrogen atoms of the bridged 2,2'-bipyridinium cation.

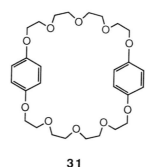

31

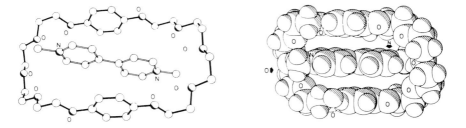

Figure 4.9   *X-ray crystal structure of the complex of the bisparaphenylene-34-crown-10 derivative* **31** *and the paraquat dication*
(*Reproduced with permission from reference 65. Copyright 1987 Royal Society of Chemistry, London.*)

In acetone solution, the diquat and paraquat dications also form 1:1 complexes with crown macrorings incorporating two 1,5-naphthalenediol units. However, a different stoichiometry (1:2 host–guest) was revealed by X-ray analysis for the solid state complex formed by compound **32** and paraquat.[66] Figure 4.11 shows the crystal structure of this 1:2 complex which provides an impressive example of polymolecular π-donor–π-acceptor stacking. Naphthalene and bipyridinium rings alternate in continuous stacks and show a high degree of overlap at short interannular distances.

---

[66] J.-Y. Ortholand, A. M. Z. Slawin, N. Spencer, J. F. Stoddart, and D. J. Williams, *Angew. Chem.*, 1989, **101**, 1402; *Angew. Chem. Int. Ed. Engl.*, 1989, **28**, 1394.

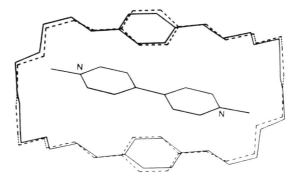

Figure 4.10 *Best least-squares fit of the X-ray crystal structures of the bisparaphenylene-34-crown-10 derivative 31 and its paraquat dication complex*
(Reproduced with permission from reference 65. Copyright 1987 Royal Society of Chemistry, London.)

32 n = 3

4.6.3 From Paraquat Complexation and Paraquat-cyclophanes to a [2]Catenane

The macroring **31** complexes paraquat (Chapter 4.6.2),[65] and the bis(paraquat)-cyclophane **33** binds hydroquinone dimethyl ether (Chapter 3.3).[67,68] These findings, schematically summarized in Figure 4.12, together with the observation that paraquat and electron-rich aromatic rings can form continuous, alternating π–π-stacks (Figure 4.11),[66] led the Stoddart group to conclude that the two cyclophanes **31** and **33** should be capable of forming the new [2]catenane **34**.[69-71]

[67] B. Odell, M. V. Reddington, A. M. Z. Slawin, N. Spencer, J. F. Stoddart, and D. J. Williams, *Angew. Chem.*, 1988, **100**, 1605; *Angew. Chem. Int. Ed. Engl.*, 1988, **27**, 1547.
[68] P. R. Ashton, B. Odell, M. V. Reddington, A. M. Z. Slawin, J. F. Stoddart, and D. J. Williams, *Angew. Chem.*, 1988, **100**, 1608; *Angew. Chem. Int. Ed. Engl.*, 1988, **27**, 1550.
[69] G. Schill, 'Catenanes, Rotaxanes, and Knots', Academic Press, New York, 1971.
[70] C. O. Dietrich-Buchecker and J. P. Sauvage, *Chem. Rev.*, 1987, **87**, 795.
[71] C. O. Dietrich-Buchecker and J. P. Sauvage, *Angew. Chem.*, 1989, **101**, 192; *Angew. Chem. Int. Ed. Engl.*, 1989, **28**, 189.

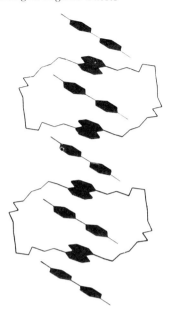

Figure 4.11 *X-ray crystal structure of the complex of the cyclophane **32** and the paraquat dication*
(Reproduced with permission from reference 66. Copyright 1989 VCH.)

When the *p*-xylylene-bisparaquat salt **35** (schematically shown in Scheme 4.5) was stirred at room temperature in acetonitrile with 1 equivalent of 1,4-bis(bromomethyl)benzene (**36**) and 2.5 equivalent of **31**, the [2]catenane **34** was obtained as tetrakis(hexafluorophosphate) salt in the excellent yield of 70%.[72] This high yield undoubtedly must be assigned to a favorable template effect in the final cyclization step. This effect, schematically shown in Scheme 4.5, originates from the tendency of the paraquat and hydroquinone diether units to form stacks with alternating π-donor and π-acceptor rings. The identity of the [2]catenane was clearly revealed by its *X*-ray crystal structure (Figure 4.13). The hydroquinone unit in the cavity takes exactly the same orientation that the hydroquinone dimethyl ether does in complex **33** (Chapter 3.3). The intra-cavity hydroquinone in **34** undergoes face-to-face interactions with the paraquat units and edge-to-face interactions with the *p*-phenylene units. The [2]catenane molecules are assembled in regular, continuous stacks in the crystal with distances of ≈ 3.5 Å between the alternating donor and acceptor units.

The specific orientation of aromatic donor and acceptor units seen in the molecular structure of the solid catenane as well as the bonding interactions at

[72] P. R. Ashton, T. T. Goodnow, A. E. Kaifer, M. V. Reddington, A. M. Z. Slawin, N. Spencer, J. F. Stoddart, C. Vicent, and D. J. Williams, *Angew. Chem.*, 1989, **101**, 1404; *Angew. Chem. Int. Ed. Engl.*, 1989, **28**, 1396.

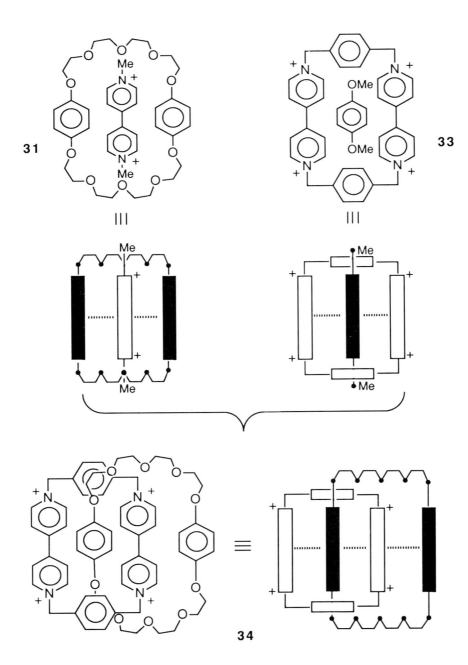

Figure 4.12 *Design concept leading to the [2]catenane 34*

Cyclophane Complexes of Charged Organic Guests 173

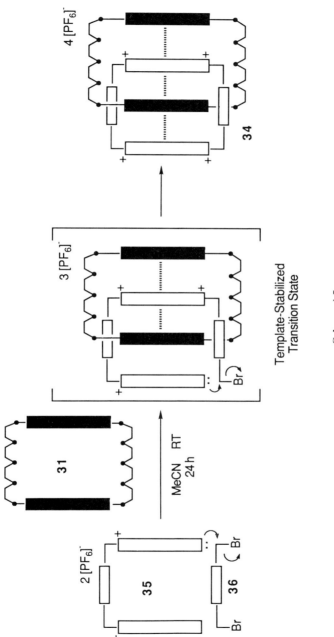

Scheme 4.5

the origin of this ordered structure are also preserved in solution. Changes from this geometry proceed *via* dynamic processes with considerable activation barriers. The [2]catenane **34** possesses 4 degenerate structures characterized by the four stacking arrays ABA'B', AB'A'B, A'BAB', and A'B'AB; A and A' labeling the two paraquat units, B and B' the two hydroquinones (Scheme 4.6). According to dynamic ^{1}H NMR studies in acetone and acetonitrile, two dynamic processes interconvert these degenerate structures of **34**. In a higher-energy pathway (I, Scheme 4.6), the 34-crown-10-macrocycle **31** rotates around one of the paraquat units in paraquat-cyclophane **33**, leading to the exchange of the hydroquinone rings B and B' between 'inside' and 'outside'. In acetonitrile, this process proceeds with an activation barrier of $\Delta G^{\neq}_{354\ K} = 14.0$ kcal mol^{-1}. In the second, energetically more favorable rearrangement (II in Scheme 4.6; $\Delta G^{\neq}_{250\ K} = 12.2$ kcal mol^{-1} in acetone-d_6), the paraquat units of **33** exchange between 'inside' and 'outside'. This exchange presumably occurs in a sweep of one hydroquinone unit in **31** around the external periphery of the paraquat-cyclophane by pirouetting about the O–C$_6$H$_4$–O axis of the other hydroquinone ring.

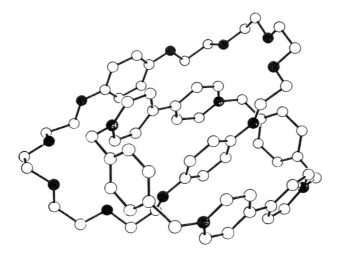

Figure 4.13 *X-ray crystal structure of the complex of the [2]catenane **34** (Reproduced with permission from reference 72.Copyright 1989 VCH.)*

By cyclic voltammetry, it was shown that the reduction potentials of the paraquat rings in the [2]catenane **34** are non-equal and differ considerably from the corresponding potentials in the free cyclophane **33** (Figure 4.14). The 1e$^-$-reduction of both paraquat units in the catenane is less facile than in **33** since the hydroquinone donor rings stabilize the e$^-$-accepting paraquat dications. This stabilization is more pronounced for the intra-cavity paraquat unit which is sandwiched between two hydroquinones. Therefore, the 1e$^-$-reduction potential of the encapsulated paraquat dication is more negative

Cyclophane Complexes of Charged Organic Guests 175

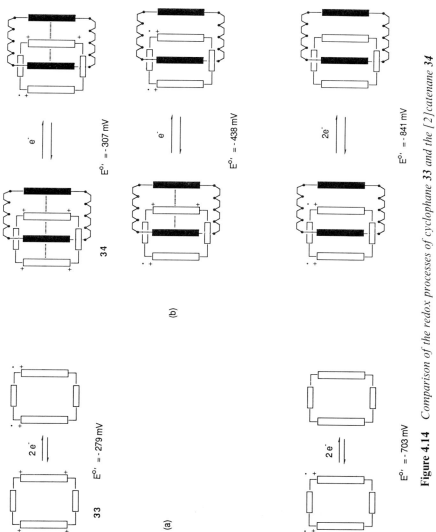

Figure 4.14 Comparison of the redox processes of cyclophane **33** and the [2]catenane **34**

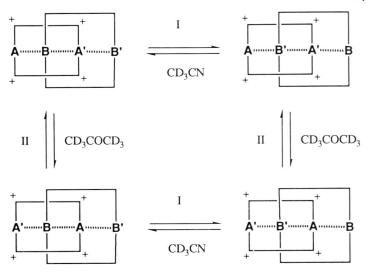

A, A' : paraquat units
B, B' : hydroquinones

Scheme 4.6

($E^{o\prime} = -438$ mV) than the reduction potential of the external paraquat unit ($E^{o\prime} = -307$ mV). Figure 4.14 shows that the subsequent reduction of the bis(radical cations) to neutral species is also more facile in the simple cyclophane **33** than in the catenane **34**.

The novel [2]catenane **34** is characterized by having: (i) a short, high-yielding template-directed synthesis, (ii) a highly ordered structure in solution and in the solid state as a result of strong EDA interactions with a distinct charge-transfer component, (iii) the tendency to form continuous π–π stacks of alternating donor and acceptor units, and (iv) unique redox properties. It is not difficult to predict that [2]catenane **34** is the first member of a new class of exciting organic materials. If cyclophane components are prepared that allow the inclusion of two aromatic rings, the formation of [3]catenanes and, ultimately, oligomeric and polymeric structures should become possible.

4.7 Azaparacyclophanes at the Interface between Molecular Receptors and Micellar Aggregates

The group of Murakami *et al.* was among the first involved in the design and preparation of water-soluble cyclophanes to bind apolar solutes (Chapter 2.2).[73] Based on the tetraaza[3.3.3.3]paracyclophane framework, these

---

[73] Y. Murakami, *Top. Curr. Chem.*, 1983, **115**, 107.

researchers constructed a variety of receptor types.[74] Although many of these compounds associate with neutral substrates, most studies describe the formation of complexes that are stabilized by Coulombic charge–charge interactions.

In compounds **37** and **38**, long alkyl chains are attached to the four nitrogens of the tetraaza[3.3.3.3]paracyclophane frame.[73,75] Apolar binding interactions in aqueous solutions were studied by several techniques. ESR spectroscopic measurements showed that the lipophilic nitroxide spin probe **40** is placed into a less polar environment upon association with **37** places (Chapter 2.4). Optical binding studies were performed with dyes whose electronic transitions are strongly dependent on the polarity of the environment. These investigations showed that Coulombic interactions provide a strong contribution to the substrate binding energy by **37** and **38**. The anionic dye, orange G, associates with cationic **37** but not with anionic **38**. Similarly, the cationic dye rhodamine 6G only interacts with the anionic compound **38**.

The geometries of the host–guest associates formed by the flexible systems **37** and **38** are not well defined. Murakami *et al.* proposed the substrate binding mode schematically shown in Figure 4.15A.[75] The tetraaza[3.3.3.3]paracyclophane is too small for deep cavity inclusion of aromatic rings or alicyclic guests (Chapter 2.2). Rather, it serves as a matrix for the four side-arms which shape a hydrophobic space for substrate incorporation. The binding mode schematically shown in Figure 4.15A can be viewed as substrate incorporation into a unimolecular micelle.[76,77]

To further enhance the hydrophobic binding efficiency, the octopus cyclophane **39**[78] with eight hydrophobic chains was prepared. Similar to **37** and **38**, the eight side-chains of **39** shape a unimolecular micelle-type binding site for apolar substrates (Figure 4.15B). Fluorescence titrations demonstrated the formation of stable 1:1 complexes by the anionic substrates ANS ($K_a = 2.8 \times 10^5$ L mol^{-1}) or TNS ($K_a = 3.0 \times 10^5$ L mol^{-1}) as well as by the neutral compound *N*-phenyl-1-naphthylamine (PNA) ($K_a = 1.3 \times 10^6$ L mol^{-1}). These association constants at 303 K in water/methanol (95:5, v/v) are 1–3 magnitudes larger than those measured for the complexes of **37** with the same guests. The eight side chains in **39** obviously form a more efficient apolar binding site than the four side chains in **37**.[79]

In the next logical step, the lipophilic side arms departing from the tetraaza[3.3.3.3]paracyclophane ring were connected at the ends to form a macrocyclic cap. In **41**, this cap is a tetraazacyclotetradecane ring.[80] A CPK model of **41** shows a very large open cavity only when the C_{11}-chains adopt

---

[74] Y. Murakami and J. Kikuchi, *Pure Appl. Chem.*, 1988, **60**, 549.
[75] Y. Murakami, A. Nakano, K. Akiyoshi, and K. Fukuya, *J. Chem. Soc., Perkin Trans. 1*, 1981, 2800.
[76] F. M. Menger, M. Takeshita, and J. F. Chow, *J. Am. Chem. Soc.*, 1981, **103**, 5938.
[77] F. M. Menger, *Top. Curr. Chem.*, 1986, **136**, 1.
[78] Y. Murakami, J. Kikuchi, M. Suzuki, and T. Matsuura, *J. Chem. Soc., Perkin Trans. 1*, 1988, 1289.
[79] Y. Hisaeda, T. Ihara, T. Ohno, and Y. Murakami, *Tetrahedron Lett.*, 1990, **31**, 1027.
[80] Y. Murakami, J. Kikuchi, and H. Tenma, *J. Chem. Soc., Chem. Commun.*, 1985, 753.

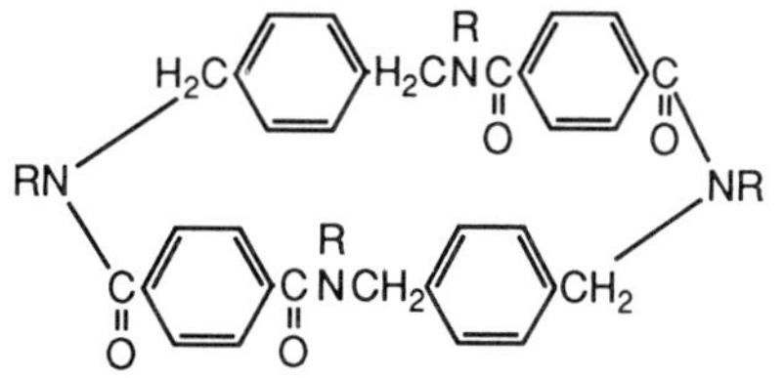

orange G

40

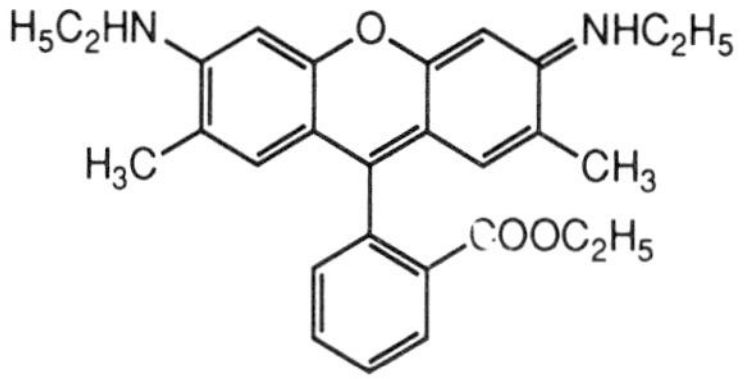

rhodamine 6G

PNA

37 R = (CH$_2$)$_{10}$N$^+$NMe$_3$ Br$^-$

38 R = (CH$_2$)$_{10}$COO$^-$

39 R = (CH$_2$)$_2$CNHCHCN[(CH$_2$)$_{13}$Me]$_2$ Cl$^-$
              (CH$_2$)$_4$NH$_3^+$

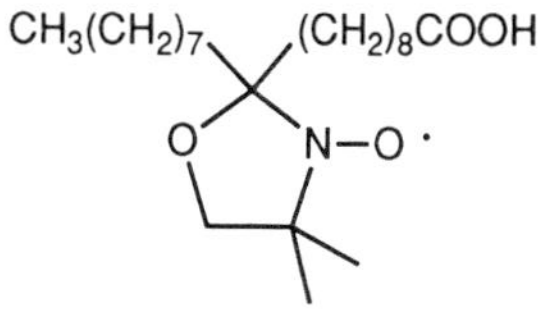

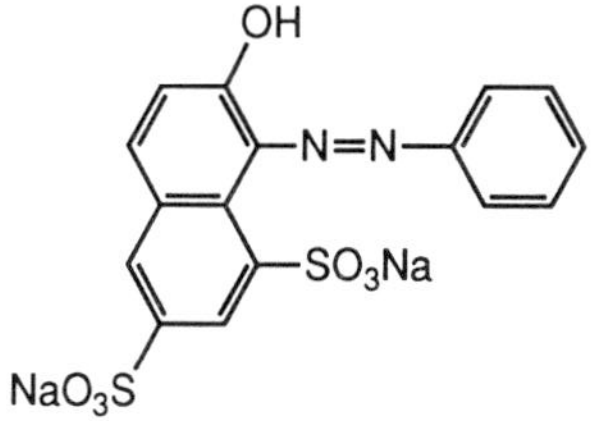

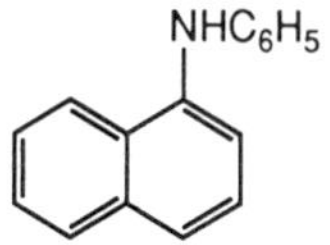

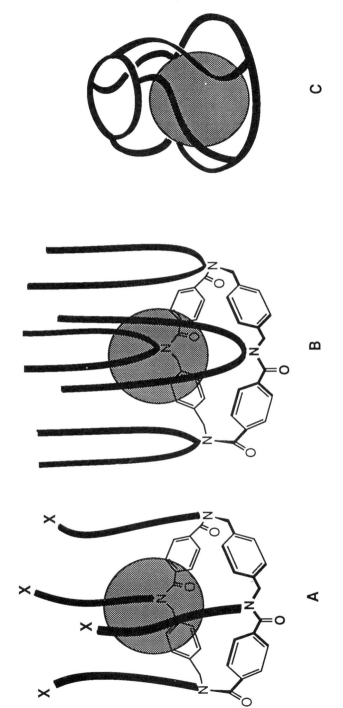

Figure 4.15 Schematic representations of substrate binding modes of (A) receptors **37** and **38**, (B) octopus cyclophane **39**, and (C) capped tetraazaparacyclophane **41**

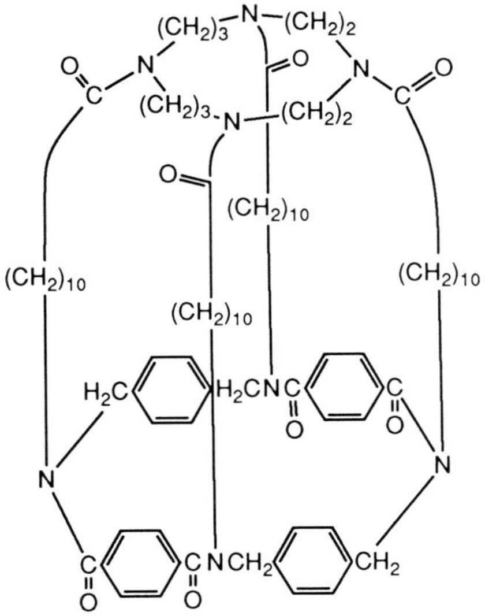

extended conformations characterized by a sequence of mostly *anti* torsional angles. Upon entrapment of naphthalene-sized guests, *e.g.* ANS or TNS, an intramolecular induced-fit type motion may lead to the collapse of the extended host into a more folded conformation capable of tighter guest encapsulation (Figure 4.15C). Such a collapse would establish closer van der Waals contacts between the two binding partners and reduce the exposure of their apolar surfaces to water; two major driving forces for apolar association. The binding between **41** and compounds like ANS and TNS was studied by fluorescence titrations performed in water/ethanol (9:1, v/v). An analysis of the complexation-induced shifts of the ^{1}H NMR resonances of both binding partners, which would support the proposed cavity inclusion (Figure 4.15C), has not been reported. Based solely on fluorescence studies, substrate incorporation into micelle-type aggregates formed by **41** or cavity binding cannot be determined. A large neutral molecule like **41** probably does not show monomeric molecular-dispersed water solubility under the conditions of the reported fluorescence binding titrations ($[\mathbf{41}]_0 = 1 \times 10^{-5}$ to 1×10^{-4} mol L^{-1}) (Chapter 2.4).

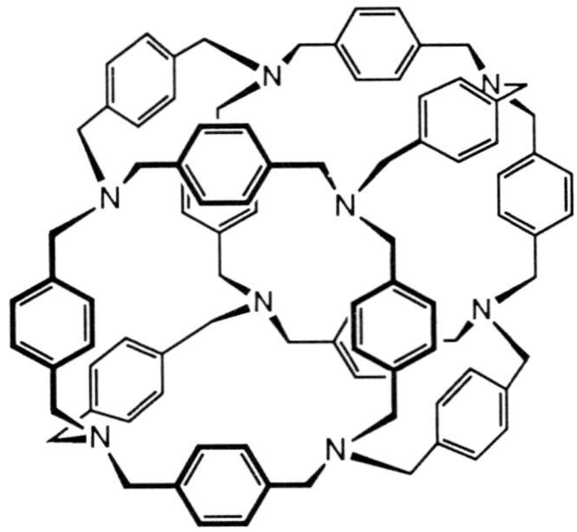

The use of tetraaza[3.3.3.3]paracyclophanes by the Murakami group culminated in the report of the 'cubic azaparacyclophane' **42** in which a relatively rigid, spacious apolar cavity is shaped by a total of six [3.3.3.3]paracyclophane faces.[81-83] According to CPK model examinations,

[81] Y. Murakami, J. Kikuchi, and T. Hirayama, *Chem. Lett.*, 1987, 161.
[82] Y. Murakami, J. Kikuchi, T. Ohno, and T. Hirayama, *Chem. Lett.*, 1989, 881.
[83] Y. Murakami, J. Kikuchi, T. Ohno, T. Hirayama, and H. Nishimura, *Chem. Lett.*, 1989, 1199.

42 possesses a large cavity with a maximum inner diameter of ≈ 9 Å. However, only guests capable of passing through the tetraaza[3.3.3.3]paracyclophane ring, with a hole size of *ca.* 5.5–6.5 Å, can be incorporated into the inner cavity of **42**. Although fluorescence studies support the association between protonated **42** and neutral or anionic substrates in acidic aqueous solution, ^{1}H NMR binding studies, which would provide convincing evidence for cavity inclusion complexation, remain to be reported.

CHAPTER 5
Hydrogen-bonded Complexes of Cyclophanes and Small Neutral Molecules

5.1 Introduction

Substrate specificity in the majority of biological recognition events is established through formation of unique hydrogen bonds between receptor and substrate. Recent developments demonstrate that hydrogen-bonding arrays and networks also provide suitable driving forces for the specific complexation of small, biologically relevant molecules by synthetic receptors in organic solvents. Prior to these studies, the role of hydrogen-bonding in the association between complementary DNA bases[1-3] and between these bases and other heterocycles[4,5] had been investigated in great detail. The interactions between small neutral molecules with H-donor centers and crown ethers had already been discussed by Pedersen in 1971.[6] Also, hydrogen-bonding interactions had been recognized as a major specificity factor in chromatographic enantiomer resolutions on chiral stationary phases.[7-10] However, the development of synthetic molecular receptors with organized hydrogen-bonding sites specifically for the binding of neutral molecules had not been vigorously pursued until the middle of the 1980s. At that time, Rebek and coworkers pioneered a fascinating new area of research with their successful design and comprehensive studies of cleft-type receptors aligned with converging functional groups for hydrogen-bonding of neutral mole-

---

[1] Y. Kyogoku, R. C. Lord, and A. Rich, *Proc. Natl. Acad. Sci. USA*, 1967, **57**, 250.
[2] G. G. Hammes and A. C. Park, *J. Am. Chem. Soc.*, 1968, **90**, 4151
[3] N. G. Williams, L. D. Williams, and B. R. Shaw, *J. Am. Chem. Soc.*, 1989, **111**, 7205.
[4] Y. Kyogoku, R. C. Lord, and A. Rich, *Nature (London)*, 1968, **218**, 69.
[5] N. T. Yu and Y. Kyogoku, *Biochim. Biophys. Acta*, 1973, **331**, 21.
[6] C. J. Pedersen, *J. Org. Chem.*, 1971, **36**, 1690.
[7] G. Blaschke, *Angew. Chem.*, 1980, **92**, 14; *Angew. Chem. Int. Ed. Engl.*, 1980, **19**, 13.
[8] Y. Dobashi and S. Hara, *J. Am. Chem. Soc.*, 1985, **107**, 3406.
[9] B. Feibush, M. Saha, K. Onan, B. Kargar, and R. Giese, *J. Am. Chem. Soc.*, 1987, **109**, 7531.
[10] W. H. Pirkle and T. C. Pochapsky, *Chem. Rev.*, 1989, **89**, 347.

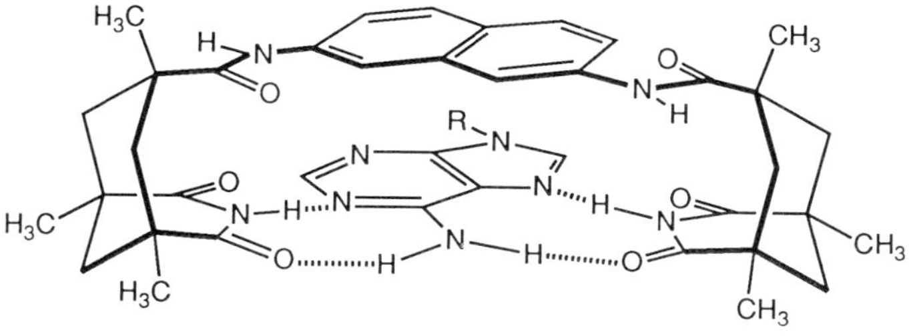

Scheme 5.1

cules.[11,12] Scheme 5.1 shows the stable complex formed in chloroform between adenine derivatives and one of these elegant clefts.

The development of nonmacrocyclic, cleft-type receptors has rapidly advanced into a major area of interest in molecular recognition research.[13-16] In addition, a variety of cyclophanes were prepared for small molecule recognition through hydrogen-bonding in organic solvents. These cyclophanes and their complexes with phenolic substrates, DNA-bases, barbiturates, and other small heterocycles like imidazole are the primary focus of Chapter 5. The studies with both clefts and cyclophanes over the past five years already have considerably advanced the understanding of multiple hydrogen-bonding interactions as a basis for selective recognition. Today, the binding ability of these receptors is still limited to organic solvents which do not compete with the substrate for hydrogen-bonding. Exciting developments in specific drug delivery and detoxification will be produced if future generations of hydrogen-bonding receptors show selective strong complexation in aqueous solution.

5.2 Small Neutral Molecule Binding to Crowns and Benzo-annelated Derivatives

5.2.1 Complexation by Unfunctionalized Macrocyclic Polyethers

Two of the pioneers of crown-cation complexation, C. J. Pedersen and D. J. Cram, also described first examples of the complexation between neutral molecules and crown ethers. In 1971, Pedersen observed the formation of solid-state complexes with thiourea and its analogues[6] and in 1974, Cram et al.

---

[11] J. Rebek, Jr., *Science (Washington, DC)*, 1987, **235**, 1478.
[12] J. Rebek, Jr., *Angew. Chem.*, 1990, **102**, 261; *Angew. Chem. Int. Ed. Engl.*, 1990, **29**, 245.
[13] S. C. Zimmerman and W. Wu, *J. Am. Chem. Soc.*, 1989, **111**, 8054.
[14] J. C. Adrian, Jr. and C. S. Wilcox, *J. Am. Chem. Soc.*, 1989, **111**, 8055.
[15] T. W. Bell and J. Liu, *J. Am. Chem. Soc.*, 1988, **110**, 3673.
[16] T. R. Kelly and M. P. Maguire, *J. Am. Chem. Soc.*, 1987, **109**, 6549.

published the formation of a crystalline complex between 18-crown-6 and acetonitrile.[17] In the years that followed, the X-ray crystal structures of a variety of complexes formed by crown ethers or benzo-annelated derivatives and neutral small molecules possessing C–H, N–H, or O–H acidic functionalities were determined.[18–22] Examples for guests that form complexes in the solid state with crown hosts are acetonitrile, nitromethane, malononitrile, dimethyl sulfoxide, hydrazine derivatives, aromatic amines, phenols, water, and alcohols. The X-ray crystallographic studies show that hydrogen-bonding between the acidic protons of the guests and the O,S,N-donor atoms of the crown hosts represents a major component of the host–guest interactions in addition to dipole–dipole and van der Waals dispersion forces. Computer modeling studies also support the conclusion that hydrogen-bonding is a major driving force for formation of these complexes.[23–25] Figure 5.1 illustrates the host–guest interactions in the crystal structure of the 1:2 complex formed between 2,6-pyrido-18-crown-6 (a *meta*-pyridinophane) and malononitrile.[26]

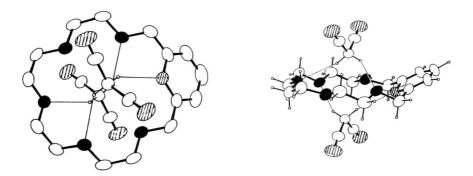

Figure 5.1 *X-ray crystal structure of the 1:2 host–guest complex formed between 2,6-pyrido-18-crown-6 and malononitrile (Reproduced with permission from reference 26. Copyright 1986 ACS.)*

---

[17] G. W. Gokel, D. J. Cram, C. L. Liotta, H. P. Harris, and F. L. Cook, *J. Org. Chem.*, 1974, **39**, 2445.
[18] F. Vögtle and W. M. Müller, *J. Incl. Phenom.*, 1983, **1**, 369.
[19] F. Vögtle, H. Sieger, and W. M. Müller, *Top. Curr. Chem.*, 1981, **98**, 107.
[20] F. Vögtle, H. Sieger, and W. M. Müller, in 'Host–Guest Complex Chemistry, Macrocycles (Synthesis, Structures, Applications)', eds. F. Vögtle and E. Weber, Springer, Berlin, 1985, pp. 319.
[21] F. Vögtle, W. M. Müller, and W. H. Watson, *Top. Curr. Chem.*, 1984, **125**, 131.
[22] A. Elbasyouny, H. J. Brügge, K. von Deuten, M. Dickel, A. Knöchel, K. U. Koch, J. Kopf, D. Melzer, and G. Rudolph, *J. Am. Chem. Soc.*, 1983, **105**, 6568.
[23] J. R. Damewood, Jr., W. P. Anderson, and J. J. Urban, *J. Comput. Chem.*, 1988, **9**, 111.
[24] J. R. Damewood, Jr., J. J. Urban, T. C. Williamson, and A. L. Rheingold, *J. Org. Chem.*, 1988, **53**, 167.
[25] P. D. J. Grootenhuis and P. A. Kollman, *J. Am. Chem. Soc.*, 1989, **111**, 4046.
[26] C. J. van Staveren, V. M. L. J. Aarts, P. D. J. Grootenhuis, J. van Eerden, S. Harkema, and D. N. Reinhoudt, *J. Am. Chem. Soc.*, 1986, **108**, 5271.

Complexation between C–H acidic guests like malononitrile, nitromethane, or acetonitrile and crown derivatives also occurs in apolar solvents like benzene and chloroform.[26-28] The free energies for formation of 1:1 complexes in these solvents range from $\Delta G° \approx -0.7$ to ≈ -3.2 kcal mol^{-1} at 298 K, and complexation is enthalpy-driven.[26] For example, the free energy for the 1:1 complexation between malononitrile and benzo-18-crown-6 in chloroform is $\Delta G° = -2.3$ kcal mol^{-1} with $\Delta H° = -5.4$ kcal mol^{-1} and $T\Delta S° = -3.1$ kcal mol^{-1}. These association constants are magnitudes smaller than those measured in apolar solvents for the complexes of crowns with alkali metal cations or primary ammonium ions. To increase the binding of neutral hydrogen-bonding molecules to macrocyclic polyethers, Reinhoudt et al. built additional recognition sites into their receptors. These developments are described in the following sections.

5.2.2 Complexation Assisted by Intraannular Acidic Groups Attached to Aromatic Rings of the Receptor

In 1977, Cram et al. reported on compounds 1–3 with carboxylate groups converging into the polyether binding site.[29] They noticed that the pK_a-values of these groups are increased compared to the pK_a of 2,6-bis(methoxymethyl)-benzoic acid. This pK_a increase was explained by the stabilization of the conjugate acid through intramolecular hydrogen-bonding to oxygen donor atoms of the macrorings. Support for this interpretation was found in the X-ray crystal structure of acid 2. By increasing the size of the macrorings beyond 4, the intramolecular hydrogen-bonding becomes less favorable, and the macrocyclic pK_a approaches those of acyclic comparison compounds.

Reinhoudt et al. proposed that an increase in pK_a, similar to those measured for 1–3 as a result of intramolecular hydrogen-bonding, should also occur if a water molecule complexes to larger macrocyclic polyether with an intraannular carboxylate and the conjugate acid becomes stabilized through hydrogen-bonding to the bound water molecule.[30] Evidence for the complexation of water molecules to larger crowns with or without convergent acidic sites was provided by a variety of X-ray crystal structures.[31,32] Figure 5.2 shows the X-ray crystal structure of the water complex of 2-carboxyl-1,3-xylyl-24-crown-7 (4).[30] The carboxylic acid proton is not dissociated and assists in the complexation of the water molecule which forms a total of three hydrogen

---

[27] P. A. Mosier-Boss and A. I. Popov, *J. Am. Chem. Soc.*, 1985, **107**, 6168.
[28] J. A. A. de Boer, D. N. Reinhoudt, S. Harkema, G. J. van Hummel, and F. de Jong, *J. Am. Chem. Soc.*, 1982, **104**, 4073.
[29] M. Newcomb, S. S. Moore, and D. J. Cram, *J. Am. Chem. Soc.*, 1977, **99**, 6405.
[30] C. J. van Staveren, V. M. L. J. Aarts, P. D. J. Grootenhuis, W. J. H. Droppers, J. van Eerden, S. Harkema, and D. N. Reinhoudt, *J. Am. Chem. Soc.*, 1988, **110**, 8134.
[31] P. D. J. Grootenhuis, J. W. H. M. Uiterwijk, D. N. Reinhoudt, C. J. van Staveren, E. J. R. Sudhölter, M. Bos, J. van Eerden, W. T. Klooster, L. Kruise, and S. Harkema, *J. Am. Chem. Soc.*, 1986, **108**, 780.
[32] G. R. Newkome, H. C. R. Taylor, F. R. Fronczek, T. J. Delord, D. K. Kohli, and F. Vögtle, *J. Am. Chem. Soc.*, 1981, **103**, 7376.

bonds to the host. Similarly, in water complexes of 2,6-pyridinium crown ether perchlorates, the pyridinium proton is hydrogen-bonded to the complexed water molecule and, according to X-ray crystallography, proton transfer to give a hydronium ion does not occur.[31] In contrast, complete proton transfer occurs in water complexes of macrocyclic polyethers bearing more acidic sulfonic acid residues as converging groups.[33] Here, the acidic proton is transferred to the complexed water molecule or, in the case of larger macrocycles, to a cluster of two or three bound water molecules.

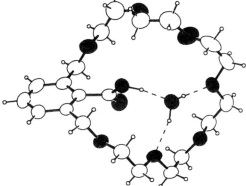

Figure 5.2 *X-ray crystal structure of the complex formed between water and 2-carboxyl-1,3-xylyl-24-crown-7 (4)*
(Reproduced with permission from reference 30. Copyright 1988 ACS.)

---

[33] J. van Eerden, M. Skowronska-Ptasinska, P. D. J. Grootenhuis, S. Harkema, and D. N. Reinhoudt, *J. Am. Chem. Soc.*, 1989, **111**, 700.

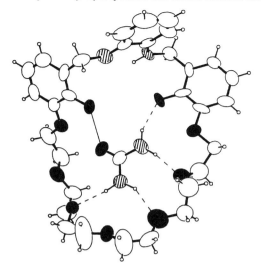

Figure 5.3 *X-ray crystal structure of the complex formed between urea and the phenolic crown ether* **5**
(Reproduced with permission from reference 30. Copyright 1988 ACS.)

An intraannular acidic functionality of a macrocyclic polyether can also stabilize complexes of other small polar molecules through hydrogen bonding. One of the most interesting small guest molecules is urea. Reinhoudt *et al.*[30,34] and other researchers[15] pursued the development of specific receptors capable of solubilizing urea in apolar solvents and mediating the selective transport of urea through liquid immobilized membranes. They showed that carboxylic and phenolic hydroxy groups converging into a crown ether ring assist the complexation of urea at a macrocyclic polyether binding site. Similar to bound water, complexed urea alters the pK_a of the converging group. If accurately measured, such a pK_a-change can provide evidence for the complexation of a neutral molecule by the crown.[34]

The cyclophane carboxylic acid **4** forms a stoichiometric complex with urea in aqueous solution.[30] The *X*-ray crystal structure shows that this complex is stabilized by a hydrogen bond between the nondissociated acid proton and the urea oxygen atom along with four hydrogen bonds between proximate crown oxygens and the urea hydrogens. Convergent phenolic groups are also effective in strengthening urea binding. Figure 5.3 shows the *X*-ray crystal structure of the urea complex formed by the bis(phenol) macroring **5**.[30] Urea is completely encapsulated within the macrocyclic cavity and acts as a donor in hydrogen bonds to three ether oxygens and to one of the phenolic oxygens. At 2.58 Å, the urea carbonyl oxygen is located at a very short distance to the other

---

[34] D. N. Reinhoudt and H. J. den Hertog, Jr., *Bull. Soc. Chim. Belg.*, 1988, **97**, 645.

phenolic oxygen. This short distance is indicative of a strong H-bond from the urea oxygen to the phenolic hydrogen. The stability of the urea complexes formed in water by **4** and analogous hosts is weak; association constants K_a vary around 1 L mol^{-1}. However, the reader should keep in mind that in aqueous solution, alkali metal cation binding by crown ethers is also weak. For the complex of 18-crown-6 with Na$^+$, a K_a-value of 6 L mol^{-1} is measured at ambient temperature.[35]

5.2.3 Urea Binding to Macrocyclic Polyethers Assisted by Co-complexation of Electrophilic Cations

In the complexes discussed above, the acidic proton of a functional group converging into the macroring strengthens urea binding by functioning as an electrophilic center which forms a hydrogen bond to the urea oxygen atom. In a continuation of their studies, Reinhoudt *et al.* explored whether co-complexed metal cations would also act as efficient electrophiles to assist urea binding.[36] They designed ditopic macrorings such as **6** having an ordered, highly polarizable binding site for soft cations and a more flexible crown binding site for urea. Upon deprotonation of the phenolic residues in **6**, very stable complexes form with soft cations, *e.g.* Ni^{2+}, Ba^{2+}, Cu^{2+}, or UO$_2^{2+}$. These cations bind to the two Schiff base nitrogens and the two phenoxides, leaving an open crown ether site for urea. Upon treating a solution of the **6**·UO$_2^{2+}$ complex with one equivalent of urea, a 1:1:1 complex precipitated. The X-ray crystal structure of this complex (Figure 5.4) shows that urea is encapsulated in the molecular cavity and binds to the vacant coordination site of the UO$_2^{2+}$ ion. Once again, hydrogen-bonding occurs between the urea hydrogens and the oxygen atoms of the flexible polyether macroring. Co-complexation is not limited to urea; solid ternary complexes also form with other small hydrogen-bonding molecules like formamide.

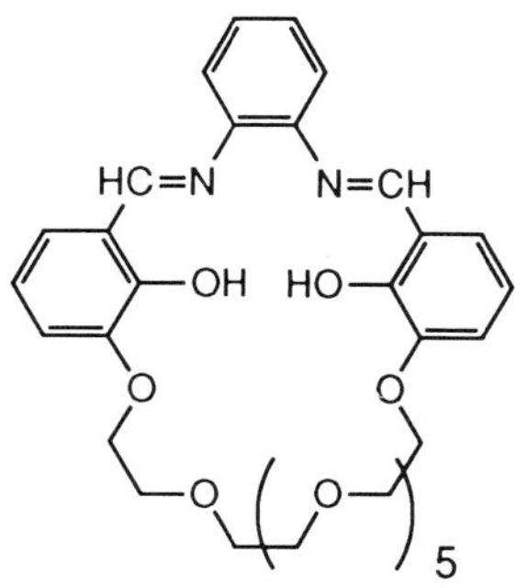

6

[35] R. M. Izatt, J. S. Bradshaw, S. A. Nielsen, J. D. Lamb, J. J. Christensen, and D. Sen, *Chem. Rev.*, 1985, **85**, 271.
[36] C. J. van Staveren, J. van Eerden, F. C. J. M. van Veggel, S. Harkema, and D. N. Reinhoudt, *J. Am. Chem. Soc.*, 1988, **110**, 4994.

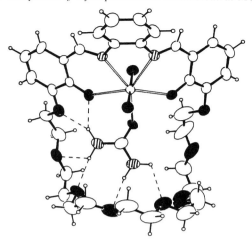

Figure 5.4 *X-ray crystal structure of the complex formed between urea and* $6 \cdot UO_2^{2+}$ *(Reproduced with permission from reference 36. Copyright 1988 ACS.)*

5.3 Concave Functionality: Cyclophanes for Phenol Complexation

In the following paragraphs, examples are presented where concave convergent functionality at a structurally well defined cyclophane cavity binding site is responsible for generating very high specificity for guest molecules. Following a design strategy previously developed for water-soluble diyne-bridged cyclophanes (Chapter 2.2),[37,38] Whitlock *et al.* prepared the highly preorganized naphthalenophanes **7** and **8** (Figure 5.5).[39,40] In these cyclophanes, the basic pyridine nitrogen of a dimethylaminopyridine residue is oriented towards the cavity binding site allowing favorable hydrogen-bonding interaction with H-donor centers of encapsulated guests. Such a hydrogen bond was found to be crucial for complexation to take place.

Table 5.1 gives the stability constants for inclusion complexes of hosts **7** and **8** with phenolic guests formed in CDCl$_3$. Figure 5.5 shows the X-ray crystal structures of the *p*-nitrophenol complexes formed by the two cyclophanes. The binding specificity shown by **7** and **8** is determined both by the complementarity of the guest to the cavity and by the strength of the hydrogen bond to the pyridine nitrogen. For steric reasons, cavity inclusion of *p*-nitrophenol derivatives only occurs if they are unsubstituted at the carbons *ortho* and *meta* to the phenolic residue. Suitably sized benzene derivatives that are not capable of hydrogen-bonding, *e.g. p*-nitroanisole and *p*-nitrochlorobenzene, do not

---

[37] E. T. Jarvi and H. W. Whitlock, *J. Am. Chem. Soc.*, 1982, **104**, 7196.
[38] S. P. Miller and H. W. Whitlock, Jr., *J. Am. Chem. Soc.*, 1984, **106**, 1492.
[39] R. E. Sheridan and H. W. Whitlock, Jr., *J. Am. Chem. Soc.*, 1986, **108**, 7120.
[40] R. E. Sheridan and H. W. Whitlock, Jr., *J. Am. Chem. Soc.*, 1988, **110**, 4071.

form stable complexes. There exists a very interesting relationship between complexation strength and the acidity of the H-donor functionality of the guest. Phenol only binds weakly to **7** ($K_a = 20$ L mol^{-1}), whereas the more acidic *p*-nitrophenol and *p*-cyanophenol form stable complexes of $K_a > 10^3$ L mol^{-1} (Table 5.1). On the other hand, even more acidic guests, *e.g.* benzoic acid, do not form cavity complexes with **7**. Rather, simple proton transfer occurs with no apparent incavitation of the anion. All these observations are consistent with the requirement of an optimum ΔpK_a between acid and base in the complexes of **7** and **8**.

Complexes of cyclophane **8** are 4–8 times more stable than complexes of **7** (Table 5.1) This stability difference has been explained by a better hydrogen-bonding geometry in the complexes of **8**. The *X*-ray crystal structure of the **7**·*p*-nitrophenol complex shows that the pyridine nitrogen points into the cavity, but is not well aligned for a linear hydrogen bond (Figure 5.5). In the **8**·*p*-nitrophenol complex, the pyridine ring parallels the cavity walls and forms a better, more linear hydrogen bond with the phenolic hydroxyl. Phenol complexation by the 'phenol sticky hosts'[39] **7** and **8** is enthalpically controlled. The enthalpic driving force is partially compensated by negative complexation entropies. For example, $\Delta S°$ values of -2 e.u. and -22 e.u. are calculated for the formation of the **7**·[4-(4'-nitrophenylazo)phenol] and **8**·[4-hydroxy-4'-nitrostilbene] complexes, respectively.

Cyclophane **9** with one diyne and one *p*-xylylene bridge forms a more stable complex in CD$_2$Cl$_2$ with *p*-nitrophenol ($K_a = 9.6 \times 10^4$ L mol^{-1}) than hosts **7** and **8** having two diyne bridges (Table 5.1).[41] In **9**, the distance between the least-squares planes through the two naphthalenes (≈ 6.88 Å) is shorter than in **7** and **8** (≈ 7.58 Å), and the increase in binding strength was explained by a better fit of the guest in the smaller cavity. Additionally, the *X*-ray crystal structure of the **9**·*p*-nitrophenol complex shows strong edge-to-face aromatic interactions between the perpendicularly aligned phenyl rings of host and guest which should further stabilize the complex.

In contrast to **7–9**, the larger anthraquinone-cyclophanes **10** and **11** do not form complexes with phenols or naphthols.[42] These macrocycles probably do not possess preorganized binding sites and prefer conformations in which the aromatic ring in the third bridge partially fills the cavity. Support for such unfavorable conformations was provided by force field computer modeling studies and by the analysis of ^{1}H NMR data.

Phosphines represent another class of strong hydrogen bond acceptors[43,44] that have been incorporated as concave functionalities into macrocyclic cavities. Whitlock *et al.* prepared two *exo–exo* (**12** and **13**) and two *exo–endo* (**14** and **15**) bis(phosphine oxide) cages (Scheme 5.2), and demonstrated both extracavity and intracavity complexation modes with *p*-nitrophenol as

---

[41] B. J. Whitlock and H. W. Whitlock, *J. Am. Chem. Soc.*, 1990, **112**, 3910.
[42] M. E. Haeg, B. J. Whitlock, and H. W. Whitlock, Jr., *J. Am. Chem. Soc.*, 1989, **111**, 692.
[43] M. Etter and P. W. Baures, *J. Am. Chem. Soc.*, 1988, **110**, 639.
[44] F. Toda, K. Mori, Z. Stein, and I. Goldberg, *J. Org. Chem.*, 1988, **53**, 308.

Hydrogen-bonded Complexes of Cyclophanes and Small Neutral Molecules 191

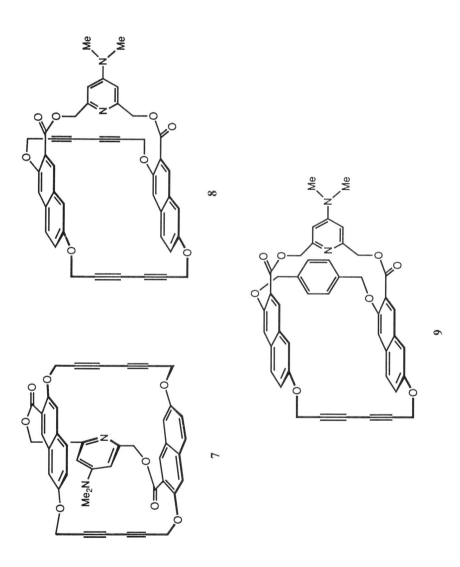

11

10

Table 5.1 *Association constants (K_a) and free energies of complexation ($-\Delta G°$) for the 1:1 complexes formed between cyclophanes 7 and 8 and phenols in $CDCl_3$, T = 298 K*

Guest	K_a [L^{-1} mol]	$-\Delta G°$ [kcal mol^{-1}]
Host 7		
p-Nitrophenol	3.0×10^3	4.7
4-(2',4'-Dinitrophenylazo)phenol	2.2×10^3	4.5
p-Cyanophenol	1.1×10^3	4.2
4-(4'-Nitrophenylazo)phenol	5.2×10^2	3.7
Host 8		
p-Nitrophenol	1.4×10^4	5.6
4-(2',4'-Dinitrophenylazo)phenol	1.3×10^4	5.6
p-Cyanophenol	9.8×10^3	5.4
4-(4'-Nitrophenylazo)phenol	3.4×10^3	4.8

guest.[45] The X-ray crystal structures of *exo–exo*-**12** and *exo–endo*-**15** are shown in Figure 5.6. The host–guest complexation shifts in ^{1}H NMR binding titrations in $CDCl_3$ indicate that the *exo–exo* hosts **12** and **13** form extracavity complexes having a 1:2 host–guest stoichiometry. Binding of the guest outside the cavity results in only small complexation-induced shifts of the NMR resonances of the host. In contrast, the *exo–endo* hosts **14** and **15** also exhibit intracavity complexation which is characterized by large upfield shifts of the aromatic host protons *ortho* to the *exo* phosphine oxide. The NMR analysis shows that the diyne host **14** with its larger, more preorganized cavity complexes the first guest at the *endo* site ($^{endo}K_a = 354$ L mol^{-1}). The complexation-induced shifts of the host resonances calculated for the 1:1 complex are large and very similar to those calculated for the 1:2 host–guest complex. The 1:2 complex ultimately formed by **14** is characterized by two almost equal association constants ($^{exo}K_a = 320$ L mol^{-1}).

Intracavity complexation by host **15** with its more collapsed cavity (Figure 5.6) is weaker ($^{endo}K_a = 223$ L mol^{-1}) than extracavity binding ($^{exo}K_a = 866$ L mol^{-1}). Here, the ^{1}H NMR binding analysis shows that extracavity binding is the initial complexation mode on the way to the formation of the 1:2 complex. For the 1:1 complex, only very weak complexation-induced shifts are calculated for the host resonances. As mentioned above, *exo*-complexation by **12**–**15** is characterized by very weak complexation-induced shifts of the host resonances.

The tris(bipyridine) ligand **16** created by Vögtle *et al.*, a macrobicyclic enterobactin mimic and a very strong binder of Fe(III),[46–48] was expanded

---

[45] B. P. Friedrichsen and H. W. Whitlock, *J. Am. Chem. Soc.*, 1989, **111**, 9132.
[46] S. Grammenudi and F. Vögtle, *Angew. Chem.*, 1986, **98**, 1119; *Angew. Chem. Int. Ed. Engl.*, 1986, **25**, 1122.
[47] P. Stutte, W. Kiggen, and F. Vögtle, *Tetrahedron*, 1987, **43**, 2065.
[48] T. J. McMurry, M. W. Hosseini, T. M. Garrett, F. E. Hahn, Z. E. Reyes, and K. N. Raymond, *J. Am. Chem. Soc.*, 1987, **109**, 7196.

194 Chapter 5

7 · 1,2-dichloroethane

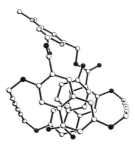

7·*p*-nitrophenol complex vertical view

7·*p*-nitrophenol complex lateral view

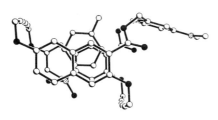

8·*p*-nitrophenol complex vertical view

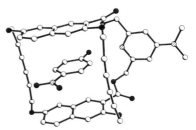

8·*p*-nitrophenol complex lateral view

Figure 5.5   *X-ray crystal structures of the 1,2-dichloroethane solvate of 7 and the p-nitrophenol complexes of 7 and 8*
 (Reproduced with permission from references 39 and 40. Copyright 1986, 1988 ACS.)

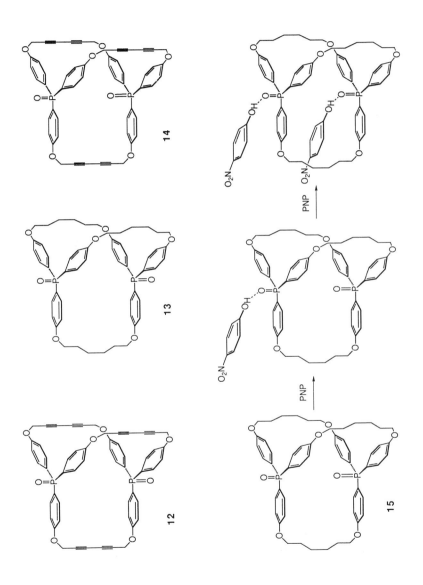

Scheme 5.2

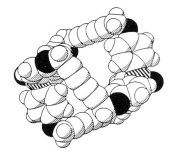

12

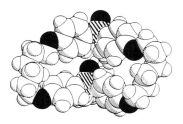

15

Figure 5.6   *X-ray crystal structures of the bis(phosphine oxide) macrocycles **12** and **15***
(Reproduced with permission from reference 45. Copyright 1989 ACS.)

producing macrobicyclic compounds such as **17** which complex trihydroxybenzenes with high selectivity in dichloromethane (Table 5.2).[49] For example, 1,3,5- and 1,2,4-trihydroxybenzene are complexed by **17**, whereas the 1,2,3-isomer is not bound. Varying the host structure demonstrated that inclusion of the guest in a suitably sized cavity is required for binding. Neither the smaller macrobicycle **16** nor open-chained derivatives of **17** having three bipyridines attached to only one of the caps are effective binders. The association constant for the phoroglucinol complex at ambient temperature was estimated in solid–liquid extraction experiments as $K_a \approx 1.1 \pm 0.2 \times 10^4$ L mol^{-1}. Multiple hydrogen-bonding between phenolic OH groups and the bipyridine acceptor centers in addition to van der Waals interactions account for this stable complex formation.

[49] F. Ebmeyer and F. Vögtle, *Angew. Chem.*, 1989, **101**, 95; *Angew. Chem. Int. Ed. Engl.*, 1989, **28**, 79.

Table 5.2 *Host–guest selectivity observed with **17** in CD_2Cl_2 at 298 K (+, complexation; −, no complexation)*

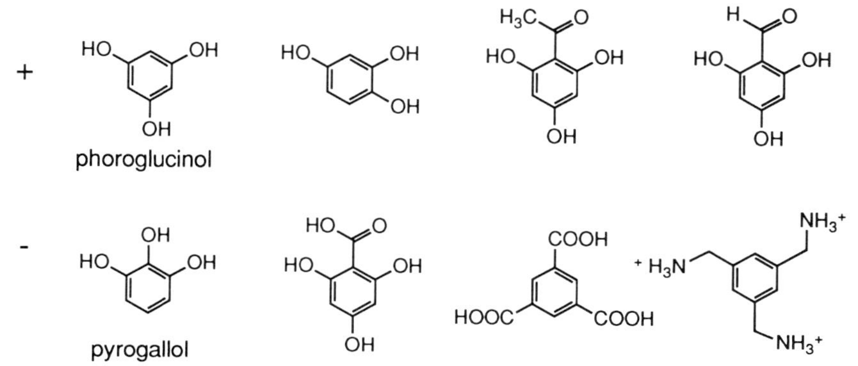

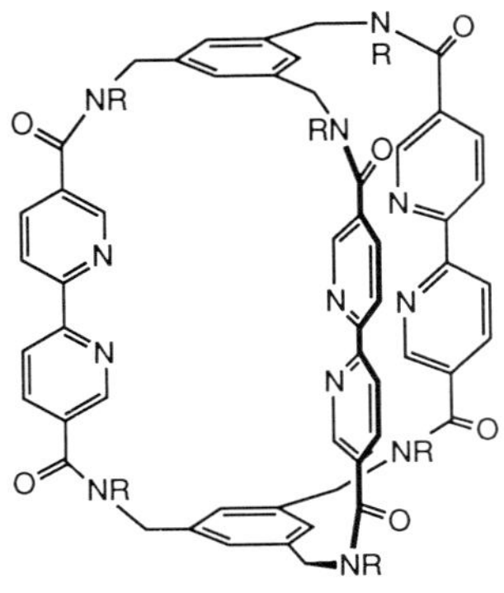

	Guests
+	phloroglucinol; 1,2,4-trihydroxybenzene; 2,4,6-trihydroxyacetophenone; 2,4,6-trihydroxybenzaldehyde
−	pyrogallol; 2,4,6-trihydroxybenzoic acid; benzene-1,3,5-tricarboxylic acid; 1,3-bis(aminomethyl)-5-(aminomethyl)benzene trication

16 R = PhCH$_2$

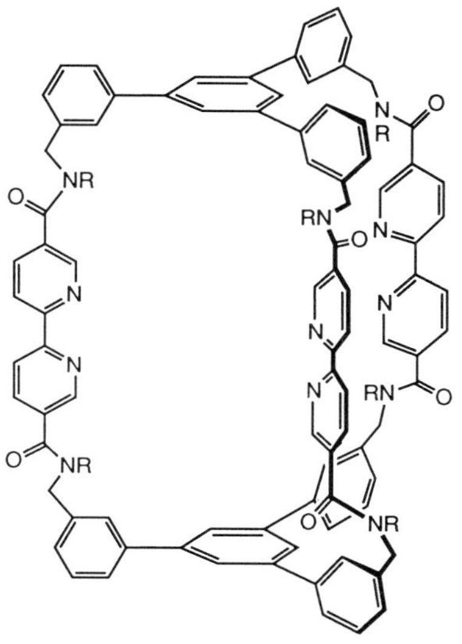

17 R = PhCH$_2$

An interesting stabilization of the phenolic guests towards oxidants was also observed. Aqueous solutions of pure phloroglucinol rapidly turn dark upon exposure to light and air. In contrast, dichloromethane solutions of phloroglucinol containing host **17** did not turn dark even upon day-long exposure to light and air.

5.4 Carbohydrate Complexation by an Octahydroxy[1.1.1.1]Metacyclophane

By acid-catalyzed reaction of resorcinol with dodecanal,[50] Aoyama et al.[51] obtained octol **18** (see also Chapters 3.5 and 4.4.1) which is readily soluble in apolar solvents. In carbon tetrachloride or benzene, this compound binds and solubilizes a variety of polar biomolecules like riboflavin (vitamin B_2), cyanocobalamine (vitamin B_{12}), hemin, and sugars. Solutions of the complexes are obtained by extracting aqueous solutions of the guests with the organic solutions of the host at 20 °C for 24 h. The complexation-induced upfield shifts observed for the ^1H NMR resonances of complexed sugar molecules indicate that the bound guests come close enough to host **18** to be subjected to the shielding anisotropy effects of its four benzene rings. Hydrogen-bonding is clearly the dominant host–guest interaction mode. This is demonstrated by the absence of any binding and solubilization of the polar guests if octol **18** is peracetylated.

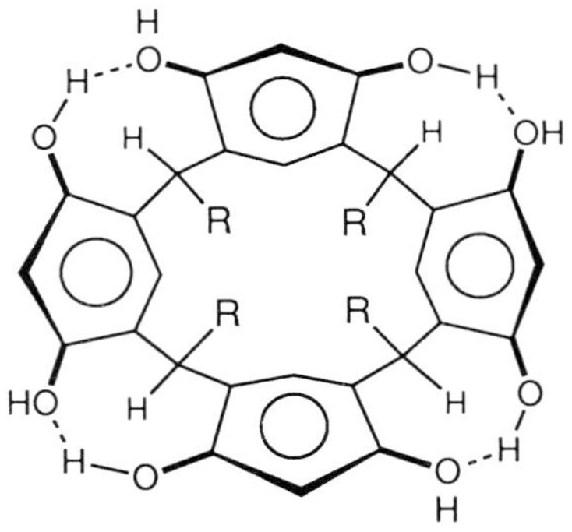

18 R = $(CH_2)_{10}CH_3$

The detailed analysis of monosaccharide recognition by **18** shows a high degree of guest selectivity.[52,53] Octol **18** possesses four major hydrogen-bonding centers (A, B, C, and D in Figure 5.7). CPK models suggest that two bonding centers, e.g. A and B separated by a *meta*-phenylene bridge, are located at a distance suitable for strong hydrogen-bonding to the two hydroxy groups of *cis*-1,4-cyclohexanediol but not to the two hydroxy groups in the *trans* isomer. The extraction experiments did indeed show a very high binding selectivity for *cis*-1,4-cyclohexanediol. The two residual hydrogen-bonding sites in the host interact with two water molecules that are also extracted into the apolar solvent (Figure 5.7).[52]

[50] A. G. S. Högberg, *J. Org. Chem.*, 1980, **45**, 4498.
[51] Y. Aoyama, Y. Tanaka, H. Toi, and H. Ogoshi, *J. Am. Chem. Soc.*, 1988, **110**, 634.
[52] Y. Aoyama, Y. Tanaka, and S. Sugahara, *J. Am. Chem. Soc.*, 1989, **111**, 5397.
[53] Y. Tanaka, Y. Ubukata, and Y. Aoyama, *Chem. Lett.*, 1989, 1905.

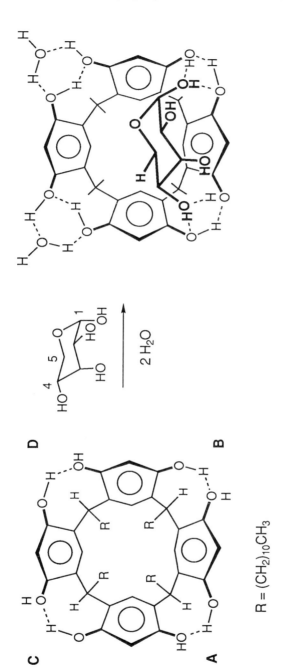

Figure 5.7 *Formation of a complex in carbon tetrachloride between octol* **18** *and D-ribose in the α-pyranose form upon extracting an aqueous solution of the sugar with a host solution in carbon tetrachloride. In addition to one ribose molecule, two water molecules are also extracted and complexed to* **18**

Similarly, sugars in the α-pyranose form with *cis*-OH groups at C-1 and C-4, *e.g.* ribose, form the most stable complexes and show the best extractabilities (Figure 5.7).[52] In comparative extraction studies, other factors have been recognized to determine sugar binding selectivity: (i) Extractabilities are highest if the OH-groups at C-3 and C-4 are *cis* oriented. (ii) The extractability increases with increasing lipophilicity of the substituent at C-5 ($CH_3 > H > CH_2OH$). (iii) Hydrogen atoms as substituents on C-2 are better than OH groups. Also, *cis*-OH groups at C-2 and C-3 lead to a more favorable extractability than *trans*-OH groups at these carbon atoms. The combination of all selectivity factors leads to the following extractability sequence for **18**: L-fucose(6-deoxy-L-galactose) > 2-deoxy-D-ribose > D-ribose > D-arabinose ≈ L-rhamnose(6-deoxy-L-mannose) >> D-galactose ≈ D-xylose ≈ D-lyxose ≈ D-mannose ≈ D-glucose (Table 5.3).

The selectivity rules reflect the tendency of the sugar hydroxy groups to interact with the hydrogen-bonding centers of the host and to avoid being exposed in an energetically unfavorable way to the apolar solvent. For example, the CH_2OH group at C-5 of an aldohexose in the α-pyranose form would not be correctly aligned for hydrogen-bonding to **18** (Figure 5.7), leaving this polar group exposed in an energetically unfavorable way to the apolar solvent. More lipophilic substituents at C-5, *e.g.* a methyl group or even a hydrogen atom, interact more favorably with the apolar solvent and, therefore, sugars with these functions at C-5 form more stable complexes and give higher extractabilities. Similarly, an OH-group at C-2 is not well aligned for hydrogen-bonding to the host and interacts with the apolar solvent. Also, the geometry of the α-ribopyranose complex (Figure 5.7) which, according to ^{1}H NMR analysis is the most favorable, indicates that an OH-group at C-3 *cis* to the OH at C-4 should better participate in the hydrogen-bonding to the host than a *trans*-OH group at C-3.

Two-point hydrogen-bonding interactions also lead to a remarkable chain-length selectivity in the binding of dicarboxylic acids to octol **18**.[54] The most stable 1:1 complex in $CDCl_3$ is formed by glutaric acid $HOOC-(CH_2)_3-COOH$ ($K_a \approx 1.2 \times 10^5$ L mol^{-1}). The complex geometry which is the most favorable, according to CPK model examinations and ^{1}H NMR data, is shown in Scheme 5.3. In its extended conformation, glutaric acid ideally spans the distance between the two hydrogen-bonding sites A and C in **18**. Very large and specific upfield shifts of the glutaric acid alkyl resonances in the ^{1}H NMR spectra support this positioning of the guest atop the shielding region of **18** (Scheme 5.3). At room temperature, the exchange between free and complexed guest is slow on the ^{1}H NMR time scale, and the signals of free and bound guest appear side-by-side in the spectrum. As expected for predominantly hydrogen-bonding interactions, complexation does not occur with dimethyl glutarate, with the octaacetyl derivative of **18**, or in hydrogen bond breaking polar solvents such as acetone or methanol.

---

[54] Y. Tanaka, Y. Kato, and Y. Aoyama, *J. Am. Chem. Soc.*, 1990, **112**, 2807.

Table 5.3 Selectivity in carbohydrate recognition by octol **18**, as measured by the extractability of the sugar from aqueous solution into a CCl$_4$ solution of the host. Highest affinity to **18**: L-fucose, lowest affinity: D-glucose. Preferential binding of all guests in the α-pyranose form is assumed. The factors determining the binding selectivity are given

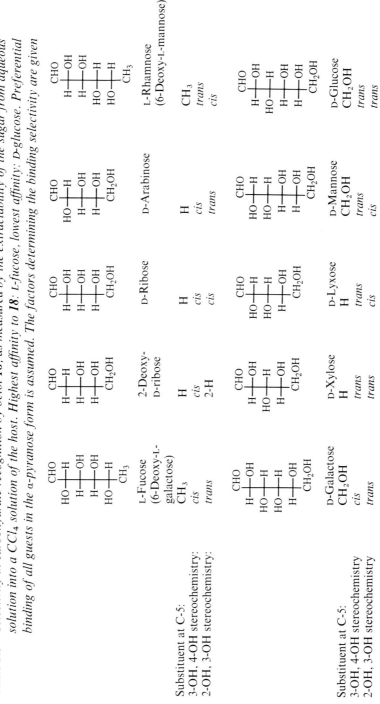

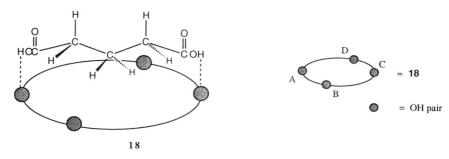

Scheme 5.3

Dicarboxylic acids with larger or smaller alkyl chains between the two carboxyl residues or monocarboxylic acids[55] form weaker complexes. The affinities of pimelic acid HOOC–(CH$_2$)$_5$–COOH ($K_a \approx 1.1 \times 10^3$ L mol^{-1}) and glutaric acid monomethyl ester ($K_a \approx 31$ L mol^{-1}) are much lower. The fact that $\Delta G°$ for the complex of glutaric acid (-6.9 kcal mol^{-1}) is larger than $2 \times \Delta G°$ for the complex of its monoester (-2.0 kcal mol^{-1}) indicates that the two-point interaction is much more favorable than two independent single-point interactions.

Ideally, the free energy for complexation of a ditopic substrate (such as glutaric acid) at two interaction sites of equal affinity in a host (such as in **18**) should always be more favorable than the free energy for the separate association of two monotopic substrates (such as the monomethyl ester of glutaric acid) at these sites. The advantages of ditopic binding are entropic in nature and are known as a chelation effect. In two separate single-point binding events under formation of a 1:2 complex, an overall loss of two sets of rotational and translational entropies occurs. In the two-point binding by a ditopic substrate, however, only one set of translational and rotational entropies is lost. The relationship $-\Delta G°_{\text{two-point}} > -2\Delta G°_{\text{single-point}}$ holds if the correct alignment of the ditopic substrate for the second interaction does not generate any strain in the complex. The following section gives an example of a less ideal two-point binding by a ditopic substrate.

5.5 Cyclophanes for Nucleotide Base Recognition

Many biological receptors are rather flexible and upon approach of the correct substrate, conformational changes occur to organize the binding site for the most favorable binding interactions. This mechanism for molecular recognition, known as *induced fit*, has been the focus in studies of nucleotide base recognition by synthetic receptors in the laboratories of A. Hamilton.[56] Inspiration for this work came from the known X-ray crystal structures of

---

[55] N. Pant and A. D. Hamilton, *J. Am. Chem. Soc.*, 1988, **110**, 2002.
[56] A. D. Hamilton, N. Pant, and A. Muehldorf, *Pure Appl. Chem.*, 1988, **60**, 533.

nucleotide-binding enzymes. For example, ribonuclease T_1 cleaves RNA specifically at guanosine.[57] This specificity is a result of multi-point binding interactions between the protein binding site and the nucleotide (Figure 5.8). Two hydrogen bonds form between the peptide backbone and O(6) and NH(1) of guanine. Another, ionic hydrogen bond is seen between a peptide histidine and the nucleotide phosphate group. Finally, electron density maps show two positions for the peptide tyrosine Tyr-45 with respect to the substrate guanine plane. At one position, the tyrosine is located at a nonbonding distance to the nucleotide. In the other orientation (shown in Figure 5.8), the tyrosine ring and the substrate guanine are at van der Waals (≈ 3.4 Å) distance and undergo efficient π—π-stacking interactions. This observation suggests an induced fit type binding mechanism with the tyrosine swinging into a stacking position upon substrate complexation.

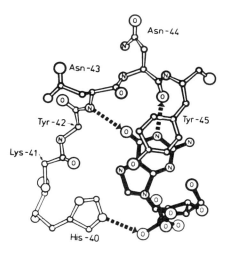

Figure 5.8 *Guanosine binding site of ribonuclease T_1. Guanosine is shown in bold (Reproduced with permission from reference 57. Copyright 1982 Nature (London).)*

To mimic such induced fit mechanisms for nucleotide recognition in a synthetic molecule, Hamilton *et al.* prepared cyclophane **19**.[58] This receptor was designed to bind thymine derivatives through triple hydrogen-bonding between the 2,6-diamidopyridine moiety of the receptor and the cyclic imide moiety of the substrate. An additional stabilization of the complex should occur by π—π-stacking between the naphthalene and thymine rings. Figure 5.9A shows the X-ray crystal structure of pure **19**. With an interplanar angle of 127.5° between the pyridine and naphthalene rings, the free host adopts an open conformation. A stoichiometric 1:1 complex forms between **19** and 1-

---

[57] U. Heinemann and W. Saenger, *Nature (London)*, 1982, **299**, 27.
[58] A. D. Hamilton and D. Van Engen, *J. Am. Chem. Soc.*, 1987, **109**, 5035.

butylthymine in $CDCl_3$, and the X-ray structure of the solid state complex was obtained (Figure 5.9B). Both triple hydrogen-bonding and π–π-stacking are operative as attractive interactions in the solid state complex. The angle between the pyridine and naphthalene rings in **19** has widened from 127.5° in the free receptor to 161.6° in the complex. Thus, upon substrate complexation, **19** acts as a molecular hinge and swings the naphthalene unit to within van der Waals distance of the bound thymine ring.[56] The closest interplanar contact between the two rings is 3.37 Å.

This *induced fit* behavior of **19**, which elegantly mimics the above-described recognition of guanosine by ribonuclease T_1, is not only observed in the solid state but also in chloroform solutions. The ^1H NMR spectrum of the complex solution supports a binding geometry that is similar to that observed in the solid state. Strong downfield shifts of the NH-protons on both host and guest (2.25 and 2.6 ppm, respectively) reflect the formation of a triple-hydrogen-bonded complex. In the complex, the thymine resonances of proton 6-H, the CH_3-group, and the $N-CH_2-$ group move upfield (≈ 0.3 ppm) while the signal of the butyl chain methyl group, distant from the thymine ring, is not significantly shifted. These selective upfield shifts are consistent with π–π-stacking interactions between the naphthalene and the thymine ring. The association constant for the **19**·1-butylthymine complex was determined as $K_a = 290$ L mol^{-1} (293 K). For comparison, 2,6-dibutyramidopyridine forms a weaker complex with 1-butylthymine, $K_a = 90$ L mol^{-1}, which shows that the π–π-stacking interaction provides additional stability to the complex of **19** (Table 5.4).

The isoalloxazine moiety in flavo-coenzymes contains a pyrimidine ring similar to the one in thymine and, therefore, should also be a suitable substrate for cyclophane **19**. It was indeed found that **19** forms a stable complex (Scheme 5.4) with 10-*n*-hexylisoalloxazine in chloroform ($K_a = 3.5 \times 10^3$ L mol^{-1}, 298 K).[59] As a result of π–π-stacking, the isoalloxazine fluorescence is quenched upon complex formation. Also, complexation inhibits the flavin-mediated photo-oxidation of 1,4-butanedithiol to the corresponding disulfide in chloroform.

As previously mentioned in Chapters 2.6 and 5.3, aromatic–aromatic interactions comprise both π–π-stacking as well as edge-to-face interactions. Both interactions are relevant to the stabilization of proteins.[60] In the edge-to-face orientation, electrostatic attraction occurs between the positively polarized hydrogen atoms on one ring with the negatively polarized region of the second. Whereas guanine binding to ribonuclease T_1 occurs in a π–π-stacking mode, its binding to the human c-H-*ras* oncogene protein involves edge-to-face interactions between the nucleotide base and a phenylalanine ring of the receptor.[61] In their studies with synthetic receptors, Hamilton *et al.*

[59] S. Shinkai, G.-X. He, T. Matsuda, A. D. Hamilton, and H. S. Rosenzweig, *Tetrahedron Lett.*, 1989, **30**, 5895.
[60] S. K. Burley and G. A. Petsko, *Science (Washington, DC)*, 1985, **229**, 23.
[61] A. M. de Vos, L. Tong, M. V. Milburn, P. M. Matias, J. Jancarik, S. Noguchi, S. Nishimura, K. Miura, E. Ohtsuka, and S.-H. Kim, *Science (Washington, DC)*, 1988, **239**, 888.

Hydrogen-bonded Complexes of Cyclophanes and Small Neutral Molecules

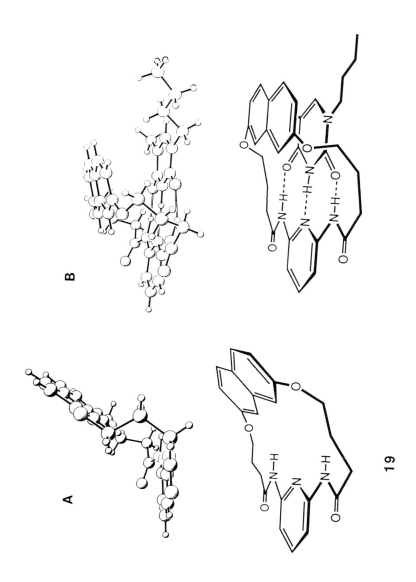

Figure 5.9 *X-ray crystal structure of receptor **19** (A) and of the complex formed between **19** and 1-butylthymine (B) (Reproduced with permission from reference 58. Copyright 1987 ACS.)*

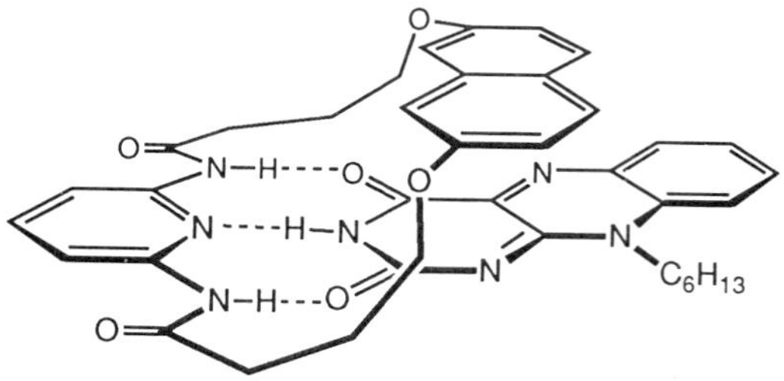

Scheme 5.4

demonstrated that thymine binding can either be promoted *via* face-to-face π–π-stacking or *via* edge-to-face interactions, depending on the electronic properties of the aromatic rings of the receptors that are involved.[62]

The diester macrocycle **20** binds 1-butylthymine in CDCl$_3$ with K_a = 570 L mol^{-1}, a much higher stability constant than the one measured for the complex of 2,6-dibutyramidopyridine (Table 5.4).[62] The X-ray crystal structure of the complex formed by **20** (Figure 5.10A) shows a π–π-stacking geometry with the cyclophane-naphthalene almost parallel to the pyrimidine ring. According to the analysis of ^{1}H NMR complexation-induced shifts, the complex adopts a similar stacking geometry in solution. On the other hand, the similar tetraether macrocycle **21** forms a weaker complex (K_a = 138 L mol^{-1}) which has a distinctively different geometry. In its crystal structure (Figure 5.10B), the cyclophane-naphthalene is almost perpendicular (interplanar angle of 77°) to the pyrimidine plane. This edge-to-face interaction also seems to provide a small stabilization for the cyclophane complex compared to acyclic 2,6-dibutyramidopyridine (Table 5.4). The ^{1}H NMR spectra of the **21**-complex indicate that a stacking geometry is also not favored in solution. The resonances around the thymine ring do not show the complexation-induced upfield shifts that are indicative of π-stacking interactions. Rather, upfield shifts specifically of the naphthalene-1,8-proton resonances (0.13 ppm) suggest that the 1,8-edge of the naphthalene is closer to the H-bonding plane than the 4,5-edge.

The thymine binding studies with **20** and **21** provide a nice example for the relationship between the electronic characteristics of binding components and the geometry of aromatic–aromatic interactions.[62] MNDO calculations suggest that the special stabilization of the stacking complex formed by **20** originates from the electrostatic component of electron donor–acceptor interactions (Chapter 1.5.1). Figure 5.11 shows a downward view on the electronic charge distributions of 2,7-dimethoxynaphthalene-3,6-dicarboxylic acid, a model for **20**, and thymine in a cofacial orientation. According to this model, the precise alignment of five pairs of oppositely charged atoms should

---

[62] A. V. Muehldorf, D. Van Engen, J. C. Warner, and A. D. Hamilton, *J. Am. Chem. Soc.*, 1988, **110**, 6561.

lead to an electrostatic stabilization of parallel stacking. Similar MNDO calculations on 2,3,6,7-tetramethoxynaphthalene, a model for **21**, predict a reversal of sign at some atoms which would lead to electrostatic repulsion if the two chromophores adopt a stacking geometry. More favorable electrostatic interactions seem to occur in the edge-to-face geometry which the butylthymine complex of **21** adopts. This interpretation is supported by the X-ray crystal structure (Figure 5.10B) which shows that the positively polarized naphthalene-1,8-protons point into the negatively charged region of the thymine comprised by the imide oxygen and nitrogen atoms.

By modifying the hydrogen-bonding edge in their cyclophanes, Hamilton *et al.* also obtained receptors that are specific for other nucleotide bases. Host **22** binds 2',3',5'-tri-*O*-pentanoylguanosine in CDCl$_3$.[63] The analysis of the ^{1}H NMR complexation shifts suggests that the complex adopts the stacking geometry **23**. In ^{1}H NMR binding titrations, the stability constant of **23** in CDCl$_3$ was determined as $K_a = 712$ L mol^{-1} (293 K). This represents a more than fivefold increase in association strength compared to the complex formed by the simple naphthyridine **24** ($K_a = 126$ L mol^{-1}) which lacks the stacking unit. Hence, the naphthalene-guanine stacking interaction contributes approximately 1 kcal mol^{-1} to the overall free binding energy (Table 5.4).

Cyclophane **25** is a specific receptor for adenine derivatives.[64] Its X-ray crystal structure is shown in Figure 5.12. The complex **26**, that forms between 9-butyladenine and **25** in CDCl$_3$, is stabilized by a combination of: (i) Watson–Crick base pairing, (ii) Hoogsteen base pairing, and (iii) π–π-stacking interactions. Watson–Crick base pairing describes the double hydrogen-bonding that occurs in duplex DNA between the pyrimidine of 2'-deoxythymidine (T) and the pyrimidine-N, NH of 2'-deoxyadenosine (A) (Figure 5.13). Hoogsteen base pairing describes the double hydrogen-bonding that occurs in the DNA major groove between the imidazole-N, NH of a 2'-deoxyadenosine of the duplex and the pyrimidine part of 2'-deoxythymidine located in the major groove. Hoogsteen base pairing leads to triple helix formation; the association between duplex DNA and a single stranded oligonucleotide in the major groove (Figure 5.13).[65-67]

Complex **26** is much more stable than the nucleotide base complexes of **19**–**22** that are stabilized by triple hydrogen-bonding and π–π-stacking. By ^{1}H NMR titrations, a stability constant K_a of 3.2×10^3 L mol^{-1} was measured for **26** in CDCl$_3$ (298 K). In contrast, only weak association occurs between 9-butyladenine and 1-butylthymine ($K_a < 23$ L mol^{-1}) which either undergo Hoogsteen or Watson–Crick base pairing only.

Recently, an attractive model has been developed by Jorgensen *et al.* to

[63] A. D. Hamilton and N. Pant, *J. Chem. Soc., Chem. Commun.*, **1988**, 765.
[64] S. Goswami, A. D. Hamilton, and D. Van Engen, *J. Am. Chem. Soc.*, 1989, **111**, 3425.
[65] (*a*) S. A. Strobel, H. E. Moser, and P. B. Dervan, *J. Am. Chem. Soc.*, **1988**, **110**, 7927; (*b*) H. E. Moser and P. B. Dervan, *Science (Washington, DC)*, 1987, **238**, 645; (*c*) S. A. Strobel and P. B. Dervan, *Science (Washington, DC)*, 1990, **249**, 73.
[66] D. Praseuth, L. Perrouault, T. L. Doan, M. Chassignol, N. Thuong, and C. Helene, *Proc. Natl. Acad. Sci. USA*, 1988, **85**, 1349.
[67] V. Sklenár and J. Feigon, *Nature (London)*, 1990, **345**, 836.

208   Chapter 5

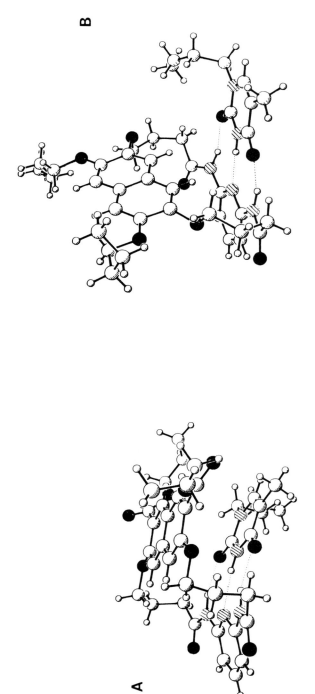

Figure 5.10 *Side views of the X-ray crystal structures of the complexes formed between 1-butylthymine and receptors **20** (A) and **21** (B) (Reproduced with permission from reference 62. Copyright 1988 ACS.)*

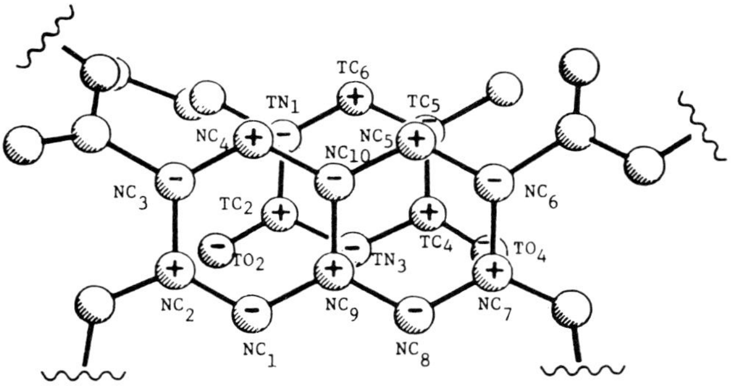

Hydrogen-bonded Complexes of Cyclophanes and Small Neutral Molecules 209

Figure 5.11 Top view of the naphthalene moiety N of receptor **20** and 1-butylthymine T in a π-stacking arrangement. The electronic charge distributions shown were obtained in MNDO calculations
(Reproduced with permission from reference 62. Copyright 1988 ACS.)

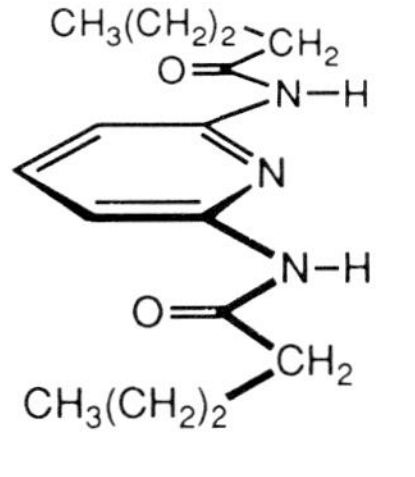

2,6-Dibutyramidopyridine

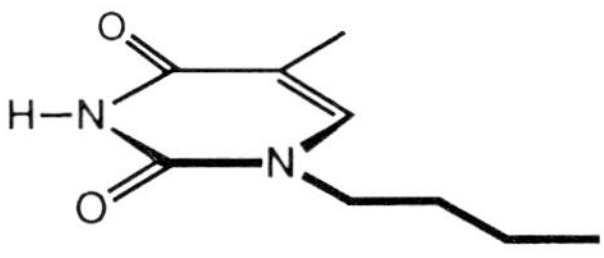

1-Butylthymine

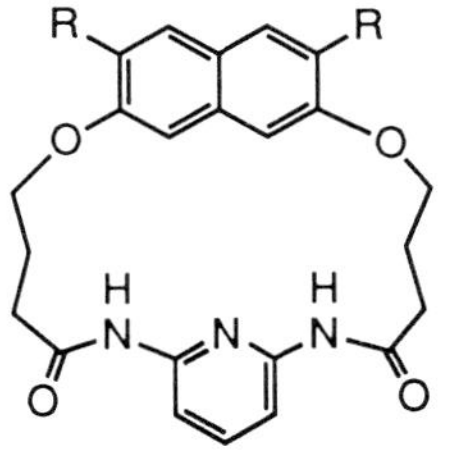

19 R = H
20 R = COO(CH$_2$)$_2$CH$_3$
21 R = O(CH$_2$)$_3$CH$_3$

Table 5.4 *Association constants (K_a) and free energies of complexation ($-\Delta G°$) for the 1:1 complexes formed by nucleotide bases and hydrogen-bonding receptors in $CDCl_3$, $T = 298$ K*

Substrate	Receptor	K_a [L^{-1} mol]	$-\Delta G°$ [kcal mol^{-1}]
1-butylthymine	19	290	3.4
	2,6-dibutyramidopyridine	90	2.7
	20	570	3.8
	21	138	2.9
2′,3′,5′-tri-O-pentanoyl-	22	712	3.9
guanosine	24	126	2.9
9-butyladenine	25	3200	4.8
9-butyladenine	1-butylthymine	<23	<1.9

explain differences in stability observed in chloroform between various hydrogen-bonded complexes.[68] Scheme 5.5 shows that triply hydrogen-bonded complexes vary greatly in their stability. According to Monte-Carlo statistical mechanics simulations using perturbation theory, the hydrogen-bonding interactions are very similar in the four complexes **27–30**. Therefore, these primary forces for association cannot be at the origin of the observed differences in stability. Rather, a secondary interaction, the electrostatic attraction or repulsion between the partially charged atoms involved in the hydrogen-bonding ($H^{\delta+}$, $O^{\delta-}$, and $N^{\delta-}$) presumably is the origin of the observed large differences. These secondary interactions are shown in Figure 5.14 for triply hydrogen-bonded systems. System A, in Figure 5.14, is predicted to be the most favorable since it contains exclusively attractive secondary interactions. Currently, there exists no synthetic complex with this specific hydrogen-bonding pattern. In system B, two attractive and two repulsive interactions compensate each other. The absence of net unfavorable secondary interactions leads to the formation of stable complexes, *e.g.* **27** and **28** in Scheme 5.5. In system C, four unfavorable secondary electrostatic interactions considerably weaken the primary hydrogen-bonding association. Therefore, the overall association strength measured for such a triply hydrogen-bonded system is weak as illustrated by the reduced stability of the complexes **29** and **30** shown in Scheme 5.5. For these complexes, the benefit of a third hydrogen bond is no longer apparent in view of association constants similar to those measured for complexes between nucleic acids with two hydrogen bonds ($K_a = 40$–130 L mol^{-1}).[1]

In a continuation of their studies, Hamilton *et al.* extended their cyclophanes to host systems containing two receptor sites.[71,72] Compound **31**

---

[68] W. L. Jorgensen and J. Pranata, *J. Am. Chem. Soc.*, 1990, **112**, 2008.
[69] Y. Kyogoku, R. C. Lord, and A. Rich, *Biochim. Biophys. Acta*, 1969, **179**, 10.
[70] T. R. Kelly, C. Zhao, and G. J. Bridger, *J. Am. Chem. Soc.*, 1989, **111**, 3744.
[71] A. D. Hamilton and D. Little, *J. Chem. Soc., Chem. Commun.*, 1990, 297.
[72] S. C. Hirst and A. D. Hamilton, *Tetrahedron Lett.*, 1990, **31**, 2401.

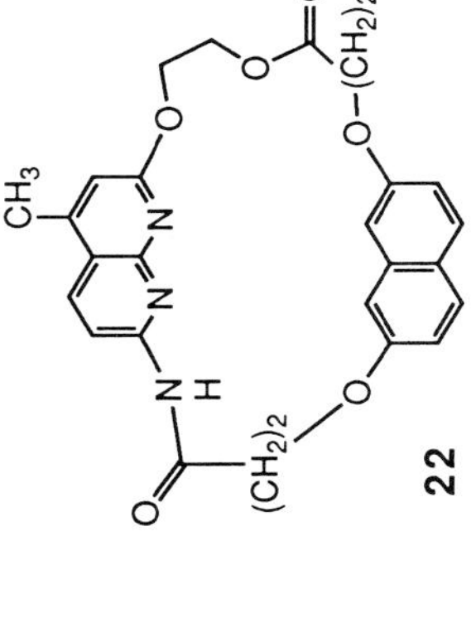

Hydrogen-bonded Complexes of Cyclophanes and Small Neutral Molecules 211

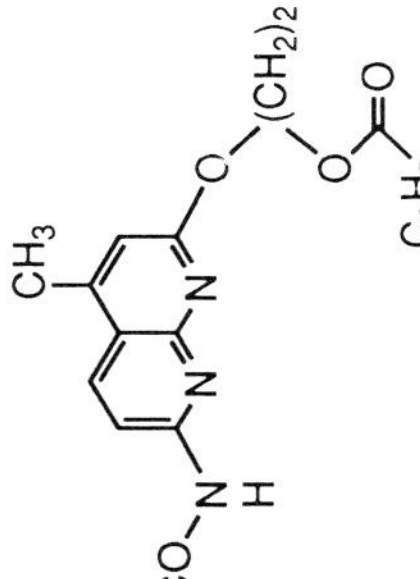

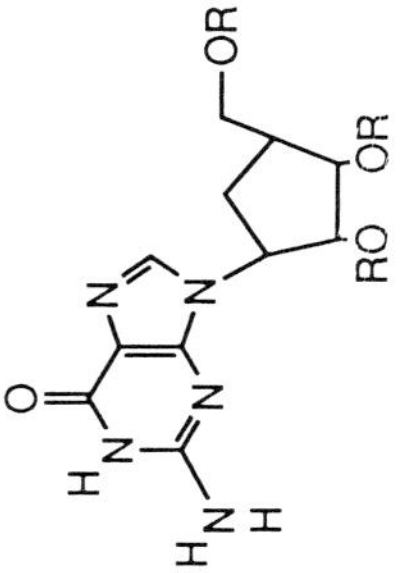

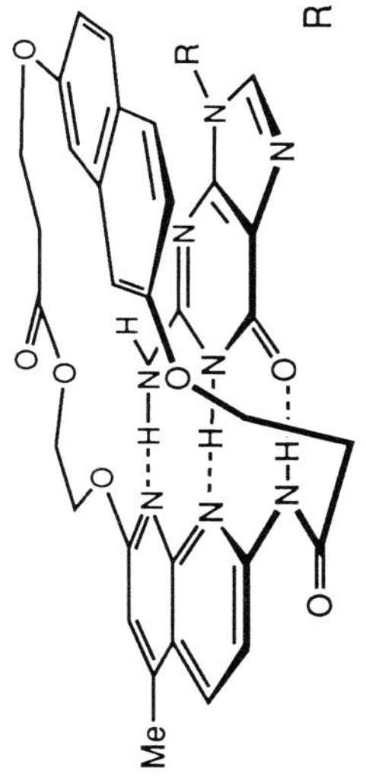

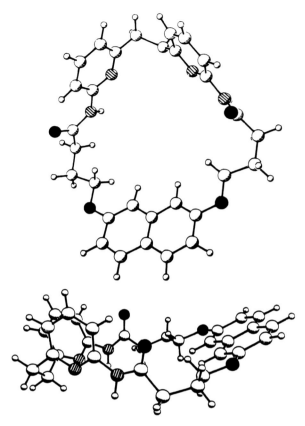

Figure 5.12 *X-ray crystal structure of cyclophane 25 in a top view (above) and a side view (below)*
(Reproduced with permission from reference 64. Copyright 1989 ACS.)

with two receptor sites separated by a diyne spacer and the bis(thymine) derivative **32** form a complex in $CDCl_3$ for which an association constant of $K_a = 2.03 \times 10^4$ L mol^{-1} ($\Delta G° = -5.9$ kcal mol^{-1}) was measured by ^{1}H NMR ($T = 298$ K).[71] The analysis of the complexation-induced shifts of the ^{1}H NMR resonances supports that the macrocyclic structure **33** represents the complex in solution. The complex **33** has an approximately 10 times larger association constant than complex **34** ($K_a = 1.6 \times 10^3$ L mol^{-1}, $\Delta G° = -4.4$ kcal mol^{-1}), which represents a model for the complexation at one binding site of **31**. This comparison shows that the double thymine binding in **33** is not optimal. If the binding of the bis(thymine) derivative **32** by **31** would be ideal, the formation free energy of the resulting complex **33** would be more than twice as large than the formation free energy of **34** (see Chapter 5.4). Possibly, some strain is involved in the binding of the bis(thymine) **32** to the dimerized receptor **31**.

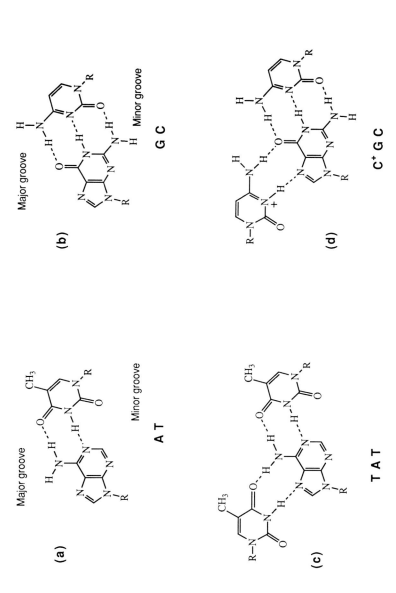

Figure 5.13 *Inserts (a) and (b) show the Watson–Crick base pairs AT and GC in double helical DNA and the orientation of their hydrogen-bonding centers towards the major and minor groove. Inserts (c) and (d) show how additional pyrimidine bases located in the major groove are bound by Hoogsteen hydrogen bonds to the complementary purine strands in the Watson–Crick duplex.*[65c]

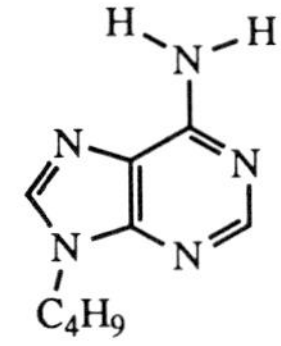

9-Butyladenine

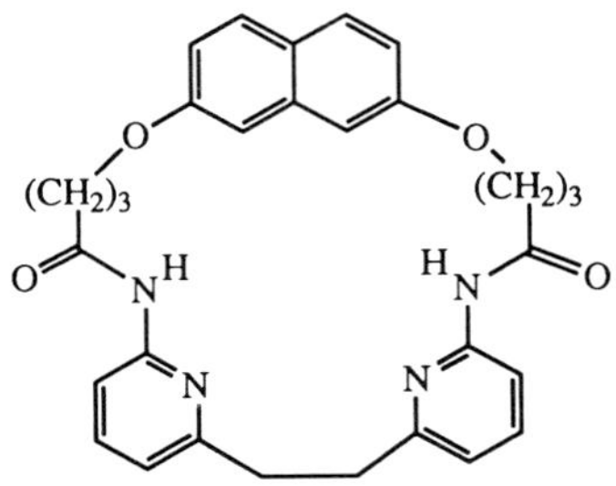

25

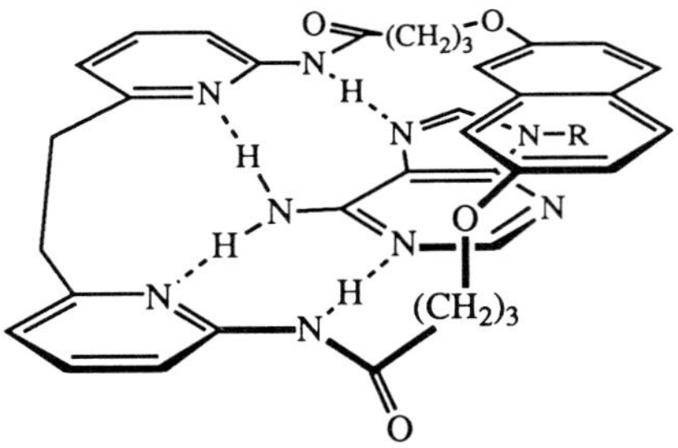

26

5.6 Barbiturate Receptors

By taking advantage of the triple hydrogen bond complementarity between 2,6-diamidopyridines and cyclic imides, Hamilton et al. developed an efficient receptor for barbiturates.[73] Barbiturates, a family of drugs derived from barbituric acid (**36a**) have widespread use as sedatives and anticonvulsants. A water-soluble artificial receptor for this large class of compounds could be useful for detoxification in cases of drug overdoses. In receptor **35**, two 2,6-diamidopyridine units are incorporated into a macrocyclic structure and provide binding sites to the six accessible hydrogen-bonding centers in 5,5-disubstituted barbiturates like barbital (**36b**). An isophthaloyl group provides the necessary organization and rigidity to the hexadentate binding site in **35**, and the diphenylmethane unit forms a lipophilic niche to accommodate suitable 5,5-substituents on the drugs. In $CDCl_3$, barbital (**36b**) forms a stable complex with cyclophane **35** which, according to ^{1}H NMR analysis, possesses the structure **37**. Large downfield complexation-induced shifts are observed for all hydrogen-bonding NH-resonances, and upfield shifts support the proximity of the ethyl groups to the diphenylmethane unit in **35**. The

---

[73] S.-K. Chang and A. D. Hamilton, *J. Am. Chem. Soc.*, 1988, **110**, 1318.

(association constants in CDCl₃)

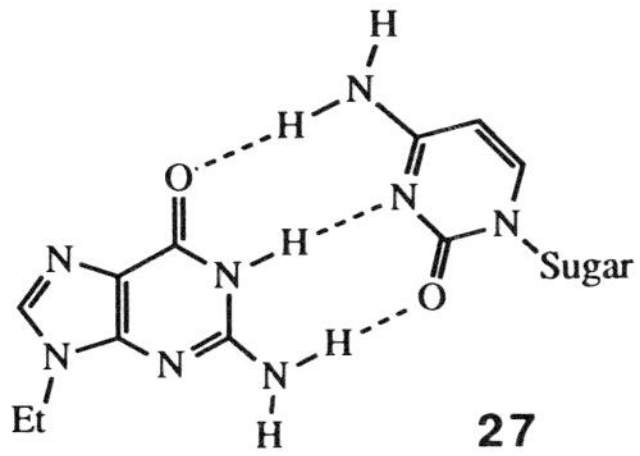

27 $K_a \approx 10^4 - 10^5$ L mol^{-1} (ref. 69)

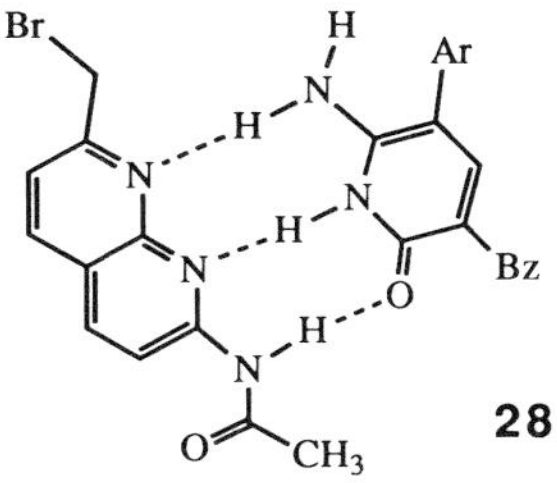

28 $K_a = 1.7 \times 10^4$ L mol^{-1} (ref. 70)

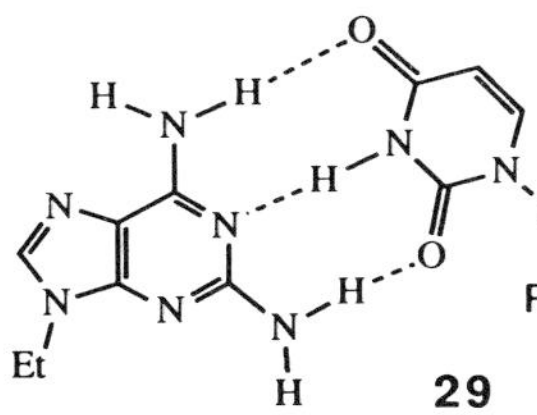

29   R = cyclohexyl $K_a \approx 170$ L mol^{-1} (ref. 1)

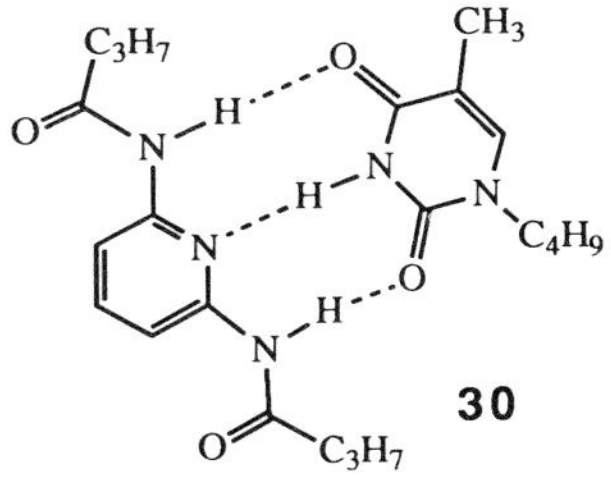

30 $K_a \approx 90$ L mol^{-1} (ref. 58)

Scheme 5.5

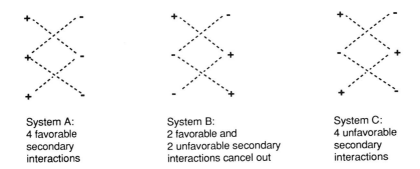

System A: 4 favorable secondary interactions

System B: 2 favorable and 2 unfavorable secondary interactions cancel out

System C: 4 unfavorable secondary interactions

Figure 5.14 *Secondary electrostatic interactions in triply hydrogen-bonded complexes*[68]

216   Chapter 5

isophthaloyl proton 2-H of the receptor is forced to lie in the deshielding region of the barbital C(2)-carbonyl group, and this resonance is shifted downfield by 0.4 ppm. The solution geometry, derived from the ^{1}H NMR data, is in very good agreement with an X-ray crystal structure of the complex.[74]

The complex that barbital and **35** form in CDCl$_3$ has a high association constant of $K_a = 1.37 \times 10^6$ L mol^{-1} (298 K, Table 5.5). The analysis of the complexation strength in a series of barbiturates and comparative studies with the nonmacrocyclic receptor **38** provides interesting insights into the origins of the high stability of complex **37**. Interruption of the macrocyclic structure and removal of the lipophilic niche for the 5,5-substituents in **38** significantly reduces complexation strength. Binding also decreases strongly for barbiturates that are not capable of forming as many hydrogen bonds as barbital (**36b**) or that lack high steric complementarity to the binding site of **35**. With their bulky phenyl rings, phenobarbital (**36c**) and mephobarbital (**36d**) do not fit correctly into the binding site. In addition, mephobarbital (**36d**) only forms three hydrogen bonds to one of the diamidopyridine edges in **35**. Hence, the complexes of these derivatives are much less stable than the barbital complex **37** (Table 5.5).

5.7 Imidazole Recognition: A Remarkable Relation between Solvent Size and Complexation Strength

Hydrogen-bonding is the major binding mode in inclusion complexes formed with the macrobicyclic receptor **39** prepared by Still *et al.*[75] The X-ray crystal structure of **39** demonstrates a large enforced cavity lined with convergent but spatially separated functionalities for hydrogen-bonding to incorporated suitably sized organic guests. Small polar organic molecules, mostly imidazole and pyridine derivatives, form complexes in chloroform by hydrogen-bonding to the amide and urea moieties of the receptor. NOESY experiments provided strong support for cavity inclusion. Table 5.6 shows the binding data for selected substrates in CDCl$_3$ (298 K). To form stable complexes, guests must fit into the cavity of **39** and be capable of both accepting and donating hydrogen bonds. In agreement with hydrogen-bonding being the major driving force for substrate incorporation, no significant binding was observed in solvents having a greater polarity than chloroform, *e.g.* in acetonitrile or dimethylsulfoxide.

To explore the effects of solvent structure on binding, Chapman and Still measured the association constant for the **40** imidazole complex in a range of organic solvents differing both in functionality and dimensions.[76] Table 5.7 documents the observed solvent dependency of the association strength. The following conclusions can be drawn:

---

[74] A. D. Hamilton, personal communication.
[75] J. D. Kilburn, A. R. MacKenzie, and W. C. Still, *J. Am. Chem. Soc.*, 1988, **110**, 1307.
[76] K. T. Chapman and W. C. Still, *J. Am. Chem. Soc.*, 1989, **111**, 3075.

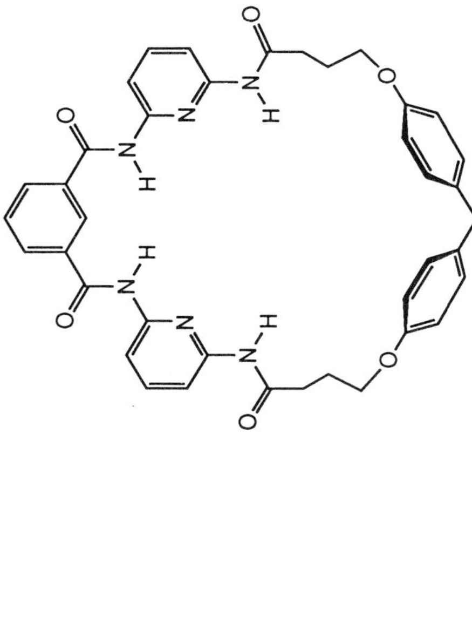

	R_1	R_2	R_3
36 a	H	H	H
b	Et	Et	H
c	Et	Ph	H
d	Et	Ph	CH_3

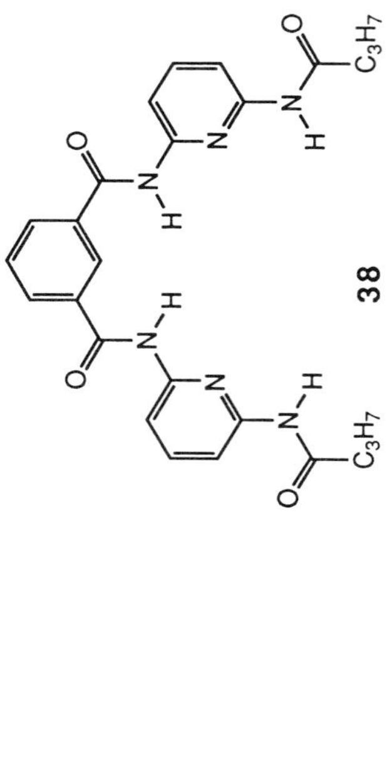

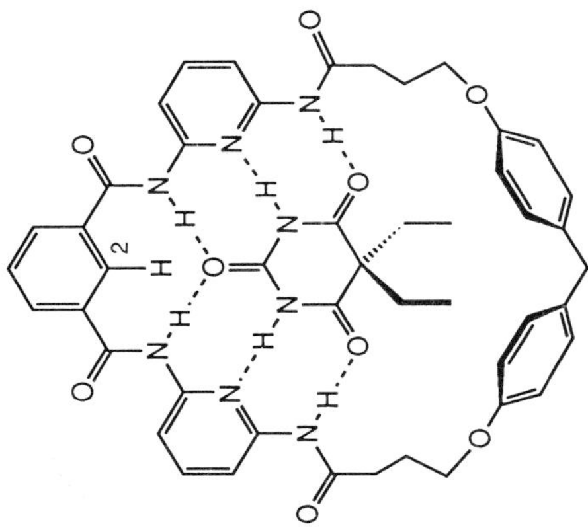

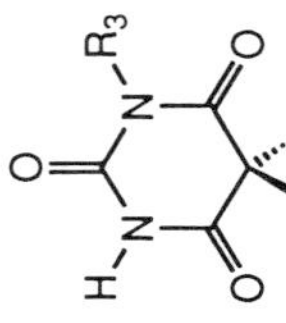

Table 5.5 *Association constants (K_a) and free energies of complexation ($-\Delta G°$) for the 1:1 complexes formed by barbiturates and receptors 35 and 38 in $CDCl_3$, T = 298 K*

Receptor	Substrate	K_a [L^{-1} mol]	$-\Delta G°$ [kcal mol^{-1}]
35	36b	1.37×10^6	8.4
	36c	1.97×10^5	7.2
	36d	6.8×10^2	3.9
38	36b	2.08×10^4	5.9

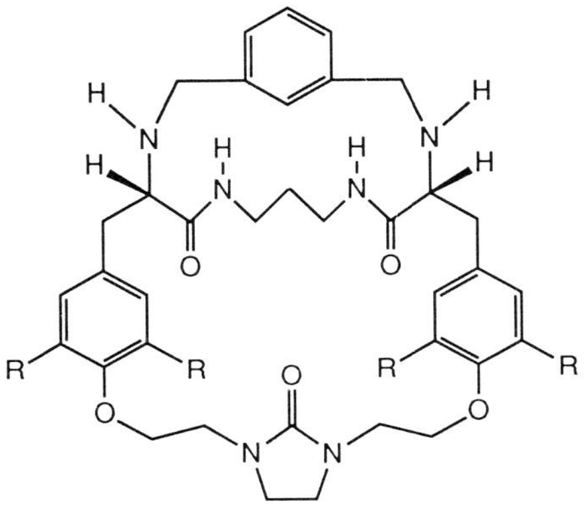

39 R = I
40 R = NO_2

Table 5.6 *Association constants (K_a) and free energies of complexation ($-\Delta G°$) for selected complexes formed by small heterocycles and receptor 39 in $CDCl_3$, T = 298 K*

Substrate	K_a [L^{-1} mol]	$-\Delta G°$ [kcal mol^{-1}]
imidazole	1420	4.3 ± 0.4
1-methylimidazole		0
2-methylimidazole	310	3.4 ± 0.3
benzimidazole	1420	4.3 ± 0.5
pyrrole		0
4-pyridone	4650	5.0 ± 1.0
4-aminopyridine	160	3.0 ± 0.3
aniline		0

(1) The extent of binding is strongly dependent upon the nature of the solvent. Even within the same functional class, significant variations are found. Thus, the association energy in the class of chlorinated hydrocarbon solvents increases by > 3.5 kcal mol^{-1} upon changing from CH_2Cl_2 to $Cl_2HC-CHCl_2$.

(2) Among solvents of a single class, there is a good correlation with solvent size. Figure 5.15A shows that binding strength increases roughly in proportion to the van der Waals surface areas of the solvent molecules.

Table 5.7 *Association constants (K_a) and free energies of complexation ($-\Delta G°$) for the complex formed between **40** and imidazole in various organic solvents, T = 298 K*

Solvent	K_a [L^{-1} mol]	$-\Delta G°$ [kcal mol^{-1}]
CH_2Cl_2	240 ± 24	3.25 ± 0.06
$CHCl_3$	490 ± 70	3.67 ± 0.09
CH_3CCl_3	8161 ± 370	5.33 ± 0.03
$CHCl_2CHCl_2$	128,000 ± 9000	6.96 ± 0.10
THF	29.0 ± 4.0	2.00 ± 0.08
2-Me-THF	77.0 ± 9.0	2.57 ± 0.07
2,5-Me$_2$-THF	185 ± 27	3.09 ± 0.10
2,2-Me$_2$-THF	156 ± 19	2.99 ± 0.08
2,2,5,5-Me$_4$-THF	1067 ± 110	4.13 ± 0.30
tetrahydropyran	104 ± 18	2.75 ± 0.10
1,4-dioxane	87 ± 10	2.65 ± 0.08
tert-butyl methyl ether	566 ± 80	3.75 ± 0.10
iso-propanol	13 ± 2	1.53 ± 0.06
tert-butyl alcohol	66 ± 13	2.48 ± 0.13
acetonitrile	no association	< 1.5

(3) Imidazole undergoes substantial desolvation as it is encapsulated by **40** during binding. Therefore, a decrease in the extent of solvation of the free substrate should favor its binding to **40**. To test this hypothesis, the solvation free energies of imidazole in various solvents relative to CH_2Cl_2 were measured by solubility and partition experiments. Figure 5.15B shows the relative free energies of imidazole solvation plotted against the free energies of formation for the **40**·imidazole complex. A strong inverse correlation was found for many solvents within the ether and halocarbon classes; the host–guest association strength decreases with increasingly favorable imidazole solvation.

(4) For some solvents, in particular 1,1,2,2-tetrachloroethane, the correlations in Figure 5.15B are not valid. Binding strength in 1,1,2,2-tetrachloroethane is much higher than in the smaller halocarbon solvents. Good correlations such as those shown in Figure 5.15B can only be expected if all solvents are either capable of, or, for steric reasons, not capable of solvating the cavity of **40**. Only in these two cases is the differential imidazole solvation the primary factor responsible for shifting the binding equilibrium. According to CPK model examinations, the solvents that correlate (Figure 5.15B) are capable of solvating the binding site, whereas 1,1,2,2-tetrachloroethane, which lies well off the correlation line, is too bulky to penetrate the cavity. In this solvent, the cavity of **40** is largely desolvated. Hence, the incorporation of the guest imidazole does not involve an unfavorable energy term for desolvation of the binding site, and a higher overall binding free energy

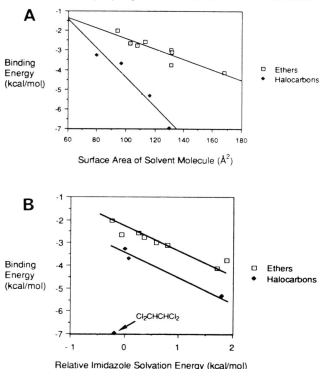

Figure 5.15 *Correlation between the stability of the 40-imidazole complex and (A) the surface area of solvents listed in Table 5.7 and (B) the relative imidazole solvation energy in these solvents
(Reproduced with permission from reference 76. Copyright 1989 ACS.)*

is measured. The studies with **40** together with other investigations[39,77] suggest the use of solvents that are too bulky to solvate the host binding cavity if strong host–guest complexation is desired.

In summary, Chapter 5 demonstrates that, within a few years, enormous progress has been made in developing cyclophane receptors to bind small biologically relevant molecules by using an array of hydrogen bonds as the primary interaction mode. Today, this complexation mode is still limited to solvents that do not compete with the substrate for hydrogen-bonding to the receptor. Research in this area must now focus at developing the systems which bind selectively *via* hydrogen-bonding in polar solvents, polar protic solvents, and, ultimately, in water. In Chapter 7, useful guidelines for achieving this goal will be discussed.

[77] J. Canceill, L. Lacombe, and A. Collet, *J. Am. Chem. Soc.*, 1986, **108**, 4230.

CHAPTER 6

Cyclophanes for Chiral Molecular Recognition

6.1 Introduction

Shortly after the discovery of crown ethers by Pedersen,[1] the potential of optically active crown ether hosts for forming diastereomeric complexes of differential stability with the enantiomers of chiral primary ammonium guests was recognized. First syntheses of chiral macrocyclic polyethers were described by Wudl and Gaeta[2] and Cram et al.[3] In the following, a great variety of optically active macrocyclic polyethers were prepared for the resolution of racemic cationic guests, e.g. chiral primary ammonium salts, protonated α-aminoalcohols and α-amino acid derivatives, through complexation in organic solution.[4-8] Among the highest enantiomer selectivities for chiral ammonium ions were obtained with the mono and bis(1,1'-binaphthyl)-crown ethers, e.g. **1** and **2** prepared by Cram et al.,[9-11] and the 9,9'-spirobifluorene crown ethers, e.g. **3** and **4** reported by Prelog et al.[12-14]

Free or polymer-bound cyclodextrins have been mostly used as macrocyclic

---

[1] C. J. Pedersen, *J. Am. Chem. Soc.*, 1967, **89**, 2495.
[2] F. Wudl and F. Gaeta, *J. Chem. Soc., Chem. Commun.*, 1972, 107.
[3] E. P. Kyba, M. G. Siegel, L. R. Sousa, G. D. Y. Sogah, and D. J. Cram, *J. Am. Chem. Soc.*, 1973, **95**, 2691.
[4] F. DeJong and D. N. Reinhoudt, *Adv. Phys. Org. Chem.*, 1980, **17**, 279.
[5] J. Jurczak and M. Pietraskiewicz, *Top. Curr. Chem.*, 1986, **130**, 183.
[6] J. F. Stoddart, *Prog. Macrocyclic Chem.*, 1981, **2**, 173.
[7] S. T. Jolley, J. S. Bradshaw, and R. M. Izatt, *J. Heterocycl. Chem.*, 1982, **19**, 3.
[8] E. Weber, J. L. Toner, I. Goldberg, F. Vögtle, D. A. Laidler, J. F. Stoddart, R. A. Bartsch, and C. L. Liotta, 'Crown Ethers and Analogs', eds. S. Patai and Z. Rappoport, Wiley, Chichester, 1989.
[9] D. J. Cram and J. M. Cram, *Acc. Chem. Res.*, 1978, **11**, 8.
[10] D. J. Cram and K. N. Trueblood, *Top. Curr. Chem.*, 1981, **98**, 43.
[11] C. B. Knobler, F. C. A. Gaeta, and D. J. Cram, *J. Chem. Soc., Chem. Commun.*, 1988, 330.
[12] V. Prelog, *Pure Appl. Chem.*, 1978, **50**, 893.
[13] V. Prelog and S. Mutak, *Helv. Chim. Acta*, 1983, **66**, 2274.
[14] M. Dobler, M. Dumic, M. Egli, and V. Prelog, *Angew. Chem.*, 1985, **97**, 793; *Angew. Chem. Int. Ed. Engl.*, 1985, **24**, 792.
[15] D. W. Armstrong and W. DeMond, *J. Chromatogr. Sci.*, 1984, **22**, 411.

2, R = H or Me

hosts for chiral recognition of neutral racemic guests.[15-18] Recent developments of nonmacrocyclic chiral receptors have been very successful. Such compounds[19,20] have shown great promise in novel approaches to enantiomer separations in chromatography,[21] crystallography, or transport experiments,[22] and as new chiral environments and reagents for asymmetric synthesis and catalysis.[23,24] In recent years, the first optically active cyclophanes for the enantioselective binding of neutral chiral molecules have been prepared. The chiral recognition properties of these compounds are the subject of Chapter 6.

The degree of enantioselection observed in the complexation of neutral racemic guests by optically active cyclophanes is still a moderate one. This can be explained with the considerable conformational flexibility of the macrorings in the unbound as well as in the complexed state. Previous chiral recognition studies with optically active crown ethers have demonstrated that the formation of very ordered diastereomeric complexes is required for a high degree of enantioselection.[9-11] The conformational heterogeneity of the optically active cyclophanes prepared today probably allows the guest enantiomers to form diastereomeric complexes that are geometrically and, hence, also energetically rather similar. Consequently, improving the enantiomer differentiation properties of chiral cyclophanes in future work requires a better control of macroring conformations. Chapter 6 illustrates some of the ongoing computer modeling efforts aimed at a better description of chiral macrocycle conformations.

6.2 Chiral Recognition in Organic Solvents

6.2.1 Analytical Resolution of Bromochlorofluoromethane by a Cryptophane

For nearly a century, attempts have been made to obtain the chiral haloform bromochlorofluoromethane (5) in optically active form.[25] When optically active samples were eventually prepared, no information on their optical purity was available.[26] The first analytical optical resolution of 5 was finally

---

[16] P. Fischer, R. Aichholz, U. Bölz, M. Juza, and S. Krimmer, *Angew. Chem.*, 1990, **102**, 439; *Angew. Chem. Int. Ed. Engl.*, 1990, **29**, 427.
[17] J. Szejtli, 'Cyclodextrin Technology', Kluwer Academic Publishers, Dordrecht, 1988.
[18] I. Jelinek, J. Snopek, and E. Smolková-Keulemansová, *J. Chromatogr.*, 1987, **405**, 379.
[19] J. Rebek, Jr., B. Askew, P. Ballester, and M. Doa, *J. Am. Chem. Soc.*, 1987, **109**, 4119.
[20] A. Echavarren, A. Galán, J.-M. Lehn, and J. DeMendoza, *J. Am. Chem. Soc.*, 1989, **111**, 4994.
[21] W. H. Pirkle and T. C. Pochapsky, *Chem Rev.*, 1989, **89**, 347.
[22] W. H. Pirkle and E. M. Doherty, *J. Am. Chem. Soc.*, 1989, **111**, 4113.
[23] E. N. Jacobsen, I. Marko, M. B. France, J. S. Svendsen, and K. B. Sharpless, *J. Am. Chem. Soc.*, 1989, **111**, 737.
[24] E. J. Corey, P.-W. Yuen, F. J. Hannon, and D. A. Wierda, *J. Org. Chem.*, 1990, **55**, 784.
[25] M. K. Hargreaves and B. Modarai, *J. Chem. Soc. C.*, 1971, 1013.
[26] S. H. Wilen, K. A. Bunding, C. M. Kascheres, and M. J. Wieder, *J. Am. Chem. Soc.*, 1985, **107**, 6997.

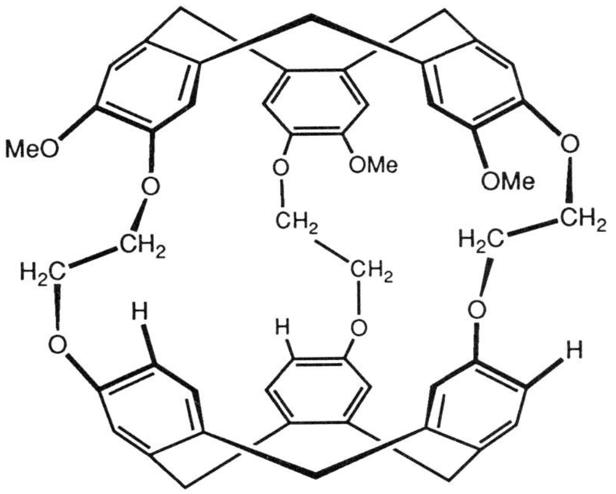

(-) - 6

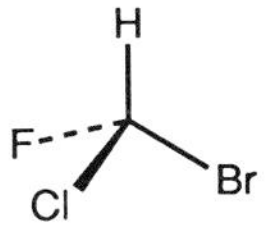

accomplished by Collet et al. through inclusion complexation of the haloform within the cavity of cryptophane 6 (Chapter 3.4).[27]

The formation of diastereomeric complexes of differential stability between cryptophane (−)-6 and the racemic haloform (±)-5 was demonstrated by ^{1}H NMR spectroscopy in $CDCl_3$. The proton resonances of the guest enantiomers in the two diastereomeric complexes move dramatically upfield ($\Delta\delta \approx 4.8$ ppm at saturation binding), and differential complexation shifts are observed. At saturation binding, the two guest resonances are separated by ≈ 10 Hz (at 200 MHz). The stability of the diastereomeric complexes was investigated at fast exchange conditions.[27] For this purpose, a solution of (±)-5 and (−)-6 was progressively diluted, and the resulting variation of the chemical shift of the guest resonance was analyzed in an iterative procedure as a function of the overall host and guest concentrations. At 332 K, the two diastereomeric complexes have only slightly different stability ($K_a = 0.30$ L mol^{-1}, $\Delta G° = 0.80$ kcal mol^{-1} versus $K_a = 0.22$ L mol^{-1}, $\Delta G° = 1.0$ kcal mol^{-1}). Since at fast exchange in the presence of a large excess of host, the ^{1}H NMR resonances of the complexed (+) and (−) enantiomers of 5 are totally separated, the enantiomeric composition of a partially enriched sample of 5 with $\alpha_D + 0.129°$ (neat)[26] could be determined from the difference in peak intensities. The e.e. (enantiomeric excess) of this sample was determined as $4.3 \pm 1\%$ which allowed an estimate of the maximum molar rotation of enantiomerically pure (+)-5 to be in the range of 1.7 ± 0.5 deg·cm^2 dmol^{-1}. The difference in peak intensities in the spectra of the complexed, partially enriched sample also allowed an assignment of the differentially shifted haloform resonances to the (+) and (−) enantiomers. This showed that the

[27] J. Canceill, L. Lacombe, and A. Collet, J. Am. Chem. Soc., 1985, 107, 6993.

diastereomeric complex between host and guest having the same sign is more stable.

6.2.2 Enantioselective Complexation of Chiral Amides by a Macrobicyclic Hydrogen-bonding Receptor

With the macrobicyclic system **7**, related to hydrogen-bonding receptors discussed in Chapter 5.7, Still *et al.* prepared an optically pure host with C_2 symmetry that binds chiral amides enantioselectively in benzene.[28] The *X*-ray crystal structure of **7** shows an internal cavity occupied by a molecule of CH_2Cl_2. The bridgehead hydrogens (H_a) point into this cavity (Figure 6.1). Simple amides form inclusion complexes with **7** (Table 6.1). In the ^1H NMR spectra, the amide N-H's of both host and guest shift downfield by more than 1 ppm upon complexation. The acetyl methyl resonance of complexed *N*-methylacetamide moves upfield by 0.5 ppm. By NOE difference spectroscopy, it was shown that this methyl group establishes proximity to both the inwards pointing bridgehead hydrogens (H_a) and the amide N-H's (H_c) of the host. A strong intramolecular NOE was also observed between H_a and H_c. Computer modeling studies provided additional support for the spatial proximity of those protons of the host and guest that show strong NOE's. The optimized complex geometry shown in Figure 6.1 was calculated starting from the conformation of **7** that was obtained in the *X*-ray crystal structure. In the simulation, the benzyl groups of **7** were replaced by methyl groups. After docking of the guest, energy minimizations were performed using the OPLS/AMBER force field.[29] The generated low energy conformations were subsequently subjected to molecular dynamics simulations.[30]

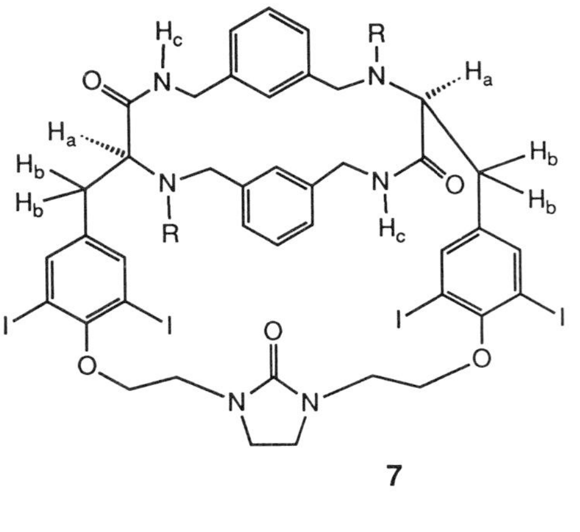

7

[28] P. E. J. Sanderson, J. D. Kilburn, and W. C. Still, *J. Am. Chem. Soc.*, 1989, **111**, 8314.
[29] W. L. Jorgensen and J. Tirado-Rives, *J. Am. Chem. Soc.*, 1988, **110**, 1657.
[30] A. E. Howard and P. A. Kollman, *J. Med. Chem.*, 1988, **31**, 1669.

Table 6.1 Free energies of complexation ($-\Delta G°$) for the 1:1 complexes formed by receptor 7 and amides in C_6D_6, $T = 298$ K

Substrate	$-\Delta G°$ [kcal mol^{-1}]		
MeNHCOMe	3.17		
MeNHCOBn[a]	3.18		
BnNHCOH	3.24		
BnNHCOMe	2.84		
BnNHCOCF$_3$	no complex observed		
BnNHCOEt	2.33		
	$-\Delta G°$ (S) [kcal mol^{-1}]	$-\Delta G°$ (R) [kcal mol^{-1}]	Enantioselection $\Delta(\Delta G°)$, kcal mol^{-1}
PhCHMeNHCOMe	3.04	2.62	0.42
PhCHMeNHCOH	3.18	2.85	0.33
PhCHMeNHCOEt	1.80	1.55	0.25
1-NpCHMeNHCOMe[b]	2.56	2.31	0.25
BnOAlaNHCOMe	2.29	1.81	0.48
MeOPGlyNHCOMe[c]	1.91	2.06	−0.15

[a] Bn = Benzyl; [b] Np = Naphthyl; [c] PGly = Phenylglycine.

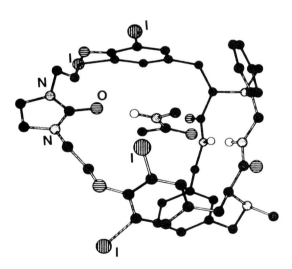

Figure 6.1 Computer model obtained by molecular dynamics simulations of the complex formed between N-methylacetamide and a receptor similar to 7. In the modeling, the benzyl groups of 7 were replaced by methyls (Reproduced with permission from reference 28. Copyright 1989 ACS.)

Host **7** shows enantioselective binding of certain chiral amides in benzene (Table 6.1). The differences in stability between diastereomeric complexes, $\Delta(\Delta G°)$, vary between ≈ 0.2 and ≈ 0.5 kcal mol^{-1}. The origins of the observed enantioselectivity are not obvious. In addition to differential hydrogen-bonding interactions, steric effects also seem to be important. Although the degrees of enantioselection given in Table 6.1 are moderate ones, they are among the highest observed with cyclophane receptors. The absence of even higher degrees of chiral recognition can be explained with the conformational flexibility of host **7** and its complexes. A further rigidification of the receptor should provide enhanced enantioselection.

6.3 Chiral Recognition in Aqueous Solutions

6.3.1 Diastereomeric Complex Formation by an Optically Active Tetraaza[6.1.6.1]paracyclophane

The first optically active paracyclophane with an apolar cavity, capable of forming diastereomeric inclusion complexes in water, was prepared by Koga *et al.*[31] In macrocycle **8**, two diphenylmethane units are bridged by two chiral C_4-chains derived from L-tartaric acid. In acidic D_2O (pD 1.2), **8** forms stable diastereomeric inclusion complexes with the enantiomers of chiral aromatic compounds like mandelic acid, atrolactic acid, and 2-phenylpropionic acid. This was demonstrated by the differential upfield shifts observed in the ^{1}H NMR spectra for the resonances of (*R*) and (*S*)-guest in the two diastereomeric complexes. For example, in the solution of **8** (0.06 M) and atrolactic acid (0.01 M), the benzylic resonance of the (*S*)-guest exhibits an upfield shift $\Delta\delta$ of 0.62 ppm, whereas this resonance in the complex of the (*R*)-guest is only shifted by 0.55 ppm. These differential upfield shifts indicate the formation of diastereomeric complexes of different predominant structures but not necessarily having different stability constants.

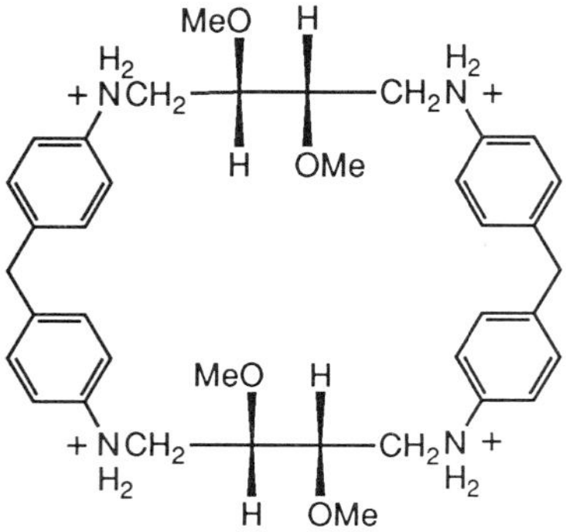

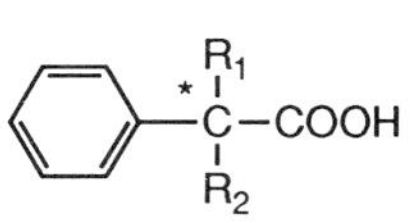

R_1	R_2	
H	OH	mandelic acid
CH_3	OH	atrolactic acid
H	CH_3	2-phenylpropionic acid

[31] I. Takahashi, K. Odashima, and K. Koga, *Tetrahedron Lett.*, 1984, **25**, 973.

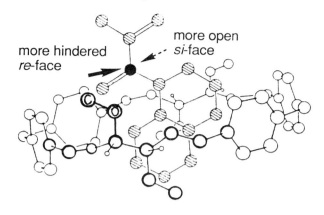

Scheme 6.1

Complexation by **8** leads to the differential stabilization of diastereomeric reaction transition states as exemplified in the following study.[32] In aqueous HCl (pH 1.2), arylglyoxylic acids are reduced with sodium borohydride to give the corresponding arylglycolic acids. These are subsequently treated with diazomethane, isolated, and analyzed as the methyl esters (equation 6.1). When the reduction of 1-naphthylglyoxylic acid (0.02 M) was performed in the presence of **8** (0.04 M), the methyl ester of (R)-1-naphthylglycolic acid (**9**) was isolated, after esterification, in 9.7% e.e. (chemical yield: 45%). The formation of an excess of the (R)-1-naphthylglycolic acid was explained by the preferred attack of the hydride reagent at the si-face of 1-naphthylglyoxylic acid incorporated pseudoaxially in the chiral binding site of **8** (Scheme 6.1). In this inclusion geometry, the methoxy groups of the chiral bridges sterically hinder the attack from the re-face.

6.3.2 Enantiospecific Binding of Chiral Onium Guests

In Chapter 4.3, it was shown that cyclophane **10** displays a strong affinity for quaternary ammonium ions. This macrocycle is chiral, and the two pure enantiomers (+)-**10** and (−)-**10** were prepared in an elegant synthetic route which takes advantage of an asymmetric Diels–Alder reaction.[33] By ^{1}H NMR, it was found that the optically active myrtanyl trimethylammonium ion (myrtanyl-TMA) binds differentially in aqueous borate buffer (pH 9) to both host enantiomers. The upfield complexation shifts at saturation binding, $\Delta\delta_{sat}$, calculated for the guest resonances in the two diastereomeric complexes are shown in Figure 6.2. At $\Delta(\Delta G°) = 0.6$ kcal mol^{-1}, a substantial

---

[32] I. Takahashi, K. Odashima, and K. Koga, *Chem. Pharm. Bull.*, 1985, **33**, 3571.
[33] M. A. Petti, T. J. Shepodd, R. E. Barrans, Jr., and D. A. Dougherty, *J. Am. Chem. Soc.*, 1988, **110**, 6825.

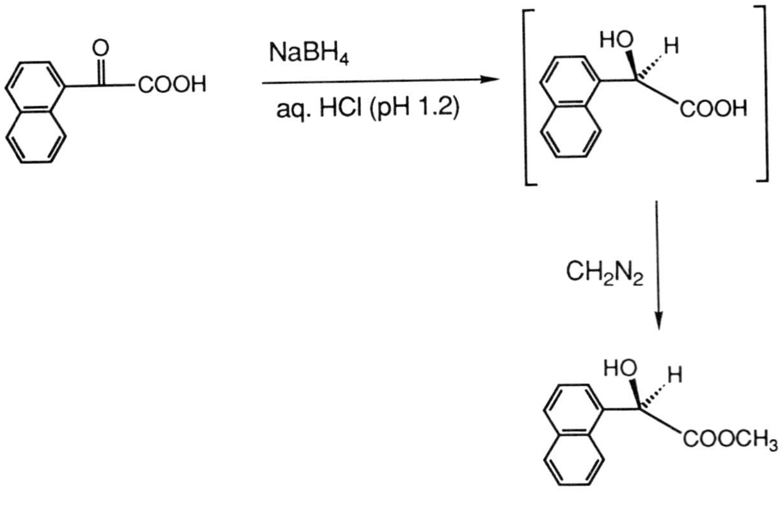

230   Chapter 6

$$\text{Naphthyl-CO-COOH} \xrightarrow[\text{aq. HCl (pH 1.2)}]{\text{NaBH}_4} \left[\text{Naphthyl-CH(OH)-COOH} \right] \quad (6.1)$$

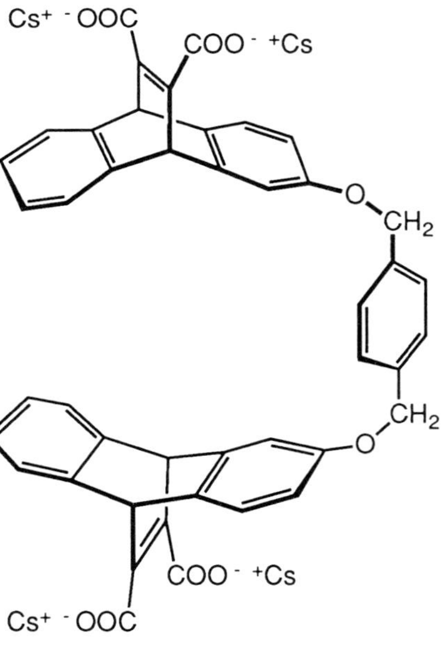

$$\downarrow \text{CH}_2\text{N}_2$$

Naphthyl-CH(OH)-COOCH$_3$

9

S,S,S,S-(+)-**10**

Table 6.2 Free energies of complexation $-\Delta G°$ (kcal mol^{-1}) at $T = 295$ K for the 1:1 complexes of enantiomerically pure cyclophane **10** and optically active guests in aqueous deuterated borate buffer, $pD = 9$

Guest	$-\Delta G°$ (kcal mol^{-1})	
	R,R,R,R-(**10**)	S,S,S,S-(**10**)
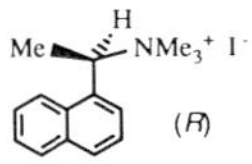 (R)-naphthyl-CHMe-NMe$_3^+$ I$^-$	5.8	6.3
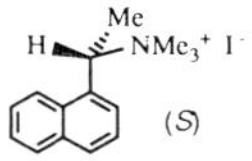 (S)-naphthyl-CHMe-NMe$_3^+$ I$^-$	6.7	5.9
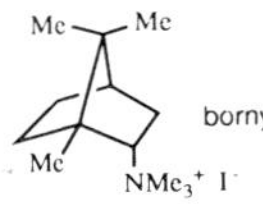 bornyl-TMA	6.4	6.5
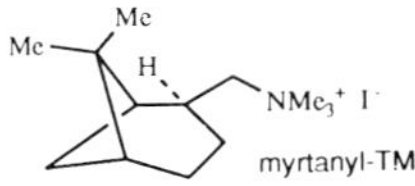 myrtanyl-TMA	5.1	4.5
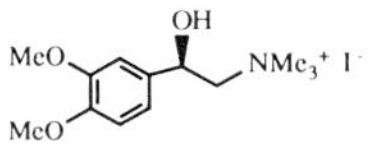 (MeO)$_2$-C$_6$H$_3$-CH(OH)-CH$_2$-NMe$_3^+$ I$^-$	5.8	5.7

enantioselection was calculated from the results of ^{1}H NMR binding titrations. Enantioselective binding was also observed for other chiral guests (Table 6.2). Interestingly, overall binding strength does not correlate with enantioselection. Bornyl-TMA forms much more stable complexes with (+) and (−)-**10** than myrtanyl-TMA. Despite the occurrence of differential ^{1}H NMR complexation shifts (Figure 6.2), no significant enantioselectivity was observed in the binding of bornyl-TMA. Clearly, differential complexation shifts, which reflect different complex geometries, do not necessarily reflect differential stabilities of diastereomeric complexes.

6.3.3 On the Origin of Cavity-filling Conformations of Cyclophanes

Naproxen [2-(6-methoxy-2-naphthyl)propionic acid, (**11a**)] is a chiral commercial anti-arthritis and anti-inflammatory agent and acts as a potent inhibitor of the enzyme cyclooxygenase which is responsible for prostaglandin synthesis.[34,35] The (S)-enantiomer of naproxen is 27 times more active than

---

[34] J.-P. Rieu, A. Boucherle, H. Cousse, and G. Mouzin, *Tetrahedron*, 1986, **42**, 4095.
[35] I. T. Harrison, B. Lewis, P. Nelson, W. Rooks, A. Roszkowski, A. Tomolonis, and J. H. Fried, *J. Med. Chem.*, 1970, **13**, 203.

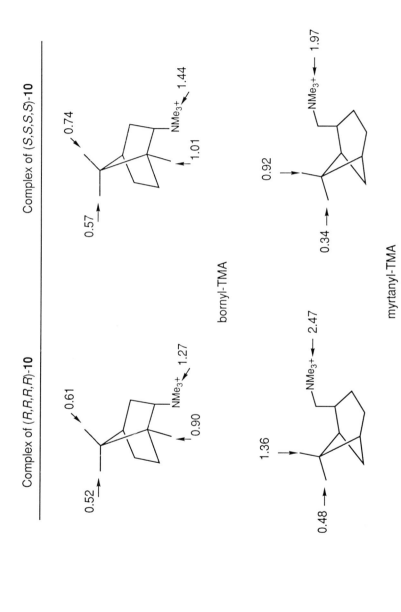

Figure 6.2 Upfield complexation shifts at saturation binding, $\Delta\delta_{sat}$, calculated for the ^1H NMR resonances of bornyl-TMA and myrtanyl-TMA in the diastereomeric complexes formed by the two enantiomers of host **10**

the (R)-enantiomer. For the optical resolution of naproxen through diastereomeric complex formation in aqueous solution, Diederich *et al.* prepared a series of water-soluble chiral cyclophanes. The first macrocycle in this series was compound **12** which incorporates a 2,2′,6,6′-tetrasubstituted biphenyl unit as an optically stable chiral spacer.[36] CPK model examinations had indicated that naproxen would fit with its naphthalene ring into the binding site of **12** and that differences in apolar and steric interactions as well as ion pairing along the chiral barrier could lead to the formation of diastereomeric inclusion complexes of differential stability.

^{1}H NMR binding studies with **13** in acidic aqueous solution, however, clearly revealed that this cyclophane does not possess an efficient binding site and, hence, does not act as a host. According to MM2 force field calculations,[37,38] the O···O distance at the biphenyl spacer of 4.10 Å is not sufficient to widen and preorganize the cavity. In addition, the bridging of two

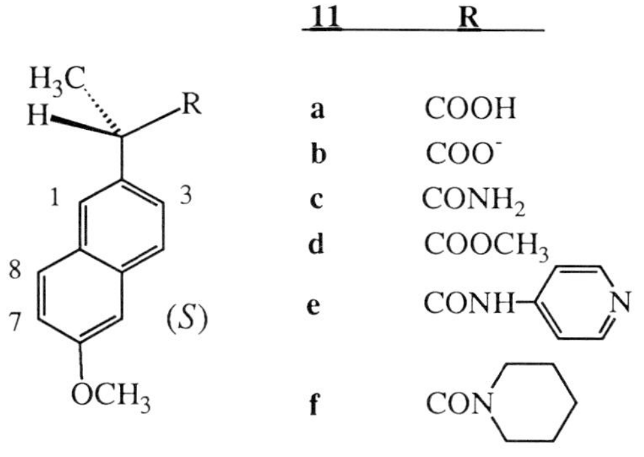

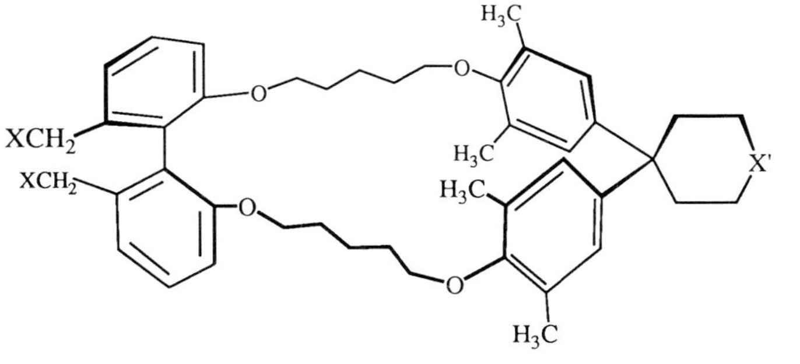

[36] Y. Rubin, K. Dick, F. Diederich, and T. M. Georgiadis, *J. Org. Chem.*, 1986, **51**, 3270.
[37] U. Burkert and N. L. Allinger, 'Molecular Mechanics', American Chemical Society, Washington, DC, 1982.
[38] N. L. Allinger, *J. Am. Chem. Soc.*, 1977, **99**, 8127.

very differently sized spacers in **13** apparently leads to an extensively twisted geometry which further reduces the size of a possible binding site.[39] The helical conformation shown in Scheme 6.2 seems to be imposed on the macrocycle by the tendency of the bridging alkane chains to maximize the number of *anti* torsional angles.

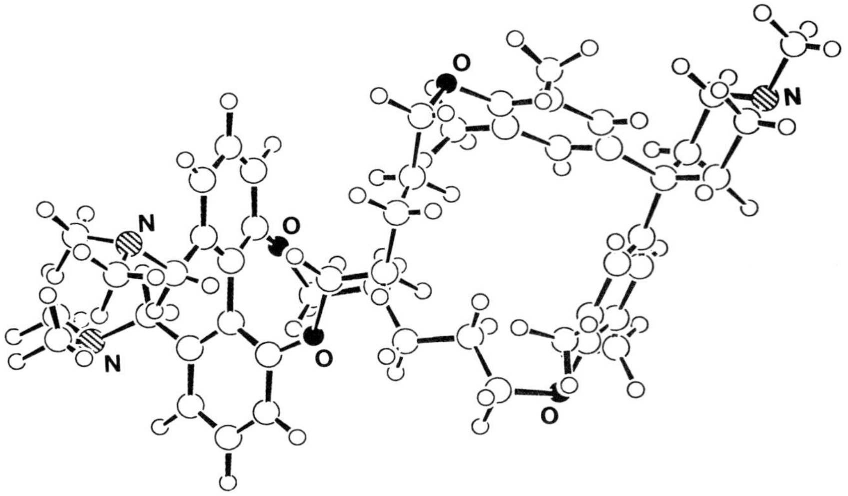

Scheme 6.2

On the way to biphenyl macrocycles related to **12** and **13**,[36,40] another result was obtained which illustrates how a potential binding site can be destroyed if the macrocycle adopts cavity-filling conformations in a process which has been described as self-complexation.[41] Starting with the ozonation of pyrene, cyclophane **14** was prepared in three steps.[36] However, a second ozonation of the 9,10-bond in the phenanthrene unit of **14** to give the chiral spacer was not successful. ^{1}H NMR spectroscopic studies indicated that **14**, invariable of solvent, temperature, and added phenanthrene, highly prefers the cavity-filling conformation **15**, in which the normally very reactive 9,10-double bond of the phenanthrene moiety is completely protected from ozonation through steric shielding. Figure 6.3 shows the peculiar chemical shifts in the ^{1}H NMR spectrum of **14** that strongly support the preference of the macroring for adopting conformation **15**.[36] Homonuclear 2D nuclear Overhauser enhancement spectroscopy (NOESY) and 1D NOE difference spectroscopy provided additional experimental evidence for the phenanthrene unit of **14** being located in close spatial proximity to the diphenylmethane unit.[40]

[39] R. Dharanipragada, S. B. Ferguson, and F. Diederich, *J. Am. Chem. Soc.*, 1988, **110**, 1679.
[40] R. J. Loncharich, E. Seward, S. B. Ferguson, F. K. Brown, F. Diederich, and K. N. Houk, *J. Org. Chem.*, 1988, **53**, 3479.
[41] P. D. J. Grootenhuis, J. van Eerden, E. J. R. Sudhölter, D. N. Reinhoudt, A. Roos, S. Harkema, and D. Feil, *J. Am. Chem. Soc.*, 1987, **109**, 4792.

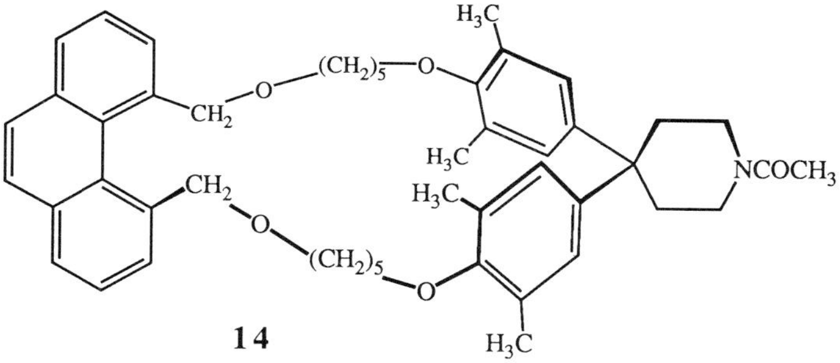

14

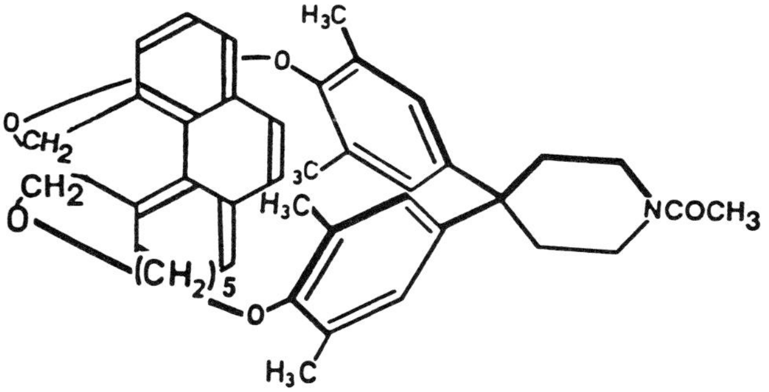

15

Following these experimental observations, a rigorous conformational analysis was undertaken which included comparative MM2 and AMBER[42,43] force field minimizations as well as molecular dynamics simulations using the SHAKE algorithm.[44] The gas-phase minimizations performed with the two different force fields were in good agreement and fully supported the ^{1}H NMR results. According to the MM2 force field, cyclophane **14** prefers the inward conformation **14-I** (Scheme 6.3) by 5.7 kcal mol^{-1} over conformations with the phenanthrene moiety outside the cavity, *e.g.* **14-O**. The inward conformation **14-I** is preferred due to very favorable van der Waals interactions with the interior of the cavity. The apolar phenanthrene unit is better stabilized by an apolar environment. This preference should be even greater in polar solvents. Molecular dynamics simulations were run on macrocycle **14** at constant pressure of 1 atm. and a temperature of 300 K for a total simulation time of 80 ps. In each simulation, the initial conformation had the phenanthrene moiety outside the cavity. After 40 ps, the phenanthrene moiety had always rotated into the interior of the cavity. For the remainder of the simulation, the phenanthrene moiety stayed within the interior of the cavity.

[42] P. K. Weiner and P. A. Kollman, *J. Comput. Chem.*, 1981, **2**, 287.
[43] S. J. Weiner, P. A. Kollman, D. A. Case, U. C. Singh, C. Ghio, G. Alagona, S. Profeta, Jr., and P. Weiner, *J. Am. Chem. Soc.*, 1984, **106**, 765.
[44] W. F. Van Gunsteren and H. J. C. Berendsen, *Mol. Phys.*, 1977, **34**, 1311.

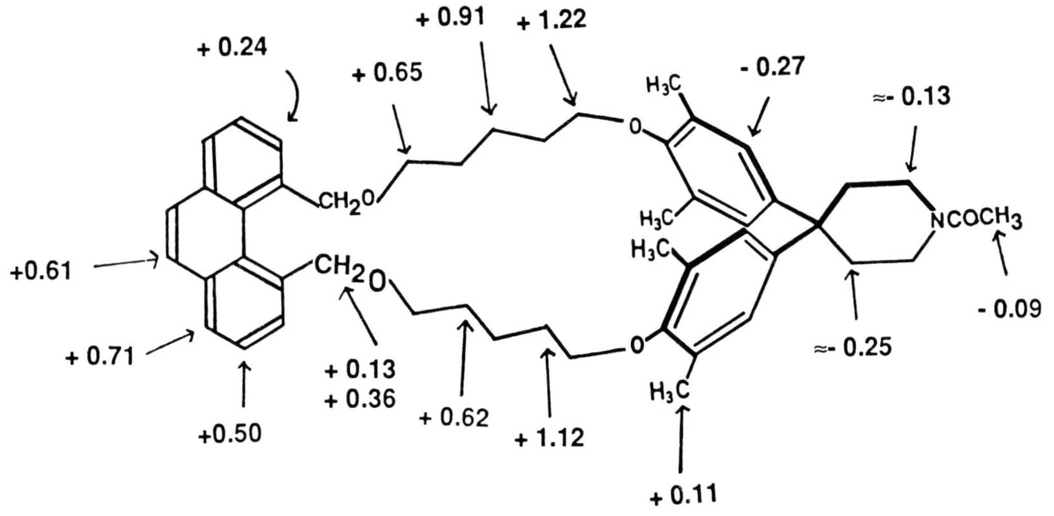

Figure 6.3 *Differences between the chemical shifts (Δδ, + = upfield shift) of comparable protons of* **14** *and the cyclization precursors 4,5-phenanthrenedimethanol and N-acetyl-4,4-bis[4-(5-chloropentoxy)-3,5-dimethylphenyl]piperidine. (Reproduced with permission from reference 40. Copyright 1988 ACS.)*

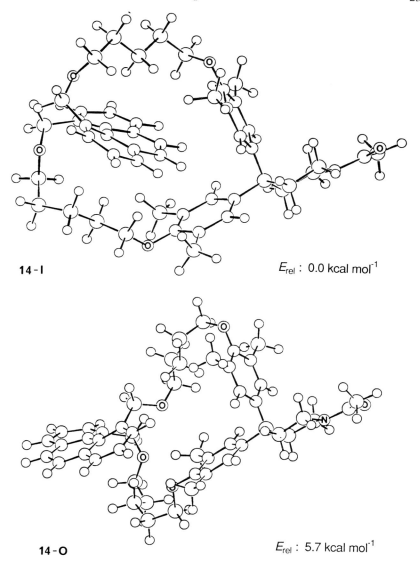

14-I E_{rel} : 0.0 kcal mol^{-1}

14-O E_{rel} : 5.7 kcal mol^{-1}

Scheme 6.3

With their two bulky 6,6′-substituents, the biphenyl macrocycles **12** and **13** are unable to adopt cavity-filling conformations with the biphenyl rotated inwards. In contrast, NMR studies as described above indicated that the biphenyl macrocycle **16** without 6,6′-substituents consists of conformations that have one phenyl ring of the biphenyl unit folded into the cavity in dynamic equilibrium with conformations consisting of the entire biphenyl unit inside or outside the cavity.[40] Both AMBER and MM2 force field calculations also support the relevance of cavity-filling conformations for the biphenyl macrocycle **16**. Scheme 6.4 shows the lowest energy conformers of **16** calculated by using the MM2 force field. In **16-⊥**, one phenyl ring of the biphenyl moiety is folded inside the cavity. The entire biphenyl unit is outside the cavity in **16-O**, and inside the cavity in **16-I**. Again, favorable van der Waals interactions within the interior of the cavity lead to a preference of inside and partially inside over outside conformations. In molecular dynamics simulations starting from outward conformations, the biphenyl moiety completely moves into the cavity within 40 ps, then moves in dynamic equilibrium between conformations with one phenyl ring inside the cavity.

16

The described molecular modeling studies of **14** and **16** completely neglect the dependence of macrocycle conformation on solvation enthalpy and entropy. Therefore, the good agreement between experimental results obtained in solution and gas-phase force field calculations seen in this study are remarkable.

The experimental and modeling studies with **13** had clearly shown that the O···O distance of ≈4.1 Å at the chiral biphenyl spacer was not sufficient to adequately widen and preorganize the cavity binding site.[36,39] MM2 force field calculations indicated that the critical O···O distance in the 4-phenyl-1,2,3,4-tetrahydroisoquinoline derivative **17** would be considerably larger and, therefore, **17** should be a more suitable chiral spacer.[39] When the optically active cyclophane (+)-**18** was prepared, its binding properties were found to be consistent with the predictions. (There exists no correlation between the arbitrarily chosen absolute configuration shown by the perspective drawing of (+)-**18** and the experimentally determined sign of optical rotation.) In water/methanol (60:40, v/v), (+)-**18** preferentially binds 2,6-disubstituted naphthalenes which take a pseudoaxial position in the cavity.[39] For example, 6-methoxy-2-naphthonitrile forms a complex with an association constant of $K_a = 336$ L mol^{-1} ($\Delta G° = -3.5$ kcal mol^{-1}) at 303 K. The (R) and (S)-enantiomers of naproxen (**11a**) and its methyl ester (**11d**) also form inclusion complexes with (+)-**18**, and differential upfield complexation shifts

Cyclophanes for Chiral Molecular Recognition 239

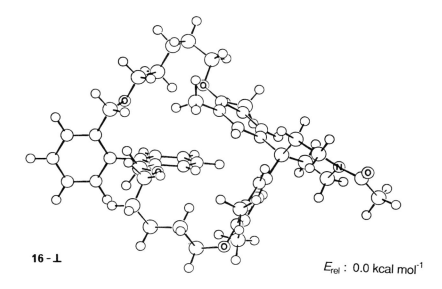

16-⊥ E_{rel} : 0.0 kcal mol^{-1}

16-O E_{rel} : 2.0 kcal mol^{-1}

Scheme 6.4

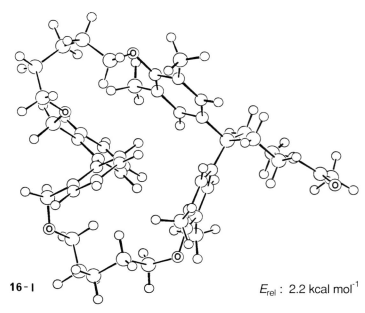

16-I E_{rel} : 2.2 kcal mol^{-1}

Scheme 6.4

of all guest resonances demonstrate the formation of diastereomeric complexes with different geometries.[39]

According to MM2 calculations on its immediate bis(tertiary amine) precursor, cyclophane **18** can adopt very favorable geometries for inclusion complexation.[39] Scheme 6.5 shows a productive low energy conformation (**19A**) of the bis(tertiary amine) precursor with an O⋯O distance at the chiral spacer of 7.58 Å, and excellent complementarity to a naphthalene substrate. As a calibration, a center-to-center distance of ≈ 6.8 Å between two aromatic rings is required for efficient π–π-stacking intercalation at van der Waals contact. However, the MM2 calculations also pointed out that, as a result of the conformational flexibility of the 4-phenyl-tetrahydroisoquinoline unit, there exists a broad range of lower energy conformations which show poorer preorganization giving a partially filled cavity (*e.g.* **19B** in Scheme 6.5). Although solvation could considerably alter the relative energies of the host conformers, the moderate binding observed experimentally with **18** suggests that low energy conformers with poorer binding ability are also prevalent in solution. The binding of naphthalene derivatives by cyclophane **18** in methanol/water (60:40, v/v) is weaker by ≈ > 1.5 kcal mol^{-1} than the binding of these substrates by bis(diphenylmethane) hosts. The 4-phenyl-tetrahydroisoquinoline unit in **18** is a conformationally more flexible spacer than a diphenylmethane unit. As a result, **18** is less preorganized than bis(diphenylmethane) cyclophanes. During complexation, a larger part of the attractive host–guest binding energy has to be spent to organize **18** from less productive

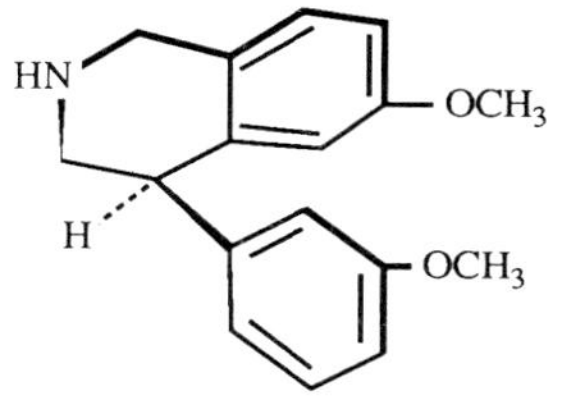

17

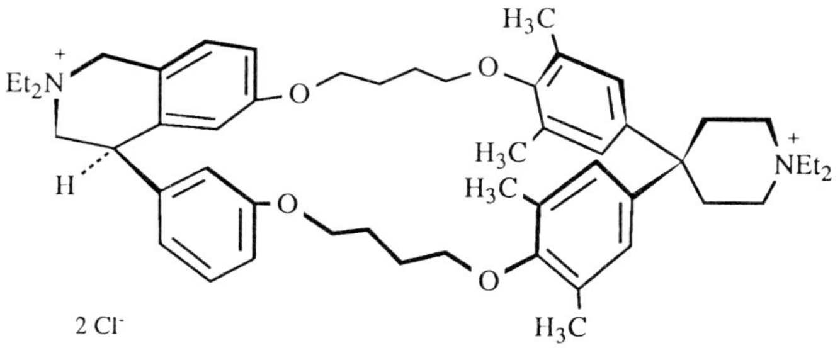

(+) - 18

lower energy conformations such as **19B** to a higher energy, productive binding conformation such as **19A**. The conformational analysis undertaken for cyclophanes **14**, **16**, and **18**, and the correlations with experimentally observed binding abilities provide a convincing example for the validity of the principle of preorganization, established by Cram (Chapter 2.3), in apolar complexation events.[45]

6.3.4 Chiral Recognition of Naproxen Derivatives

The minor groove of the 1,1'-binaphthyl unit shapes efficient chiral cation binding sites as demonstrated by Cram *et al.* in their impressive studies with optically active binaphthyl crown ethers such as **1** and **2**.[9-11] For an 88° dihedral angle Θ about the chirality axis, the O···O distance in **20** is *ca.* 3.6 Å for the minor groove. With an O···O distance of *ca.* 7.0 Å, the major groove in **20** is almost twice as wide and ideal for shaping an aromatic binding site in chiral cyclophane hosts. Starting from the regioselectively protected 2,2',7,7'-tetrahydroxy-1,1'-binaphthyl derivative **21**, cyclophane **22** was prepared in racemic form.[46,47] Scheme 6.6 shows the MM2-optimized structure[48] of the

---

[45] S. C. Zimmerman, M. Mrksich, and M. Baloga, *J. Am. Chem. Soc.*, 1989, **111**, 8528.
[46] F. Diederich, M. R. Hester, and M. A. Uyeki, *Angew. Chem.*, 1988, **100**, 1775; *Angew. Chem. Int. Ed. Engl.*, 1988, **27**, 1705.
[47] M. R. Hester, M. A. Uyeki, and F. Diederich, *Isr. J. Chem.*, 1989, **29**, 201.
[48] H. Beckhaus, *Chem. Ber.*, 1983, **116**, 86.

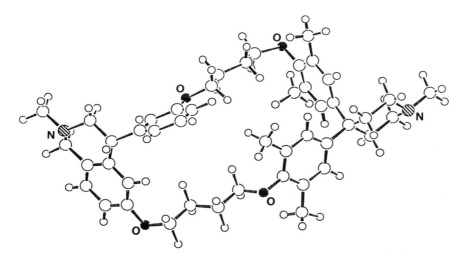

19-A E_{rel}: 6.47 kcal mol^{-1}

19-B E_{rel}: 0.00 kcal mol^{-1}

Scheme 6.5

immediate tris(tertiary amine) precursor to the quaternized cyclophane **22**. A wide open, preorganized binding site exists in all low-energy conformers generated in the modeling study. A direct comparison of the association constants for the 1:1 complexes formed by 6-methoxy-2-naphthonitrile in D$_2$O/methanol-d_4 (60:40, v/v) demonstrated that host **22** (K_a = 1990 L mol^{-1},

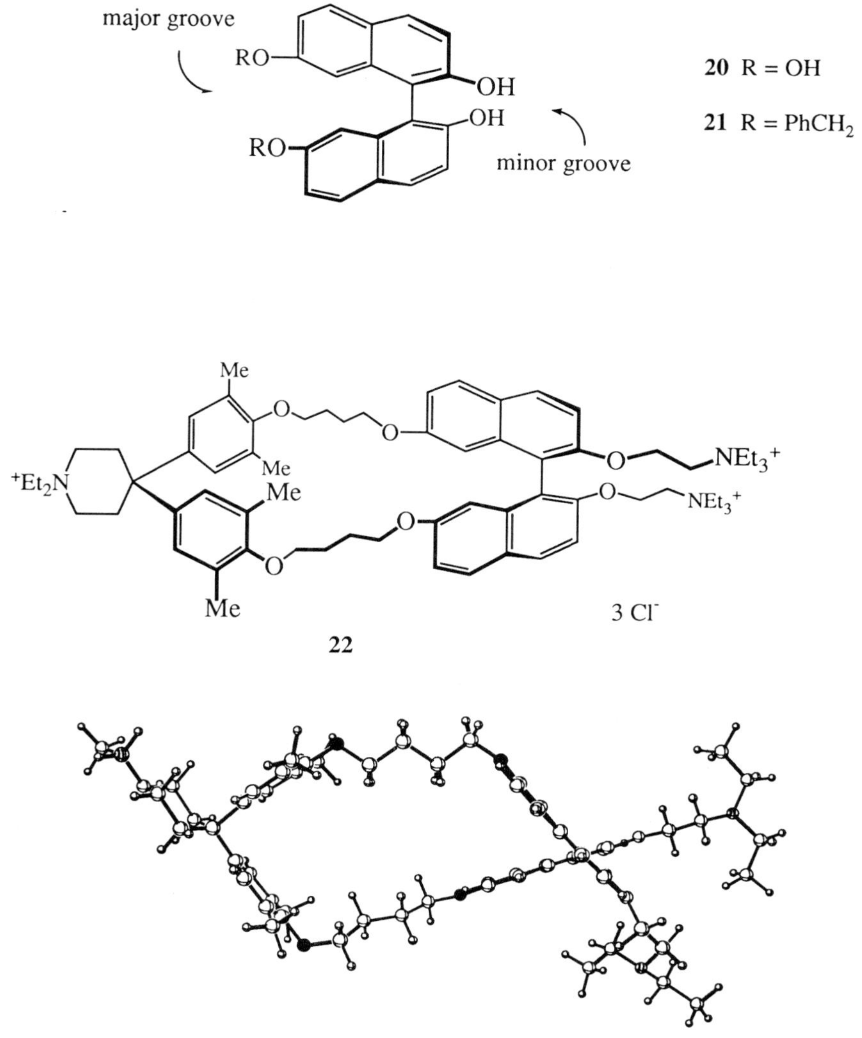

20 R = OH
21 R = PhCH$_2$

Scheme 6.6

293 K) is a much better binder than **18** ($K_a = 336$ L mol^{-1}, 303 K) which incorporates the 4-phenyl-tetrahydroisoquinoline unit as the chiral spacer.

Starting from optically pure (R)- and (S)-**21**, readily resolved through simple clathrate formation with quinine,[49] the two cyclophane enantiomers (R)-**22** and (S)-**22** were prepared.[50] ^{1}H NMR complexation studies in D$_2$O/

---

[49] C. Rosini, G. Uccello-Barretta, D. Pini, C. Abete, and P. Salvadori, *J. Org. Chem.*, 1988, **53**, 4579.

[50] P. P. Castro, T. M. Georgiadis, and F. Diederich, *J. Org. Chem.*, 1989, **54**, 5835.

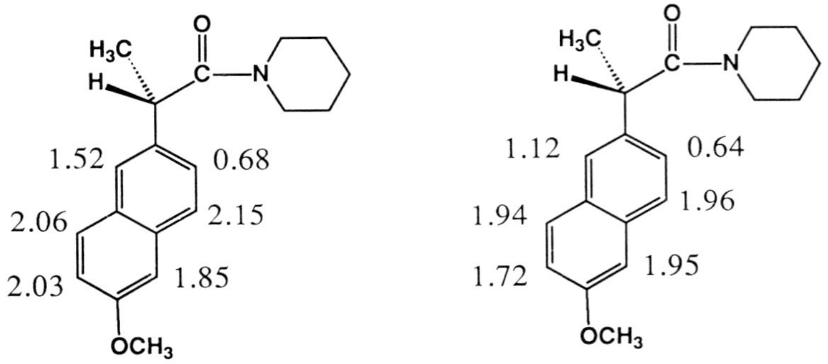

Figure 6.4 *Upfield ^1H NMR complexation shifts (ppm) calculated for saturation binding of (S)-11f by (R,S)-22 in $D_2O/MeOH$-d_4 (60:40, v/v)*

methanol-d_4 (60:40, v/v, $T=293$ K) demonstrated that the naproxen derivatives (R,S)-**11a–f** form diastereomeric 1:1 complexes of differential stability and geometry with the host enantiomers (R) and (S)-**22**. Figure 6.4 shows the large differential complexation shifts of the resonances of (S)-**11f** in the two diastereomeric, axial-type inclusion complexes.

Table 6.3 lists the association constants K_a and the free energies of formation $-\Delta G°$ (293 K) for the diastereomeric complexes formed between (R) and (S)-**22** and the naproxen derivatives (S)-**11a–f**. Corresponding data were obtained in ^1H NMR titrations with the (R)-naproxen derivatives. For all (S)-naproxen derivatives, the (S)-(S)-complexes are more stable, and the difference in stability between diastereomeric complexes $\Delta(\Delta G°)$ varies between 0.16 and 0.33 kcal mol^{-1}. The stability difference is most pronounced for the complexes of the naproxen derivatives **11e** and **11f** which have bulky amide groups. This indicates that differential steric interactions are most probably at the origin of the observed chiral recognition. The degree of enantioselection is quite substantial considering that the binding sites of (R,S)-**22** are partially shaped by an achiral diphenylmethane unit. A much higher enantioselection in the binding of naproxen derivatives is therefore expected for preorganized, optically active cyclophane hosts shaped by two 1,1′-binaphthyl major grooves.

Table 6.3 Association constants, K_a, and free energies of formation, $-\Delta G°$, of the diastereomeric complexes between (R,S)-**22** and (S)-**11a–f** in $D_2O/MeOH$-d_4 (60:40, v/v, $T=293$ K).[a] The calculated differences in stability between diastereomeric complexes $\Delta(\Delta G°)$ are given

Naproxen Derivative	(R)-**22**		(S)-**22**		$\Delta(\Delta G°)$ (kcal mol^{-1})
	K_a (L mol^{-1})	$-\Delta G°$ (kcal mol^{-1})	K_a (L mol^{-1})	$-\Delta G°$ (kcal mol^{-1})	
11a[b]	2105	4.45	2540	4.56	0.16
11b[c]	1040	4.04	1335	4.19	0.15
11c	775	3.87	1010	4.02	0.15
11d	2075	4.45	3110	4.68	0.23
11e	1760	4.35	2840	4.63	0.28
11f	1405	4.22	2490	4.55	0.33

[a] Errors in K_a: ±10%; [b] in 0.1 M DCl/MeOH-d_4 (60:40, v/v); [c] in 0.01 M K_2CO_3/MeOH-d_4 (60:40, v/v).

CHAPTER 7
Solvent Effects in Molecular Recognition

7.1 Introduction

The cyclophane complexes discussed in the previous five chapters, like all other biotic and abiotic complexes, are held together by a variety of attractive interactions between the binding partners. Without such attractions, a complex would not form. The formation of stable complexes requires steric and electronic complementarity between host and guest along with preorganization of the host (Chapter 2.3). Following these principles should guarantee a successful design of a gas-phase complex. However, all the complexes described in the previous chapters are formed in solution, and numerous examples have illustrated high solvent dependency for molecular recognition events. For example, apolar complexation is strongest in water. On the other hand, stable synthetic complexes held together by hydrogen-bonding as the main driving force have not yet been prepared in aqueous solution. This illustrates that considerations of solvent effects are of equal importance to the principles of stereoelectronic host–guest complementarity and host preorganization for the design of a complexation event in solution. This chapter describes how the nature of the solvent and specific solvation effects determine the stability of molecular complexes. Water is the essential biological fluid which promotes apolar aggregation and complexation processes necessary to sustain all functions of life. Therefore, complexation studies in water are of special interest and are the focus of Chapter 7.

7.2 Attractive Host–Guest Interactions

A variety of attractive host–guest interactions stabilize molecular complexes, and contributions of individual forces to the total interaction energy vary from complex to complex.[1] In inclusion complexes formed by incorporation of an

---

[1] C. Reichardt, 'Solvents and Solvent Effects in Organic Chemistry', 2nd ed., VCH, Weinheim, 1988.

apolar substrate or substrate moiety into a lipophilic cavity, the predominant attractive host–guest interactions are the van der Waals interactions which are comprised of a dipole–dipole, a dipole–induced dipole, and an induced dipole–induced dipole interaction term.[2] The latter interaction, also known as London dispersion interactions, is composed of an attractive and a repulsive term. For an interatomic interaction, the sum of the two terms is described by the two-parameter Lennard-Jones potential $Ar^{-12} - Br^{-6}$ with r equal to the interatomic distance, and the parameters A for the repulsive and B for the attractive term. Van der Waals forces are the major attractive host–guest interactions in the formation of complex **2** between cyclophane **1** and *p*-xylene (Scheme 7.1, Chapter 2.7.1). Additional stabilization of complexes formed by non-dipolar components like *p*-xylene results from less well defined quadrupolar and higher-multipolar interactions which reflect charge distributions that are more complicated than dipoles. Aromatic–aromatic interactions between the face-to-face and edge-to-face aromatic rings of the two binding partners in **2** represent another stabilizing interaction which usually is discussed separately from the van der Waals term.[3] The ionization potentials of *p*-benzodinitrile and the trialkoxybenzene rings of **1** differ greatly. Therefore, electron donor–acceptor interactions[4] should further stabilize the inclusion complex similar to **2** which is formed by this electron–acceptor guest.

Different types of attractive host–guest interactions stabilize the cyclophane complexes **3–5**. A strong ion–dipole interaction[1] between the included quaternary 1-adamantylammonium center and the surrounding aromatic π-systems of the host has been recognized to stabilize complex **3** (Chapter 4.3).[5] Attractive Coulombic interactions between charged centers provide a major driving force for the formation of complex **4** (Chapter 4.4).[6] Finally, the guanosine complex **5** forms as a result of triple hydrogen-bonding between host and guest (Chapter 5.5).[7] Obviously, van der Waals and aromatic–aromatic interactions that stabilize complex **2** are also effective to various degrees in complexes **3–5**.

7.3 The Formation of an Apolar Cyclophane Complex in Solution: A Complex Picture

As mentioned in the introduction, the criteria for the design of a gas phase complex consist of: (i) the electronic host–guest complementarity leading to attractive interactions, (ii) the steric complementarity of the binding partners, and (iii) the preorganization of the host. The situation is much more complicated in solution. Figure 7.1 shows a schematic representation of the

---

[2] B. Chu, 'Molecular Forces', Wiley-Interscience, New York, 1967.
[3] W. L. Jorgensen and D. L. Severance, *J. Am. Chem. Soc.*, 1990, **112**, 4768.
[4] V. Gutmann, 'The Donor–Acceptor Approach to Molecular Interactions', Plenum, New York, 1978.
[5] D. A. Stauffer and D. A. Dougherty, *Tetrahedron Lett.*, 1988, **29**, 6039.
[6] H.-J. Schneider, D. Güttes, and U. Schneider, *J. Am. Chem. Soc.*, 1988, **110**, 6449.
[7] A. D. Hamilton and N. Pant, *J. Chem. Soc., Chem. Commun.*, 1988, 765.

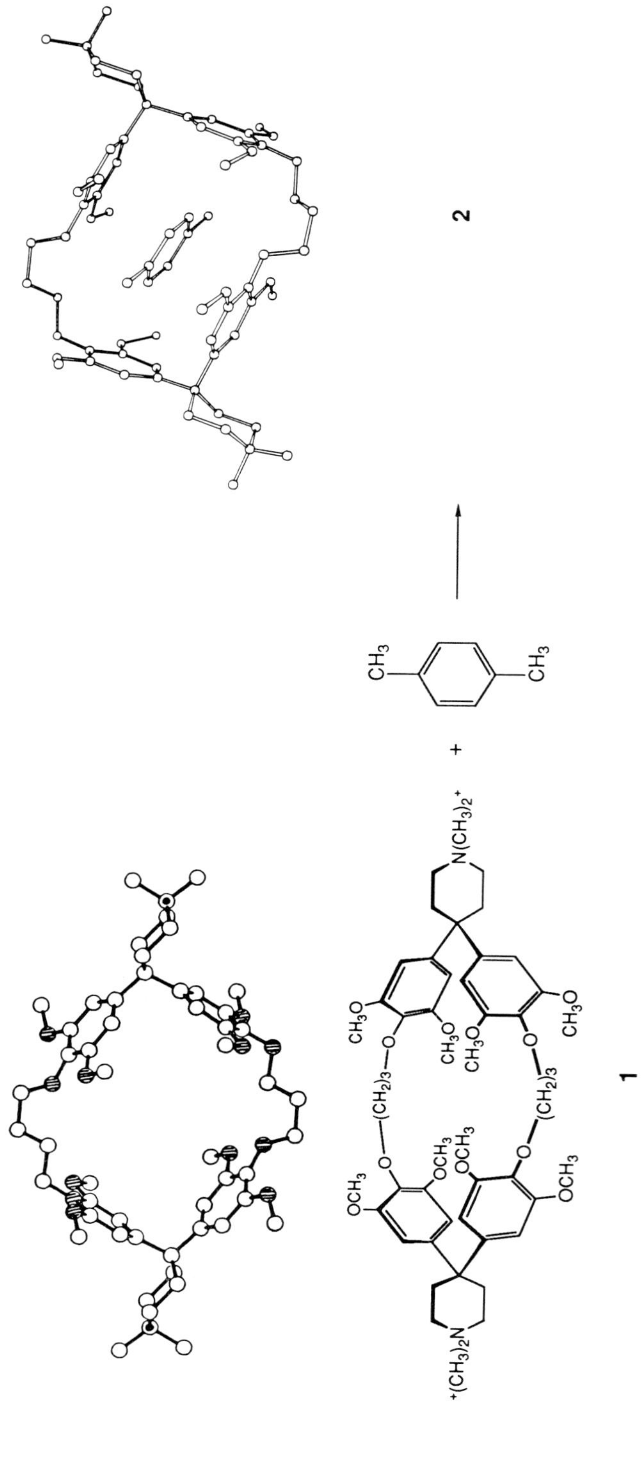

Scheme 7.1

Solvent Effects in Molecular Recognition

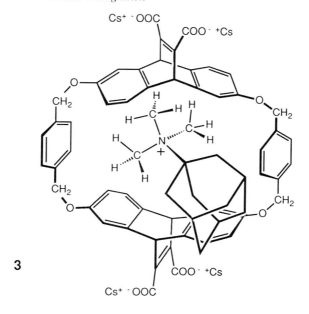

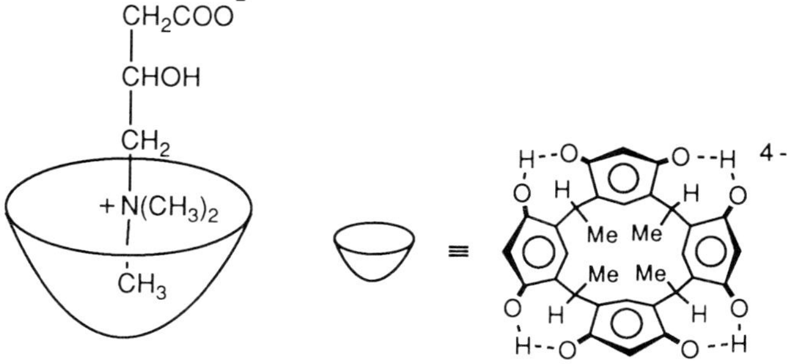

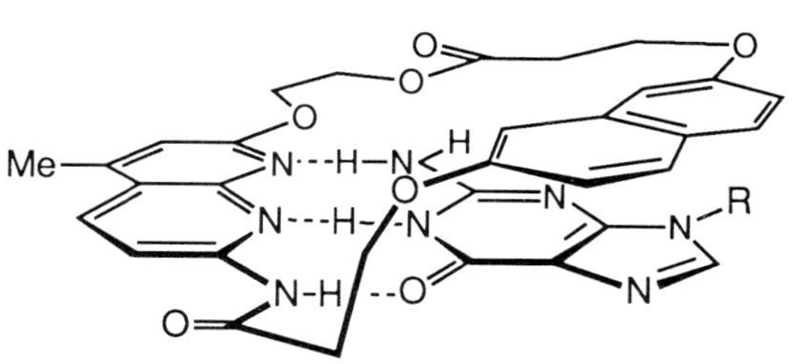

3

4

5

thermodynamic process which describes an apolar inclusion event in water, e.g. the formation of the complex between cyclophane **1** and *p*-xylene. A similar representation had previously been used by Tabushi *et al.* to describe the inclusion of organic guest molecules by cyclodextrins in water.[8,9] Several simplifications are introduced into this scheme. Changes in the geometry of the host during the entire process are ignored. Also, in a first approximation, the exterior solvation of the host in the bound and unbound states is assumed to be identical.

Figure 7.1 clearly documents the great complexity of an apolar complexation event in solution as compared to the gas phase. Host–guest complexation in the gas phase essentially is a single-step event (H + G $\rightleftharpoons$ HG). It approximately corresponds to step E in Figure 7.1 which describes the formation of a complex between the guest in the gas phase and the host, only solvated externally. To understand the complexation process in solution, steps A–D in Figure 7.1, which are all solvent-related, also need to be considered. Experimental or computational approaches which target a microscopic level description of host–guest complexation in solution, attempt to provide accurate numbers for the various enthalpic and entropic terms in thermodynamic processes such as in Figure 7.1. The sum of these terms should give the thermodynamic quantities $\Delta G°$, $\Delta H°$, and $T\Delta S°$ measured for the overall inclusion process in solution.

In the thermodynamic process shown in Figure 7.1, step A describes the desolvation of the host cavity with transfer of the cavity water molecules (S) into the gas phase. In this process, the van der Waals interactions between the host (H) cavity and the solvating water molecules ($+\Delta H^{SH}_{vdW}$) as well as the interactions between the solvating water-molecules themselves ($+\Delta H'^{SS}_{intermol}$) are lost. The transfer of the solvent molecules from the cavity into the gas phase is accompanied with an entropy gain ($+T\Delta S'^{S}_{trans, rot}$). The water molecules are subsequently transferred into the bulk (step B) which is accompanied by a loss in entropy ($-T\Delta S''^{S}_{trans, rot}$) and a gain in cohesive interactions ($-\Delta H''^{SS}_{intermol}$).

In aqueous solution, water molecules solvate apolar solutes by forming a cluster similar to an 'icelike' structure.[10–13] Such an ordered array allows the water molecules around an apolar surface to partially maintain strong hydrogen bonds. Step C describes the transfer of a guest out of this 'icelike' cluster into the gas phase. This transfer is accompanied by a loss of solvent–guest interactions ($+\Delta H^{SG}_{vdW}$) and a gain in entropy of the guest ($+T\Delta S'^{G}_{trans, rot}$). In the following step D, the water cluster breaks down with the transfer of its components into the bulk. This process is characterized by a gain in solvent cohesive interactions ($-\Delta H'''^{SS}_{intermol}$) and a gain in entropy

---

[8] I. Tabushi, Y. Kiyosuke, T. Sugimoto, and K. Yamamura, *J. Am. Chem. Soc.*, 1978, **100**, 916.
[9] I. Tabushi and T. Mizutani, *Tetrahedron*, 1987, **43**, 1439.
[10] H. S. Frank and M. W. Evans, *J. Chem. Phys.*, 1945, **13**, 507.
[11] H. A. Scheraga, *Acc. Chem. Res.*, 1979, **12**, 7.
[12] F. H. Stillinger, *Science (Washington DC)*, 1980, **209**, 451.
[13] K. Shinoda, *J. Phys. Chem.*, 1977, **81**, 1300.

Solvent Effects in Molecular Recognition

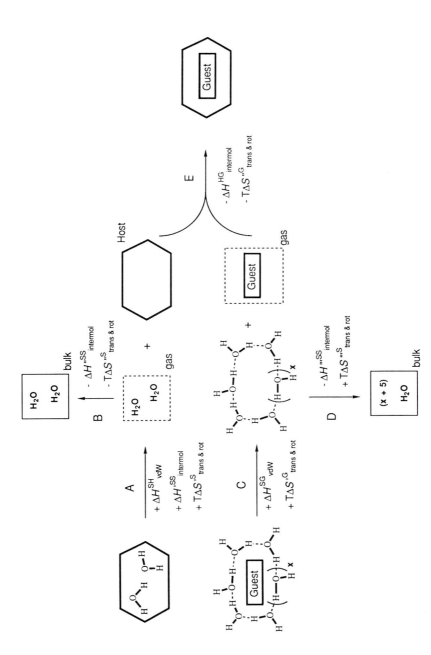

Figure 7.1 *A composite process to describe the thermodynamics of apolar inclusion complexation in water. See text Chapter 7.3 for further explanations*

($+ T\Delta S'''S_{trans, rot}$). Finally, in step E, attractive host–guest interactions ($-\Delta H^{HG}_{intermol}$) lead to complex formation under loss of the translational and rotational entropy of one of the two components ($-T\Delta S''^{G}_{trans, rot}$).

The thermodynamic process in Figure 7.1 is also valid for apolar binding processes in other solvents; only the water molecules need to be replaced by the other solvent molecules. However, the overall complexation driving force and the individual thermodynamic terms shown in the scheme should differ considerably as a function of solvent. In the following, experimental data for apolar binding in water and other solvents will be explained in terms of the composite process shown in Figure 7.1.

7.4 Enthalpically Controlled Tight Molecular Complexation in Water

As shown in Chapter 2.7.1 (Table 2.4), cyclophanes **1** and **6** form stable complexes with *para*-disubstituted benzene derivatives in water.[14] Weaker complexation occurs in methanol-d_4, and Table 7.1 contrasts the thermodynamic quantities measured for selected complexes of **6** in D_2O and methanol-d_4. The thermodynamic characteristics $\Delta H°$ and $T\Delta S°$ were obtained by van't Hoff analysis of ^1H NMR binding titrations at various temperatures. Over temperature ranges of $\Delta T \approx 20$ °C, the temperature dependency of $\Delta H°$, *i.e.* heat capacity changes $\Delta C_p°$, were sufficiently small to give linear van't Hoff plots (Table 7.1). The $\Delta H°$ values in water calculated from the van't Hoff analysis have been confirmed by directly measuring the heats of complexation with a microcalorimeter.[15]

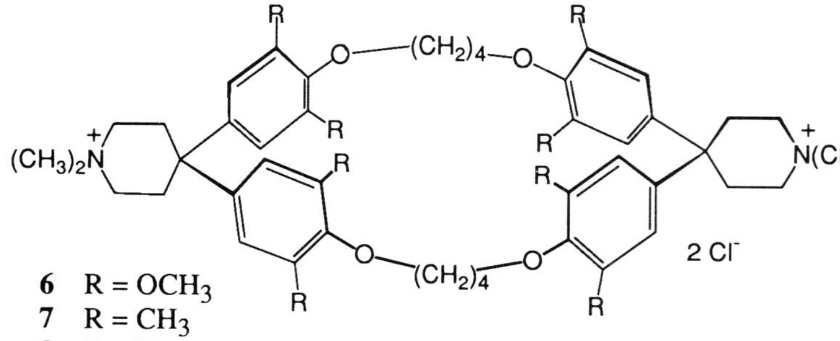

6 R = OCH_3
7 R = CH_3
8 R = H

The results in Tables 7.1 and 2.4 show that binding in water is entropically unfavorable (large negative $T\Delta S°$) and predominantly enthalpy-driven (large negative $\Delta H°$). A part of this favorable enthalpy results from attractive host–guest interactions in the formation of tight complexes ($-\Delta H^{HG}_{intermol}$ in step E of Figure 7.1). However, a comparison of the thermodynamic data for

---

[14] S. B. Ferguson, E. M. Seward, F. Diederich, E. M. Sanford, A. Chou, P. Inocencio-Szweda, and C. B. Knobler, *J. Org. Chem.*, 1988, **53**, 5593.
[15] T. Wyman and F. Diederich, unpublished results.

Table 7.1 *Association constants K_a and enthalpic ($\Delta H°$) and entropic ($T\Delta S°$) contributions to the free energies of complexation ($\Delta G°$) at 293.4 K for complexes of cyclophane **6** with 1,4-disubstituted benzene guests in D_2O and methanol-d_4*

	K_a $L\,mol^{-1}$	$\Delta G°$ $kcal\,mol^{-1}$	$\Delta H°$ $kcal\,mol^{-1}$	$T\Delta S°$ $kcal\,mol^{-1}$
Complexes of **6** in D_2O				
p-Benzodinitrile	7.8×10^3	-5.23	-9.5 ± 1.0	-4.3 ± 1.0
p-Dimethoxybenzene	1.0×10^4	-5.38	-10.2 ± 2.5	-4.8 ± 2.5
Complexes of **6** in methanol-d_4				
p-Benzodinitrile	2.4×10^1	-1.86	-4.2 ± 1.5	-2.4 ± 1.5
p-Dimethoxybenzene	8×10^0	-1.20	-4.4 ± 1.5	-3.2 ± 1.5

complexation in water and methanol shows that water intrinsically provides a larger part of the enthalpic driving force. Complexation in methanol is much weaker, mainly as a result of a less favorable enthalpic term (Table 7.1). The large difference in enthalpic driving force for complexation in water and methanol cannot be explained by differences in host–guest interactions. According to ^{1}H NMR analysis, the complexation geometries are similar in both solvents, and this should lead to similar host–guest interaction enthalpic terms $-\Delta H^{HG}_{intermol}$ (step E in Figure 7.1) in the two environments. Therefore, a large portion of the favorable enthalpic component observed in water must result from specific solvent contributions which will be discussed in the following section.

7.4.1 The Role of Solvent Cohesive Interactions

Water is characterized by strong cohesive interactions.[1] Bulk water represents a dynamic network with each water molecule participating in four strong hydrogen bonds to neighboring solvent molecules. The water molecules that solvate the apolar surfaces of the free guests and especially the cavity of the free host have reduced cohesive interactions and are enthalpically higher in energy than the water molecules in the bulk solvent.[16] Water molecules around apolar surfaces participate in fewer strong hydrogen bonds than bulk solvent molecules. Upon inclusion complexation, these water molecules are released into the bulk and become enthalpically lower in energy. In terms of the composite process in Figure 7.1, more solvent–solvent interaction energy ($-\Delta H''^{SS}_{intermol}$) is gained in step B than is lost ($+\Delta H'^{SS}_{intermol}$) in step A, and enthalpy is gained in step D upon collapse of the surface-solvating water cluster ($-\Delta H'''^{SS}_{intermol}$). Cohesive interactions in methanol are weaker than in water, and therefore less enthalpy is gained upon transfer of methanol molecules that solvate the complementary host–guest surfaces into the bulk.

[16] W. Saenger, *Angew. Chem.*, 1980, **92**, 343; *Angew. Chem. Int. Ed. Engl.*, 1980, **19**, 344.

The two terms $-\Delta H''^{SS}_{intermol}$ and $-\Delta H'''^{SS}_{intermol}$ for the process in methanol are less negative than in water.

A similar explanation for the role of solvent cohesive interactions in promoting apolar binding processes in water had previously been proposed by Sinanoglu.[17] As a result of the strong cohesive forces in water, energy is required to create a cavity in a liquid in order to accommodate the guest molecules. Such cavities are also created when water molecules enter the free host cavity upon solvation. The enthalpy terms for these cavity-creating processes are the same but of opposite sign to those for solvent–solvent interactions in steps A, B, and D of Figure 7.1.[18]

7.4.2 The Role of London Dispersion Interactions

A second contribution to the favorable enthalpy term in water results from the substitution of less favorable dispersion interactions between water molecules and the complementary host–guest surfaces (enthalpy loss $+\Delta H^{SH}_{vdW}$ in step A of Figure 7.1 and $+\Delta H^{SG}_{vdW}$ in step C) by more favorable dispersion interactions between the hydrocarbon surfaces in the complex (enthalpy gain $-\Delta H^{HG}_{intermol}$ in step E).[19] The attractive B term in the Lennard-Jones potential which describes London dispersion interactions (Chapter 7.2) is proportional to the polarizability α of the interacting atoms and groups. Dispersion energies are additive, and at constant distances between interacting atoms, the attractive forces increase with increasing atom polarizability. Oxygen atoms ($\alpha = 0.84$ Å^3) and hydroxyl residues ($\alpha = 1.20$ Å^3), the constituents of water, have low polarizability, whereas the polarizabilities of hydrocarbon residues, *e.g.* a CH_2 group ($\alpha = 1.77$ Å^3), a CH_3 group ($\alpha = 2.17$ Å^3), or an aromatic CH group ($\alpha = 2.07$ Å^3) are much larger.[20,21] Therefore, the dispersion forces between water molecules and a hydrocarbon surface are weaker than the forces between two hydrocarbon surfaces. Upon complexation by **1** or **6** with a *p*-disubstituted aromatic guest, less favorable water-solute contacts are replaced by more favorable contacts between the surfaces of the two binding partners. Methanol, on the other hand, possesses a polarizable methyl group which interacts favorably with the host and guest. Therefore, the terms $+\Delta H^{SH}_{vdW}$ and $+\Delta H^{SG}_{vdW}$ (Figure 7.1) become more positive (more unfavorable), and the differences in dispersion interactions between the nonbinding and the binding states in methanol are not as large.

7.4.3 Hydrophobic Bonding

Apolar binding and association in water is often referred to as 'hydrophobic binding'. The terms 'hydrophobic effect', 'hydrophobic interactions', and even

---

[17] O. Sinanoglu, in 'Molecular Associations in Biology', ed. B. Pullman, Academic, New York, 1968, pp. 427.
[18] J. J. Moura-Ramos, M. S. Dionísio, R. C. Gonçalves, and H. P. Diogo, *Can. J. Chem.*, 1988, **66**, 2894.
[19] A. Fersht, 'Enzyme Structure and Mechanism', Freeman, New York, 1985, pp. 293–310.
[20] A. Fersht and C. Dingwall, *Biochemistry*, 1979, **18**, 1245.
[21] J. A. McCammon, P. G. Wolynes, and M. Karplus, *Biochemistry*, 1979, **18**, 927.

'hydrophobic bond' are widely used to describe the phenomenon that the interactions between apolar surfaces of solutes and solvating water molecules are less favorable than the cohesive interactions of these water molecules in the bulk and the interactions between associated apolar surfaces.[22-25] In the past, hydrophobic association has often been uniquely viewed as an entropically driven process. The thermodynamic quantities $\Delta H° \approx 0$ and T$\Delta S° > 0$ measured for the transfer of an apolar solute from water into organic solvents or the gas phase, or for the formation of micelles and membranes in water has led to this previously commonly accepted viewpoint. Upon aggregation, the associating monomers lose less rotational and translational entropy than binding partners which form a tight molecular complex. Hence, the favorable desolvation entropies are not completely compensated for by a large unfavorable entropy term for monomer association, and thus a positive entropy is measured for the overall aggregation process. The van der Waals contacts between the monomer components in micelles and membranes are less tight than in molecular complexes, and only a small overall enthalpy term is measured for the formation of these aggregates.

The thermodynamic characteristics (large negative $\Delta H°$ and negative $T\Delta S°$) measured for the tight complexation of p-disubstituted neutral benzene derivatives by cyclophanes **1** and **6**[14] are in sharp contrast to those observed for the transfer of apolar solutes from water into organic solvents or the gas phase and for the formation of micelles and membranes. Cyclophanes represent only one class of receptors that complex apolar solutes in an enthalpy-driven way.[26] Similar thermodynamic characteristics to those in Tables 2.4 and 7.1 have been measured for the tight binding of aromatic substrates in hydrophobic pockets of enzymes and antibodies,[27,28] in the cavity of cyclodextrins,[29-33] and in AT-rich regions of the DNA minor groove which can adopt the shape of a deep narrow cleft.[34]

The different thermodynamic characteristics measured for tight complexation as compared to micellar aggregation and membrane formation primarily result from differences in the enthalpic and entropic terms for the solute

---

[22] C. Tanford, 'The Hydrophobic Effect: Formation of Micelles and Biological Membranes', 2nd ed., Wiley, New York, 1980.
[23] A. Ben-Naim, 'Hydrophobic Interactions', 2nd ed., Plenum, New York, 1983.
[24] P. L. Privalov and S. J. Gill, *Adv. Protein Chem.*, 1988, **39**, 191.
[25] N. Müller, *Acc. Chem. Res.*, 1990, **23**, 23.
[26] D. A. Stauffer, R. E. Barrans, Jr., and D. A. Dougherty, *J. Org. Chem.*, 1990, **55**, 2762.
[27] R. L. Biltonen and N. Langerman, *Methods Enzymol.*, 1979, **61**, 287.
[28] P. D. Ross and S. Subramanian, *Biochemistry*, 1981, **20**, 3096.
[29] K. Harata, *Bioorg. Chem.*, 1981, **10**, 255.
[30] K. Harata, K. Tsuda, K. Uekama, M. Otagiri, and F. Hirayama, *J. Incl. Phenom.*, 1988, **6**, 135.
[31] M. R. Eftink and J. C. Harrison, *Bioorg. Chem.*, 1981, **10**, 388.
[32] M. R. Eftink, M. L. Andy, K. Bystrom, H. D. Perlmutter, and D. S. Kristol, *J. Am. Chem. Soc.*, 1989, **111**, 6765.
[33] G. L. Bertrand, J. R. Faulkner, Jr., S. M. Han, and D. W. Armstrong, *J. Phys. Chem.*, 1989, **93**, 6863.
[34] K. J. Breslauer, R. Ferrante, L. A. Marky, P. B. Dervan, and R. S. Youngquist in 'Structure and Expression, Vol.2, DNA and its Drug Complexes', eds. R. H. Sarma and M. H. Sarma, Adenine Press, 1988, pp. 273.

association steps (*e.g.* step E in Figure 7.1). As described above, the association of monomers to form micelles or membranes is characterized by a moderate gain in enthalpy and a moderate loss in entropy. In contrast, numerous van der Waals contacts in tight complexes generate a very favorable host–guest interaction term $-\Delta H^{HG}_{intermol}$ (step E in Figure 7.1) and lead to a favorable (negative) enthalpy for the overall tight complexation process. On the other hand, this tight fit considerably reduces the translational and rotational degrees of freedom of the guest upon complexation. Therefore, in the complexation of benzene derivatives by **1** or **6**, the favorable desolvation entropies in steps A, C, and D (Figure 7.1) are masked by a large entropy loss ($-T\Delta S''^{G}_{trans, rot}$) in the complexation step E, and an overall negative entropy is measured. A more favorable entropy for tight complexation should be measured if larger host–guest surfaces are desolvated than in the binding of benzene derivatives by cyclophanes **1** and **6**.[35,36]

A study by Eftink *et al.* illustrates how the thermodynamic characteristics for two similar binding events change with the tightness of the host–guest interactions.[37] In water at pH 7.22 ($T = 298$ K), β-cyclodextrin binds 1-adamantanecarboxylate tightly in a predominantly enthalpy-driven process ($\Delta G° = -6.2 \pm 0.1$ kcal mol^{-1}, $\Delta H° = -5.4 \pm 0.1$ kcal mol^{-1}, $T\Delta S° = +0.8 \pm 0.1$ kcal mol^{-1}). In contrast, γ-cyclodextrin with its larger cavity binds the same guest under identical conditions in a predominantly entropically driven way ($\Delta G° = -5.05 \pm 0.03$ kcal mol^{-1}, $\Delta H° = +1.26 \pm 0.08$ kcal mol^{-1}, $T\Delta S° = +6.32 \pm 0.1$ kcal mol^{-1}).

Interestingly, a relationship should exist between the thermodynamic characteristics of a complexation event and binding selectivity. High substrate selectivity in a binding event normally is characterized by a tight fit of the interacting molecules. Therefore, highly selective substrate binding should be enthalpically driven.

Chapter 7 thus far has shown that apolar binding and association processes in water may either be enthalpically or entropically driven. Unfortunately, as mentioned above, hydrophobic bonding has in the past often been uniquely viewed as an entropically driven process. The characterization of apolar binding processes in water as 'hydrophobic' is acceptable as long as no specific thermodynamic quantities are implied.

For hydrophobic binding events in biological systems, *e.g.* enzyme-substrate binding,[22,38] as well as for cyclodextrin-apolar substrate binding,[39] negative changes in the heat capacity, *i.e.* the variation of $\Delta H°$ with temperature $\Delta C_p° = (\partial \Delta H°/\partial T)_p$, are invariably measured. Recently, Dougherty *et al.* performed a statistical analysis on data obtained in variable-temperature ^{1}H NMR binding titrations for the formation of **3** and other complexes of the same as well as related hosts (Chapters 2.8 and 4.3).[26] These

---

[35] R. B. Hermann, *J. Phys. Chem.*, 1972, **76**, 2754.
[36] C. Chothia, *Nature (London)*, 1974, **248**, 338.
[37] W. C. Cromwell, K. Byström, and M. R. Eftink, *J. Phys. Chem.*, 1985, **89**, 326.
[38] J. M. Sturtevant, *Proc. Natl. Acad. Sci.*, 1977, **74**, 2236.
[39] J. C. Harrison and M. R. Eftink, *Biopolymers*, 1982, **21**, 1153.

calculations revealed that significant heat capacity changes between ≈ −20 and ≈ −200 cal deg^{-1} mol^{-1} accompany the formation of these inclusion complexes in water. The evaluation of variable temperature ^{1}H NMR binding data by the statistical analysis of Dougherty *et al.* also yielded negative heat capacity changes of similar magnitude for selected complexes of **6** with *p*-disubstituted benzene derivatives.[40] These data are in good agreement with direct calorimetric measurements.[15] An accurate microscopic-level understanding of the negative heat capacity changes that accompany molecular binding phenomena in water and, to a lesser extent, also in organic solvents, is presently not available.[38]

7.5 Apolar Complexation in Organic Solvents: A Simple General Model for Solvation Effects on Apolar Binding

In the previous section, simple models were developed to explain the thermodynamic characteristics measured for tight binding in water and to explain why water is a better solvent for apolar complexation than methanol. To gain further insight into the role of solvents in apolar complexation events, Diederich *et al.* studied the stability of the pyrene complex **9** in water and in 17 organic solvents covering the entire polarity range.[41] The experimental results of this study were discussed in detail in Chapter 3.2.2. An impressive dependency of complexation strength on the nature of solvent was observed. Upon changing from the most polar solvent, water, to the least polar solvent considered in this study, carbon disulfide, complexation free energies decrease from $\Delta G° = -9.4$ kcal mol^{-1} to $\Delta G° = -1.3$ kcal mol^{-1} giving a difference in binding free energy of $\Delta(\Delta G°) = 8.1$ kcal mol^{-1}. This large difference in binding strength results almost exclusively from solvation effects. A linear free energy relationship (LFER) is valid between the free energies of complexation

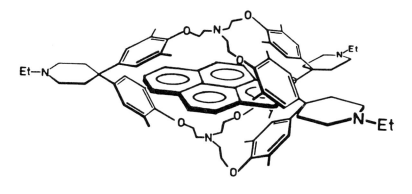

9

---

[40] D. A. Dougherty, personal communication.
[41] D. B. Smithrud and F. Diederich, *J. Am. Chem. Soc.*, 1990, **112**, 339.

and the empirical solvent polarity parameters $E_T(30)^1$ for the various solvents (Chapter 3.2.2). As shown in Chapter 3.6, LFER's with $E_T(30)$ are also useful for rationalizing solvation effects and predicting complexation strength in binary aqueous solvent mixtures.

Of particular interest to the study with **9** was the fundamental question of whether the apolar complexation promoting characteristics of water can be correlated with such characteristics of the other solvents or whether water exhibits an unusual behavior in promoting complexation.[42-53] Based on the experimental results, a simple and general macroscopic model of solvation effects on apolar complexation was developed.

In Chapter 7.4 above, it was proposed that water is a better solvent than methanol for apolar complexation since water is composed of only hydroxy groups of low polarizability and is characterized by larger solvent cohesive interactions. The analysis of the solvent dependency measured for the stability of complex **9** showed that these ideas can be expanded to include the entire solvent polarity scale. Apolar binding is generally strongest in solvents with low molecular polarizability[54] and with high cohesive interactions.[55,56] Solvent molecules with high cohesive interactions interact more favorably with bulk solvent molecules than with the complementary apolar surfaces of free host and guest, and therefore, energy is gained upon the release of surface-solvating solvent molecules into the bulk during complexation. Upon complexation, the less favorable dispersion interactions between solvent molecules of low polarizability and highly polarizable hydrocarbon surfaces are replaced with more favorable dispersion interactions between the complementary surfaces of host and guest.

No special concepts are needed to explain the great ability of water to promote apolar complexation. The linear free energy relationship (Chapter 3.2.2) between the free energies of formation of **9** and the solvent polarity parameter $E_T(30)$ also holds for water. Binding strength decreases regularly from water to polar protic solvents, to dipolar aprotic solvents, and to apolar solvents. Water does not promote apolar complexation beyond the level expected on the basis of its physical properties such as dielectric constant,

[42] M. H. Abraham, *J. Am. Chem. Soc.*, 1980, **102**, 5910.
[43] M. H. Abraham, *J. Am. Chem. Soc.*, 1982, **104**, 2085.
[44] J. H. Hildebrand, *Proc. Natl. Acad. Sci. USA*, 1979, **76**, 194.
[45] M. S. Ramadan, D. F. Evans, and R. Lumry, *J. Phys. Chem.*, 1983, **87**, 4538.
[46] R. D. Cramer, III, *J. Am. Chem. Soc.*, 1977, **99**, 5408.
[47] F. A. Greco, *J. Phys. Chem.*, 1984, **88**, 3132.
[48] D. Mirejovsky and E. M. Arnett, *J. Am. Chem. Soc.*, 1983, **105**, 1112.
[49] R. Wolfenden, *Science (Washington, DC)*, 1983, **222**, 1087.
[50] S. J. Gill, S. F. Dec, G. Olofsson, and I. Wadsö, *J. Phys. Chem.*, 1985, **89**, 3758.
[51] M.-P. Bassez, J. Lee, and G. W. Robinson, *J. Phys. Chem.*, 1987, **91**, 5818.
[52] B. Y. Zaslavsky and E. A. Masimov, *Top. Curr. Chem.*, 1988, **146**, 171.
[53] P. L. Privalov and S. J. Gill, *Pure Appl. Chem.*, 1989, **61**, 1097.
[54] K. J. Miller and J. A. Savchik, *J. Am. Chem. Soc.*, 1979, **101**, 7206.
[55] J. H. Hildebrand and R. L. Scott, 'Regular Solutions', Prentice-Hall, Englewood Cliffs, NJ, 1962.
[56] A. F. M. Barton, 'Handbook of Solubility Parameters and Other Cohesion Parameters', CRC Press, Boca Raton, FL, 1983, Chapter 8, pp. 139.

7.6 Functional Group Solvation Determines the Stability of Cyclophane Complexes

Specific solvation effects of functional groups of host and guest located at the complementary apolar surfaces of the two binding partners can have a strong influence on complexation strength. This has already been illustrated by several studies that were presented in previous chapters of this monograph. For example in Chapter 2.9, it was shown that a large cyclophane forms a much more stable complex with lithocholic acid ($K_a = 7075$ L mol^{-1}) than with deoxycholic acid ($K_a = 250$ L mol^{-1}) in D$_2$O/methanol-d_4 (50:50, v/v). Inclusion of deoxycholic acid into the apolar host cavity would require a considerable amount of energetically unfavorable desolvation of the hydroxy group at C-12α. Therefore, this bile acid prefers a different complex geometry which is destabilized by ≈ 2 kcal mol^{-1} as compared to the complex of lithocholic acid. In Chapter 3.2.1, it was shown that the perylene complex of cyclophane **10** in methanol ($T = 303$ K) is ≈ 3.2 kcal mol^{-1} less stable than the perylene complex formed by **11**. Both hosts differ only by their nitrogen functionality and form complexes that have very similar geometries which closely resemble the pyrene complex **9** formed by cyclophane **11**. The large difference in binding strength must be explained by a complexation-induced reduction in the solvation of the two amide groups in the periphery of the cavity of **10** which is more unfavorable than the partial desolvation of the tertiary amine centers in **11**. Other complexation studies also support that the reduction in the solvation of amide groups is particularly costly.

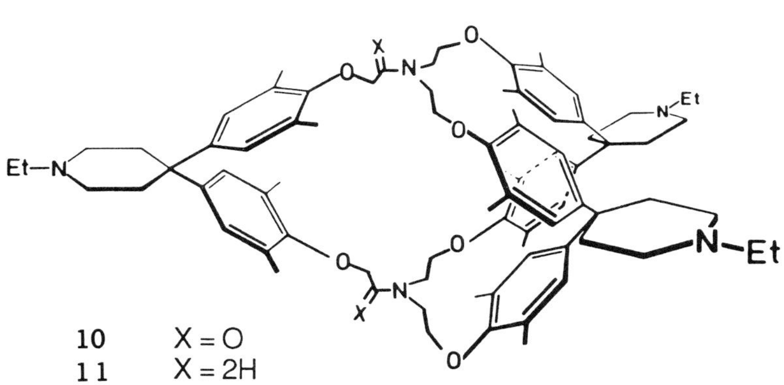

10   X = O
11   X = 2H

[57] F. Diederich, *Angew. Chem.*, 1988, **100**, 372; *Angew. Chem. Int. Ed. Engl.*, 1988, **27**, 362.

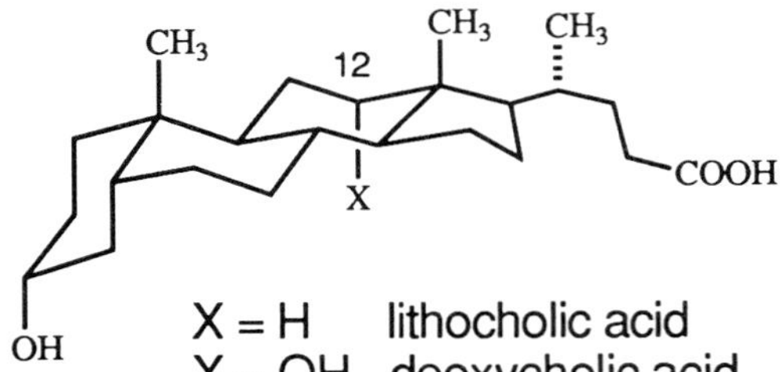

X = H lithocholic acid
X = OH deoxycholic acid

The two examples above illustrate a very general principle.[57] If the energetically favorable solvation of a polar functional group of one or both of the binding partners is reduced in the complex as compared to the free component, and if no new binding interaction compensates for this loss in solvation energy, a considerable reduction in complexation strength is observed.

Another example for the reduction of apolar binding strength due to unfavorable solvation effects is seen in the series of structurally similar cyclophanes **6–8**.[58] Table 7.2 shows that the octamethyl host **7** is the best binder, followed by the octamethoxy host **6** and unsubstituted cyclophane **8**. The methoxy groups deepen the cavity making **6** a better binder than **8**. However, hydrogen-bonding between the methoxy groups and water molecules provides some favorable solvation to parts of the cavity and therefore, the solvation-dependent driving forces for apolar binding are reduced. A partial favorable cavity solvation does not occur in the octamethyl derivative **7** giving its deep cavity a pronounced apolar character resulting in stronger complexation.

Table 7.2 *Association constants K_a and free energies of complexation $\Delta G°$ at 293 K for complexes of cyclophanes **6–8** with 1,4-disubstituted benzene guests in D_2O–CD_3OD (60:40, v/v)*

	Host	K_a $L\,mol^{-1}$	$-\Delta G°$ $kcal\,mol^{-1}$
Guest: *p*-Benzodinitrile			
	6	390	3.48
	7	1580	4.29
	8	140	2.89
Guest: *p*-Dimethoxybenzene			
	6	340	3.41
	7	580	3.72
	8	95	2.66

[58] D. B. Smithrud, E. M. Sanford, I. Chao, S. B. Ferguson, D. R. Carcanague, J. D. Evanseck, K. N. Houk, and F. Diederich, *Pure Appl. Chem.*, in press.

Finally, solvation effects are responsible for the current absence of receptors that bind their substrates *via* hydrogen-bonding in water. All known hydrogen-bonding receptors are effective only in etheral solvents, halocarbons, or aromatic solvents like benzene that do not compete with the substrate for hydrogen-bonding to the receptor (Chapter 5). In water, the hydrogen bond donor and acceptor centers of the receptor undergo strong hydrogen-bonding to the water molecules. Replacement of the hydrogen bonds to the solvent by hydrogen bonds to the substrate is an approximately neutral enthalpic process. Only chelation entropy is gained in this process if multiple hydrogen bonds are formed to one bound substrate upon release of several hydrogen-bonding water molecules into the bulk.

In molecular complexation processes in water, host–guest hydrogen-bonding seems to be responsible for substrate selectivity rather than contributing the most important driving force for binding. The association energy for a hydrogen-bonded complex of an organic molecule in water should predominantly come from the favorable desolvation of apolar surfaces and tight van der Waals contacts between host and guest. Therefore, water-soluble cyclophanes with functional groups converging into apolar binding cavities should provide ideal receptors to bind hydrogen-bonding substrates with high selectivity in water.

Figure 7.2 shows how hydrogen-bonding interactions would generate substrate selectivity in aqueous cyclophane inclusion complexes that are stabilized predominantly by van der Waals forces and by the gain in solvent cohesive interactions resulting from apolar surface desolvation. Figure 7.2A illustrates the inclusion of an apolar guest without suitable, correctly oriented hydrogen-bonding centers that are complementary to the converging hydrogen-bonding groups at the receptor binding site. This binding process would be accompanied by costly desolvation processes as discussed above. Upon substrate inclusion, the favorable solvation of the polar groups of the host would be replaced by less favorable contacts with the apolar surfaces of the included guest. This unfavorable desolvation would lead to the formation of a weaker complex.

On the other hand, the inclusion of the correct apolar guest with complementary, suitably oriented functionality should be much more favorable (Figure 7.2B). In this case, the unfavorable energy term for the desolvation of the polar groups of host and guest upon complexation is compensated by the newly gained hydrogen-bonding interactions in the complex. Hence, the correct guest forms a stable complex since functional group desolvation does not significantly reduce the most relevant driving force for complexation which is the gain in van der Waals interactions and in solvent cohesive interactions following apolar surface desolvation.

Although a large amount of experimental evidence supports the dominance of van der Waals forces and apolar surface desolvation (hydrophobic interactions) over hydrogen-bonding (hydrophilic interactions) as a driving force for complexation in water, this concept has been recently challenged in theoretical work. According to Ben-Naim, hydrophilic interactions probably

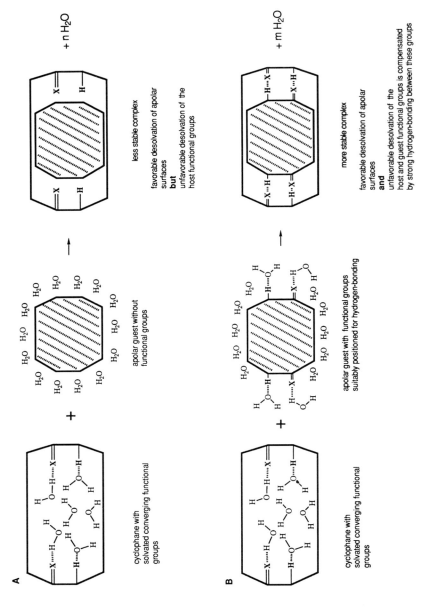

Figure 7.2 *(A) Unfavorable functional group desolvation leads to weak complexation. (B) The formation of host–guest hydrogen bonds compensates for the desolvation of the functional groups and leads to strong complexation*

are more significant in biochemical processes than hydrophobic interactions.[59,60] The design of water-soluble cyclophanes with functional groups converging into an apolar cavity is a challenging, very worthwhile objective. Such receptors would allow to test the concepts advanced in Figure 7.2 and to explore experimentally the role of hydrophilic and hydrophobic interactions for binding strength and selectivity.[61]

[59] A. Ben-Naim, *J. Chem. Phys.*, 1989, **90**, 7412.
[60] A. Ben-Naim, K. L. Ting, and R. L. Jernigan, *Biopolymers*, 1990, **29**, 901.
[61] R. U. Lemieux, *Chem. Soc. Rev.*, 1989, **18**, 347.

CHAPTER 8
Supramolecular Function, Reactivity, and Catalysis

8.1 Introduction

The six previous chapters demonstrated that cyclophanes rival biological receptors in their ability to complex organic molecules tightly and selectively. The final chapter of this monograph now focuses at cyclophane studies in which the increased understanding of host–guest binding has been applied to generate supramolecular function, reactivity, and catalysis. Examples for supramolecular function are analytical chromatographic separations of compound mixtures based on the individual molecular binding affinity to a cyclophane immobilized on a solid phase[1,2] or the acceleration of the transport of lipophilic organic molecules through aqueous solutions.[3] However, the focus of this chapter is the generation of reactivity and catalytic activity in supramolecular complexes. Inspiration for these developments is derived from enzyme research which shows that the proper organization of the reactive centers of substrate and catalyst, achieved through binding, is a prerequisite for effective and selective catalysis.[4] Catalytic mechanisms such as entropically favorable proximity and orientation of the reacting centers of catalyst and substrates or transition-state binding and stabilization are by no means limited to protein structure and material. The proper design of supramolecular architecture should provide new generations of homogeneous solution catalysts that meet the efficiency and stability requirements for use in industrial processes. These synthetic catalysts could recognize a large variety of substrates and accelerate reactions for which no biological catalysts are available. Their study should also provide insight into the nature and the

---

[1] H.-J. Schneider, W. Müller, and D. Güttes, *Angew. Chem.*, 1984, **96**, 909; *Angew. Chem., Int. Ed. Engl.*, 1984, **23**, 910.
[2] M. Bühner, W. Geuder, W.-K. Gries, S. Hünig, M. Koch, and T. Poll, *Angew. Chem.*, 1988, **100**, 1611; *Angew. Chem., Int. Ed. Engl.*, 1988, **27**, 1553.
[3] F. Diederich, *Angew. Chem.*, 1988, **100**, 372; *Angew. Chem., Int. Ed. Engl.*, 1988, **27**, 362.
[4] A. Fersht, 'Enzyme Structure and Mechanism', 2nd ed., Freeman, New York, 1985.

efficiency of catalytic mechanisms and help in determining the factors contributing to enzymatic catalysis.

Early progress in supramolecular catalysis is intimately related to cyclodextrin chemistry. Soon after his fundamental studies on inclusion complexation by cyclodextrins around 1950,[5] Friedrich Cramer investigated the enzyme-like catalytic properties of these macrocycles.[6-8] Subsequently, cyclodextrin reactivity and catalysis[9] was explored in numerous elegant studies in the laboratories of Myron Bender,[10] Ronald Breslow,[11,12] and Iwao Tabushi.[13-15] The development of synthetic supramolecular catalysts was pioneered in the 1970s in the laboratories of Jean-Marie Lehn,[16,17] Donald Cram,[18,19] and Richard Kellogg[20] (Chapter 1.4.3). Since then, an increasing number of chemists have pursued research in this fundamentally and technologically interesting area.[21,22] The major part of this chapter describes the properties of catalytic cyclophanes that possess well defined binding and catalytic sites for neutral organic molecules.

8.2 Carrier-mediated Transport of Arenes Through Water

The transport of cations through lipophilic natural or artificial membranes accelerated by natural or synthetic ionophores is a well-known process.[23-26] In contrast, only little information was available about the inverted form of this process, the transport of lipophilic substrates through an aqueous solution mediated by artificial carriers.[27] With water-soluble cyclophanes, the molecular carrier-mediated acceleration for the transport of lipophilic arenes

[5] F. Cramer, 'Einschlußverbindungen', Springer, Berlin, 1954.
[6] F. Cramer, *Chem. Ber.*, 1953, **86**, 1576.
[7] F. Cramer and W. Dietsche, *Chem. Ber.*, 1959, **92**, 1739.
[8] F. Cramer, *Angew. Chem.*, 1961, **73**, 49.
[9] W. Saenger, *Angew. Chem.*, 1980, **92**, 343; *Angew. Chem., Int. Ed. Engl.*, 1980, **19**, 344.
[10] M. L. Bender and M. Komiyama, 'Cyclodextrin Chemistry', Springer, Berlin, 1978.
[11] R. Breslow, *Science (Washington DC)*, 1982, **218**, 532.
[12] R. Breslow, *Adv. Enzymol. Relat. Areas Mol. Biol.*, 1986, **58**, 1.
[13] I. Tabushi, *Pure Appl. Chem.*, 1986, **58**, 1529.
[14] I. Tabushi and Y. Kuroda, *Adv. Catal.*, 1983, **32**, 417.
[15] I. Tabushi, *Tetrahedron*, 1984, **40**, 269.
[16] J.-M. Lehn, *Science (Washington DC)*, 1985, **227**, 849.
[17] J.-M. Lehn, *Angew. Chem.*, 1988, **100**, 91; *Angew. Chem., Int. Ed. Engl.*, 1988, **27**, 89.
[18] D. J. Cram, P. Y.-S. Lam, and S. P. Ho, *Ann. N. Y. Acad. Sci.*, 1986, **471**, 22.
[19] D. J. Cram, *Angew. Chem.*, 1988, **100**, 1041; *Angew. Chem., Int. Ed. Engl.*, 1988, **27**, 1009.
[20] R. M. Kellogg, *Angew. Chem.*, 1984, **96**, 769; *Angew. Chem., Int. Ed. Engl.*, 1984, **23**, 782.
[21] F. P. Schmidtchen, *Top. Curr. Chem.*, 1985, **132**, 101.
[22] J. Rebek, Jr., *Angew. Chem.*, 1990, **102**, 261; *Angew. Chem., Int. Ed. Engl.*, 1990, **29**, 245.
[23] Y. A. Ovchinnikov, V. T. Ivanov, and A. M. Shkrob, 'Membrane Active Complexones', Elsevier, Amsterdam, 1974.
[24] D. W. McBride, Jr., R. M. Izatt, J. D. Lamb, and J. J. Christensen in 'Inclusion Compounds', eds. J. L. Atwood, J. E. D. Davies, and D. D. MacNicol, Academic Press, London, 1984, Vol. 3, pp. 571.
[25] J. D. Lamb, R. M. Izatt, and J. J. Christensen, *Prog. Macrocyclic Chem.*, 1981, **2**, 41.
[26] P. G. Potvin and J.-M. Lehn, *Prog. Macrocyclic Chem.*, 1987, **3**, 167.
[27] A. Xenakis and C. Tondre, *J. Phys. Chem.*, 1983, **87**, 4737.

through an aqueous phase along a concentration gradient was observed for the first time.[3,28] The U-type cell shown in Figure 8.1 was used in experiments in which the relative rate of transport of arenes from one hexane phase (source) into a second hexane phase (receiving) through an aqueous phase was measured. When the aqueous phase contained cyclophane **1** as a carrier, the transportation of the stronger binding arenes was considerably accelerated (Table 8.1). For example, the rate of transport of pyrene (0.01 M in source phase) through a 5×10^{-3} M solution of **1** showed a 2100-fold acceleration over the transport through pure water. A competitive inhibition experiment demonstrated that host–guest complexation in the aqueous phase is essential for transport acceleration. 2,6-Naphthalenedisulfonate, which has a high binding affinity to **1** ($K_a > 10^6$ L mol^{-1}, $T = 293$ K, Chapter 4.2), is an efficient

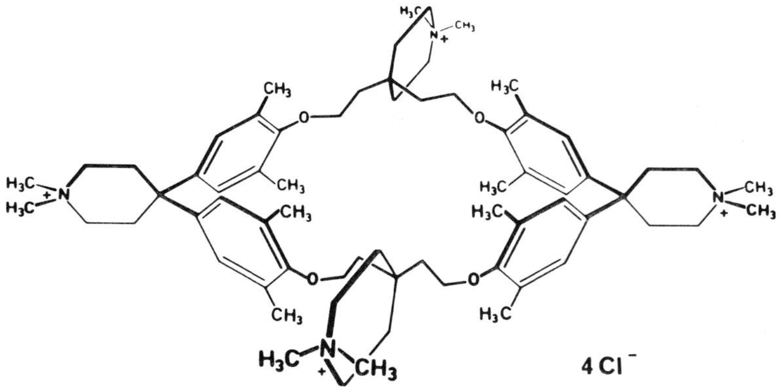

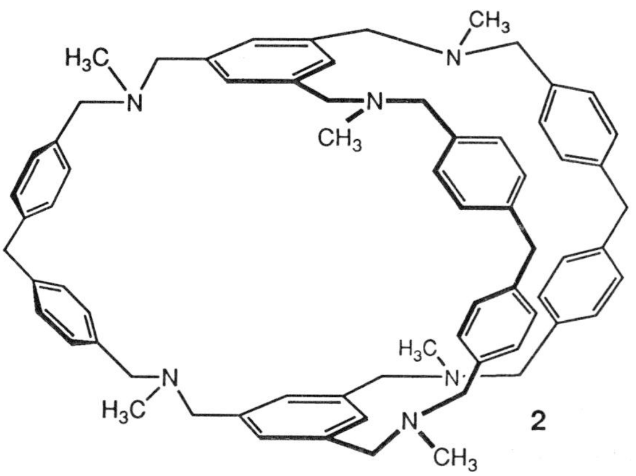

[28] F. Diederich and K. Dick, *J. Am. Chem. Soc.*, 1984, **106**, 8024.

competitive inhibitor of the complexation between **1** and pyrene ($K_a = 1.8 \times 10^6$ L mol^{-1}). At an inhibitor concentration of 0.01 M in the aqueous solution, the rate of transport of pyrene is reduced to almost the rate through pure water.

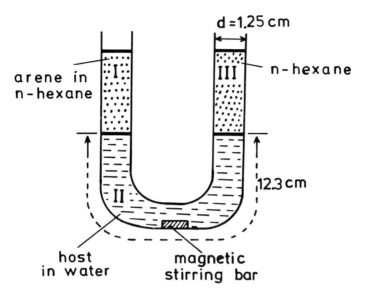

Figure 8.1 *U-type cell for studying the transport of arenes through an aqueous phase mediated by **1** as molecular carrier. The transport data of Table 8.1 were obtained with this cell*

Table 8.1 *Acceleration of the transport of arenes through a 5×10^{-4} M solution of **1** as aqueous phase II against the transport through pure water ($T = 293-295$ K). The experiments were run in the U-type cell shown in Figure 8.1. The initial concentration of the arenes in the n-hexane phase I was 10^{-2} mol L^{-1}. The association and distribution constants included in the table were obtained by liquid–liquid extractions*

Arene	Acceleration factor	K_a (L mol^{-1})	K_{dist}
Pyrene	430	1.8×10^6	8.3×10^{-6}
Fluoranthene	395	1.8×10^6	1.1×10^{-5}
Naphthalene	3.7	1.2×10^4	3.8×10^{-4}
Azulene	3.6	2.1×10^4	6.0×10^{-4}
Durene	1.8	1.9×10^3	3.8×10^{-5}

High selectivity was observed in the acceleration of arene transport through water mediated by **1** (Table 8.1). Investigations into the origin of this selectivity showed that the relative rate of transport of an arene is approximately proportional to the amount of arene present in the aqueous phase, no matter whether a carrier is present or not. Therefore, the highest acceleration factors are measured for arenes that form the most stable complexes with **1** in the aqueous phase (high K_a values) *and* have the lowest distribution constants K_{dist} (equations 8.1 and 8.2). For example, pyrene having a low distribution constant K_{dist} means that there is only a very small amount of this guest present in pure water giving a very slow non-mediated transport. However, upon strong host–guest complexation in the aqueous phase (high K_a), a large amount of pyrene is extracted, and a fast transport rate as well as a high acceleration factor are measured. Strong binding leads to high extractability since the association constant K_a for the complex of **1** in the aqueous solution is directly proportional to the extraction constant K_e (equations 8.3–8.5).

$$G_{\text{Hexane}} \xrightleftharpoons{K_{dist}} G_{H_2O} \tag{8.1}$$

$$K_{dist} = \frac{G_{H_2O}}{G_{\text{Hexane}}} \tag{8.2}$$

$$G_{\text{Hexane}} + H_{H_2O} \xrightleftharpoons{K_e} HG_{H_2O} \tag{8.3}$$

$$K_e \; (\text{L mol}^{-1}) = \frac{HG_{H_2O}}{G_{\text{Hexane}} \cdot H_{H_2O}} \tag{8.4}$$

$$K_e \; (\text{L mol}^{-1}) = K_a \cdot K_{dist} \tag{8.5}$$

Other examples of the cyclophane-mediated transport of arenes through aqueous solutions have been reported.[2] Vögtle *et al.* described the separation of naphthalene from phenanthrene in transport experiments using U-type cells and the protonated macrobicycle **2** as molecular carrier.[29]

---

[29] F. Vögtle, W. M. Müller, U. Werner, and H.-W. Losensky, *Angew. Chem.*, 1987, **99**, 930; *Angew. Chem., Int. Ed. Engl.*, 1987, **26**, 901.

The high substrate selectivity observed in all these experiments suggests that the carrier-mediated transport of neutral molecules could provide an attractive approach to non-destructive material separations. However, the described mechanism uses a 'passive transport' along a concentration gradient as the thermodynamic driving force. After transport of $\approx$ 10–15% of the total quantity of arene initially present in the source phase, back transport becomes significant. For quantitative material separations, unidirectional 'active transport' driven by the electron flow in a redox process, by the light energy in a photochemical process, or by the proton flux in a pH gradient needs to be developed.[17,30-33] In an effort to accomplish active arene transport in one defined direction, Diederich *et al.* prepared the water-soluble flavin cyclophane **3** with redox-switchable complexation properties.[34]

8.3 Redox-dependent Binding Ability of a Flavinophane

In Chapter 1.4.2, it was described that the isoalloxazine unit in flavin coenzymes is planar in the oxidized form, whereas the dihydroisoalloxazine unit in the 2e$^-$ reduced form takes a butterfly shape by bending around the two nitrogens N-5 and N-10 of the central ring in the tricyclic system. The flavinophane **3** was prepared to investigate how the different shapes of the isoalloxazine unit in the oxidized and 2e$^-$ reduced forms affect the complexation ability of the cavity binding site.[34]

In alkaline deuterated borate buffer (D$_2$O, D$_3$BO$_3$/NaOD, pD 10.4), the bright yellow oxidized flavinophane **3a** is readily and quantitatively reduced to the colorless trianionic host **3b** (Scheme 8.1). This reduction under argon can be achieved: (i) with sodium borohydride, (ii) with sodium dithionite, (iii) by irradiating with a 220 W daylight lamp in the presence of ethylenediamine tetraacetate, or (iv) electrochemically. By introduction of oxygen into the solution of **3b**, the oxidized flavin is regenerated quantitatively. The redox cycles could be repeated numerous times without affecting the molecular structure of the cyclophane, and similar results were obtained with the isoalloxazine **4**. Both CPK model examinations and ^{1}H NMR investigations provided support that the macrocyclic dihydroisoalloxazine unit prefers bending outward rather than inward, thus yielding a more elongated cavity as shown schematically in **3b**.

Host–guest binding interactions and structurally well defined self-association processes of flavinophane **3** and the nonmacrocyclic isoalloxazine **4** were investigated by ^{1}H NMR in aqueous borate buffer in concentration ranges

---

[30] M. Okahara and Y. Nakatsuji, *Top. Curr. Chem.*, 1986, **128**, 37.
[31] J. D. Winkler, K. Deshayes, and B. Shao, *J. Am. Chem. Soc.*, 1989, **111**, 769.
[32] S. Shinkai, *Pure Appl. Chem.*, 1987, **59**, 425.
[33] M. Delgado, L. Echegoyen, V. J. Gatto, D. A. Gustowski, and G. W. Gokel, *J. Am. Chem. Soc.*, 1986, **108**, 4135.
[34] E. M. Seward, R. B. Hopkins, W. Sauerer, S.-W. Tam, and F. Diederich, *J. Am. Chem. Soc.*, 1990, **112**, 1783.

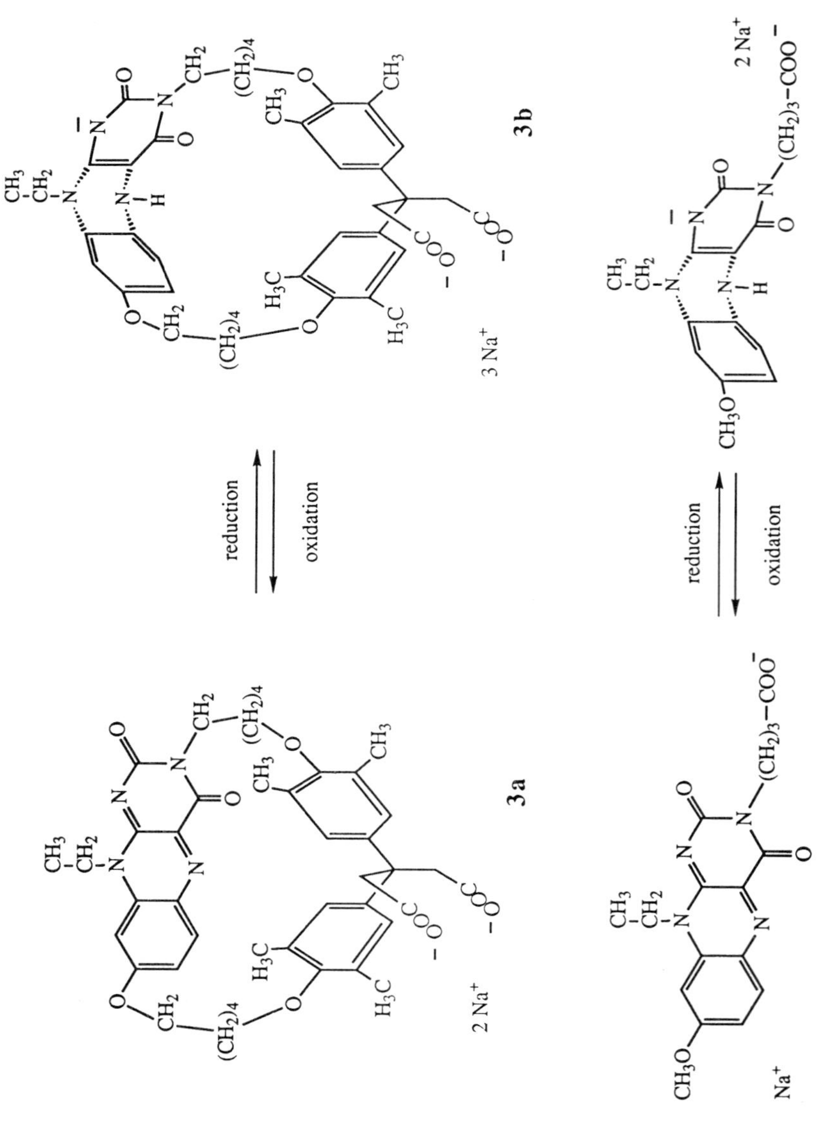

4a 4b

Scheme 8.1

below 5×10^{-3} mol L^{-1}. Under these conditions, micelle-type aggregation does not occur. Whereas the oxidized forms **3a** and **4a** show very similar binding and association characteristics, striking differences are observed between the reduced cyclophane **3b** and the nonmacrocyclic model **4b**.

The two oxidized derivatives **3a** and **4a** undergo a very similar strong self-association.[35] This process is identified by characteristic shifts of the ^{1}H NMR resonances of the two molecules. The resonances of the isoalloxazine protons as well as those of the methylene units immediately adjacent to the tricyclic units move increasingly upfield (≈ 0.1–0.2 ppm) with increasing concentration. The magnitude of the change in chemical shift encountered by the residual proton resonances in **3a** and **4a** decreases sharply with increasing distance of these protons from the isoalloxazine unit, and the resonances of the diphenylmethane protons in **3a** display almost no change in position until the cac of 4.5×10^{-3} mol L^{-1} is reached. These particular ^{1}H NMR shifts are best explained by the formation of dimers, stabilized by hydrophobic π–π-stacking interactions between the isoalloxazine units in **3a** and **4a** (Figure 8.2). Such a π–π-stacking dimerization is also supported by strong fluorescence quenching and a large hypochromicity observed for the characteristic flavin bands ($\lambda_{max} = 427$ nm for **3a**) in the electronic absorption spectra (Chapter 1.4.1).[36,37]

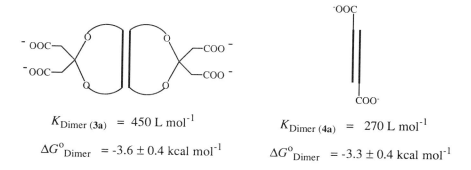

$K_{Dimer\ (3a)} = 450$ L mol^{-1}

$\Delta G°_{Dimer} = -3.6 \pm 0.4$ kcal mol^{-1}

$K_{Dimer\ (4a)} = 270$ L mol^{-1}

$\Delta G°_{Dimer} = -3.3 \pm 0.4$ kcal mol^{-1}

Figure 8.2 *Energetics and schematic representation of the geometry of the dimers formed by **3a** and **4a** in a deuterated aqueous borate buffer at pD 10.4 and T = 295 K*

Cavity inclusion complexation of naphthalene guests, *e.g.* 6-hydroxy-2-naphthonitrile, is not observed for the oxidized flavinophane **3a**. Similar to **4a**, the oxidized cyclophane **3a** prefers to bind these guests outside the cavity in a π–π-stacking mode (Figure 8.3).[38,39] This complexation mode is supported by upfield shifts (≈ 0.1–0.4 ppm at ≈ 60–70% saturation binding) of all guest

[35] E. T. Jarvi and H. W. Whitlock, *J. Am. Chem. Soc.*, 1982, **104**, 7196.
[36] G. Kotowycz, N. Teng, M. P. Klein, and M. Calvin, *J. Biol. Chem.*, 1969, **244**, 5656.
[37] R. H. Sarma, P. Dannies, and N. O. Kaplan, *Biochemistry*, 1968, **7**, 4359.
[38] E. T. Jarvi and H. W. Whitlock, Jr., *J. Am. Chem. Soc.*, 1980, **102**, 657.
[39] T. Ishida, M. Itoh, M. Horiuchi, S. Yamashita, M. Doi, M. Inoue, Y. Mizunoya, Y. Tona, and A. Okada, *Chem. Pharm. Bull.*, 1986, **34**, 1853.

¹H NMR resonances and all resonances of the isoalloxazine protons and the protons of the butanoic side chain in proximity to the tricyclic units in **3a** or **4a**. The free energies for the formation of 1:1 π–π-stacking complexes between **3a** or **4a** and naphthalene derivatives vary between 3 and 4 kcal mol^{-1}. On the basis of these findings in the association and complexation studies with the oxidized **3a** and **4a**, π–π-stacking interactions must be assumed to play a considerable role in the binding of aromatic substrates in the proximity of the isoalloxazine unit of FAD and FMN at flavoenzyme active sites.[40]

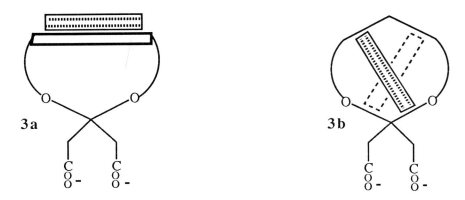

Figure 8.3 *Schematic representations of the different modes of complexation of naphthalene derivatives by the flavinophanes **3a** and **3b**. Two favorable substrate orientations in the cavity of **3b** are shown*

In sharp contrast to the oxidized isoalloxazine **4a**, the reduced model compound **4b** does not form stable complexes with naphthalene derivatives in the aqueous borate buffer. Similarly, self-association is not observed. Apparently, the extended planar surface of the oxidized tricyclic isoalloxazine moiety is required for efficient π–π-stacking interactions.

A strikingly different complexation behavior is observed with the reduced flavinophane **3b**. CPK model examinations suggest that a suitably sized cavity, similar in size and shape to the binding sites in bis(diphenylmethane) hosts (Chapter 2.3), forms if the dihydroisoalloxazine unit in **3b** bends outwards. All ¹H NMR binding titration results, indeed, support that naphthalene derivatives are encapsulated in an axial-like orientation in the cavity of **3b** as shown in Figure 8.3. In support of such a geometry, which resembles the geometry of the arene complexes of bis(diphenylmethane) hosts, the resonances of the methylene bridges of **3b** move considerably upfield upon complexation ($\approx +0.2$ to 0.5 ppm). Correspondingly, all aromatic host resonances and the signals for the acetic acid residues in **3b** move downfield in

---

[40] E. F. Pai and G. E. Schulz, *J. Biol. Chem.*, 1983, **258**, 1752.

the complex. Besides this characteristic up- and downfield shift pattern of the host resonances, the specifically large upfield shifts at saturation binding of some of the guest resonances ($\Delta\delta_{sat\ calcd} = +2.74$ ppm for 1-H in 6-hydroxy-2-naphthonitrile) further support the cavity inclusion geometry schematically depicted in Figure 8.3. Such upfield shifts are incompatible with a π–π-stacking complex geometry. The cavity-inclusion complexes formed by **3b** have similar stability compared to the external π–π-stacking complexes formed by the oxidized host.

In the development of the flavinophane **3**, the initial target of a redox-switchable host for active transport was partially achieved. The cyclophane clearly can be switched from cavity binding to non-cavity binding in a redox process. However, the hydrophobic π–π-stacking association, which occurs externally with **3a**, has the same strength as the cavity binding with **3b**. This study shows that flavinophanes resembling **3** but with an externally bridged and, hence, shielded isoalloxazine should exhibit the desired redox-dependent complexation behavior. The bridging of the isoalloxazine should prevent external π–π-stacking, and binding should only occur in the larger cavity of the reduced flavin-host.

8.4 Catalytic Cyclophanes: Supramolecular Reactivity and Catalysis

In recent years, the potential of water-soluble cyclophanes to act as enzyme-like catalysts and to accelerate reactions in supramolecular complexes has been increasingly investigated.[3] Although these designed systems are not yet comparable to enzymes in their catalytic efficiency and selectivity, a variety of interesting results have already been obtained which demonstrate the potential for fascinating future developments in the area of catalytic cyclophanes. An illustration for the work done in this field is given in the last sections of Chapter 8 which describe studies of functionalized cyclophanes that exhibit esterase activity or that activate coenzyme models at their well defined binding sites for the efficient catalysis of oxidations and of carbon–carbon bond forming reactions.

In addition to the examples discussed below in detail, a variety of other studies of supramolecular cyclophane catalysis have been reported.[41] The octopus cyclophane **5** with eight hydrocarbon chains, described by Murakami et al., provides a favorable, micellar-type reaction environment (Figure 4.15) for the alkylation of hydrophobic Vitamin B_{12} derivatives and subsequent photochemical carbon-skeleton rearrangements of the bound alkyl ligands.[42] Schneider et al. showed[43] that the tetraaza[8.1.8.1]paracyclophane **6**, like the

---

[41] (a) H.-J. Schneider and A. Junker, *Chem. Ber.*, 1986, **119**, 2815. (b) B. J. Whitlock and H. W. Whitlock, Jr., *Tetrahedron Lett.*, 1988, **29**, 6047.
[42] Y. Murakami, Y. Hisaeda, J. Kikuchi, T. Ohno, M. Suzuki, Y. Matsuda, and T. Matsuura, *J. Chem. Soc., Perkin Trans. 2*, 1988, 1237.
[43] H.-J. Schneider and N. K. Sangwan, *J. Chem. Soc., Chem. Commun.*, 1986, 1787.

larger cyclodextrins,[44,45] accelerates the Diels–Alder reaction of cyclopentadiene with diethyl fumarate in dioxane–water mixtures compared to the reaction in the pure solvent mixture. The same cyclophane also catalyzes nucleophilic substitution reactions.[46] In dioxane–water (1:1) at 30 °C, the reaction of 2-bromomethylnaphthalene (0.043 M) with an excess of sodium nitrite (0.43 M) is accelerated by a factor of 20 in the presence of **6** (0.4 M) as compared to the conversion in absence of **6** (Scheme 8.2). Concomitantly, the product ratio [R–ONO]:[R–NO$_2$] obtained with the ambident nucleophile changes from 0.50:1 to 0.16:1. The observed acceleration was interpreted in terms of an accumulation of nitrite ions at the positively charged centers of **6** in the complex. Therefore, supramolecular catalysis generates a favorable proximity between the two substrates involved in the nucleophilic substitution (Scheme 8.2).[21]

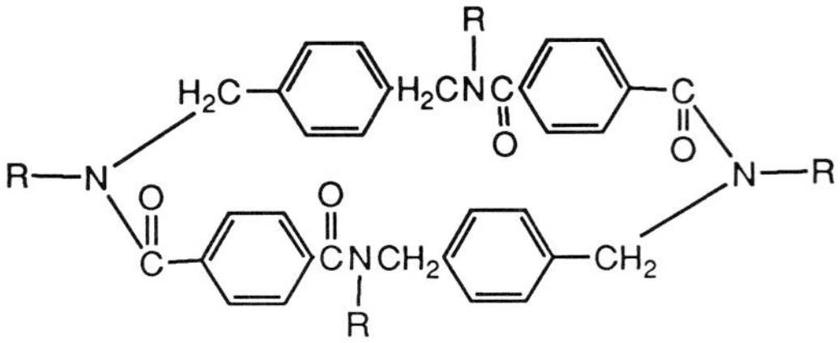

5 R = –(CH$_2$)$_2$CNH–(CH$_2$)$_4$–CHCN$\big\langle$(CH$_2$)$_{13}$CH$_3$ / (CH$_2$)$_{13}$CH$_3$

NHC-(CH$_2$)$_5$N$^+$Me$_3$ Br$^-$

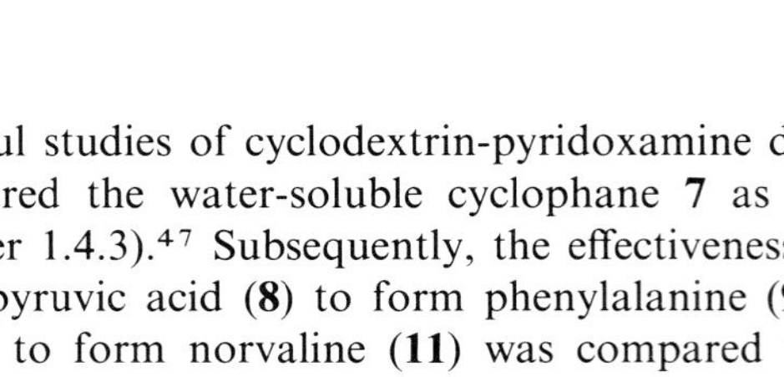

Following successful studies of cyclodextrin–pyridoxamine derivatives,[12] Breslow *et al.* prepared the water-soluble cyclophane **7** as an artificial transaminase (Chapter 1.4.3).[47] Subsequently, the effectiveness of **7** in the amination of phenylpyruvic acid (**8**) to form phenylalanine (**9**) and of α-ketovaleric acid (**10**) to form norvaline (**11**) was compared to the same

---

[44] D. C. Rideout and R. Breslow, *J. Am. Chem. Soc.*, 1980, **102**, 7816.
[45] N. K. Sangwan and H.-J. Schneider, *J. Chem. Soc., Perkin Trans. 2*, 1989, 1223.
[46] H.-J. Schneider and R. Busch, *Angew. Chem.*, 1984, **96**, 910; *Angew. Chem., Int. Ed. Engl.*, 1984, **23**, 911.
[47] J. Winkler, E. Coutouli-Argyropoulou, R. Leppkes, and R. Breslow, *J. Am. Chem. Soc.*, 1983, **105**, 7198.

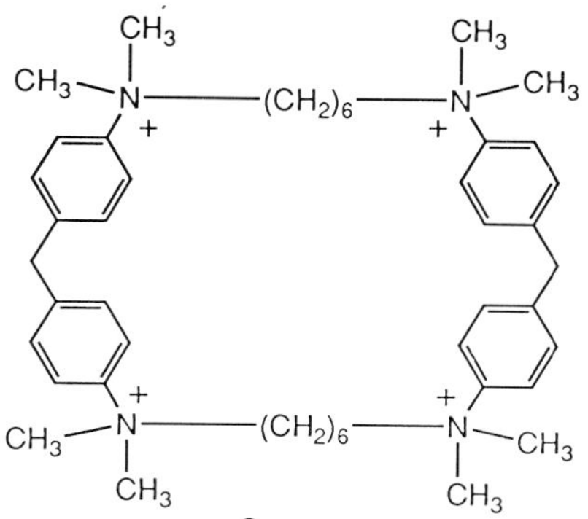

Scheme 8.2

[Structure **6**: cyclophane with four quaternary ammonium groups connected by (CH₂)₆ bridges and methylene-phenyl groups]

Reaction: 2-naphthyl-CH₂Br + NaNO₂ →(6, dioxane/water (1:1), 303 K) 2-naphthyl-CH₂ONO + 2-naphthyl-CH₂NO₂

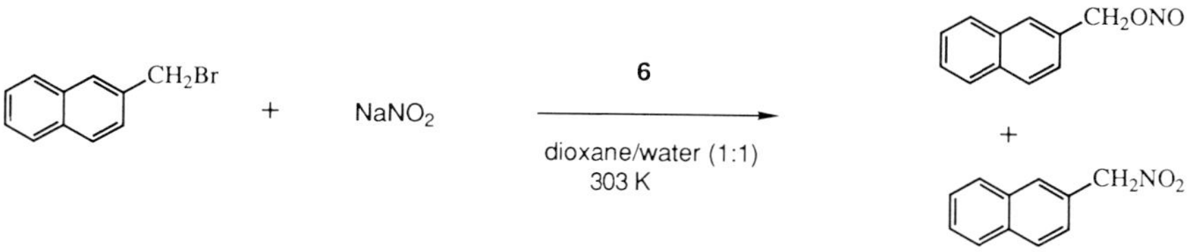

aminations performed by simple pyridoxamine and by a cyclodextrin derivative resembling **7** but with β-cyclodextrin instead of the cyclophane. For this comparison, initial reaction rates were studied at 0.5 mM concentrations of the keto acids and 0.5 mM pyridoxamine derivative in 2.7 M phosphate buffer, pH 9.3, at 299 K.

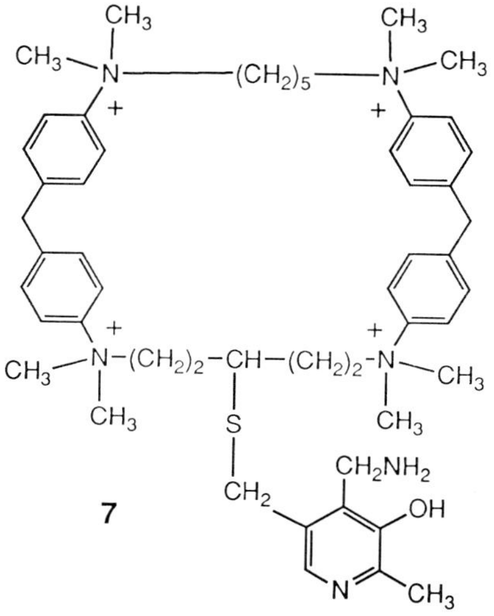

[Structure **7**: cyclophane similar to 6 but with (CH₂)₅ bridge on top and a (CH₂)₂-CH(S-CH₂-pyridoxamine)-(CH₂)₂ bridge on bottom bearing a pendant pyridoxamine group]

⟨C₆H₅⟩-CH₂-CO-COOH   →(7, 2.7 M phosphate buffer, pH 9.3, 299 K)→   ⟨C₆H₅⟩-CH₂-CH(NH₂)-COOH

phenylpyruvic acid (8)   phenylalanine (9)

CH₃CH₂CH₂-CO-COOH   →   CH₃CH₂CH₂-CH(NH₂)-COOH

α-ketovaleric acid (10)   norvaline (11)

Scheme 8.3

The pyridoxamine-cyclophane **7** converted **8** to **9** at a rate 31 ± 3 times faster as did simple pyridoxamine, and the conversion of **10** to **11** was 6 ± 1 times faster. Under the same conditions, the cyclodextrin analogue accelerated the reaction of **8** by a factor of 15 ± 2 and the reaction of **10** by a factor of 2. These data clearly shows that the binding of the phenyl group of **8** into the macrocyclic cavity of **7** contributes significantly to the rate and that, in terms of selective rate acceleration of transamination, the functionalized cyclophane **7** is comparable to its cyclodextrin analogue.

8.4.1 Cyclophanes with Esterase Activity: From Nonproductive to Productive Binding

Ever since the first studies by Cramer on the catalytic properties of cyclodextrins,[5-8] hydrolase mimics have been the focal point in the efforts to prepare bio-organic catalysts that mimic enzymes by forming stoichiometric complexes with their substrates in a reversible equilibrium prior to the reaction steps.[10,48-53] For example, some of the first water-soluble cyclophanes, *e.g.* **12**[54] and **13**[55] were prepared in the groups of Murakami and Tabushi, respectively, to accelerate the hydrolysis of bound esters. These early studies of catalytic cyclophanes have been thoroughly reviewed.[56,57]

Many macrocyclic hydrolase mimics model some aspects of α-chymotrypsin, the best studied member of the digestive serine proteases capable of hydrolyzing amides as well as esters.[4] α-Chymotrypsin mimics (E–OH) are compounds that: (i) complex their substrates (S) prior to reaction, (ii) react according to the minimum mechanism for the α-chymotrypsin-catalyzed

---

[48] R. Breslow, G. Trainor, and A. Ueno, *J. Am. Chem. Soc.*, 1983, **105**, 2739.
[49] J. M. Lehn and C. Sirlin, *New J. Chem.*, 1987, **11**, 693.
[50] D. J. Cram, P. Y.-S. Lam, and S. P. Ho, *J. Am. Chem. Soc.*, 1986, **108**, 839.
[51] S. Sasaki, M. Kawasaki, and K. Koga, *Chem. Pharm. Bull.*, 1985, **33**, 4247.
[52] J. P. Guthrie, J. Cossar, and B. A. Dawson, *Can. J. Chem.*, 1986, **64**, 2456.
[53] F. M. Menger and M. Ladika, *J. Am. Chem. Soc.*, 1987, **109**, 3145.
[54] Y. Murakami, Y. Aoyama, and M. Kida, *J. Chem. Soc., Perkin Trans. 2.*, 1980, 1665.
[55] I. Tabushi, Y. Kimura, and K. Yamamura, *J. Am. Chem. Soc.*, 1981, **103**, 6486.
[56] Y. Murakami, *Top. Curr. Chem.*, 1983, **115**, 107.
[57] I. Tabushi and K. Yamamura, *Top. Curr. Chem.*, 1983, **113**, 145.

13

12

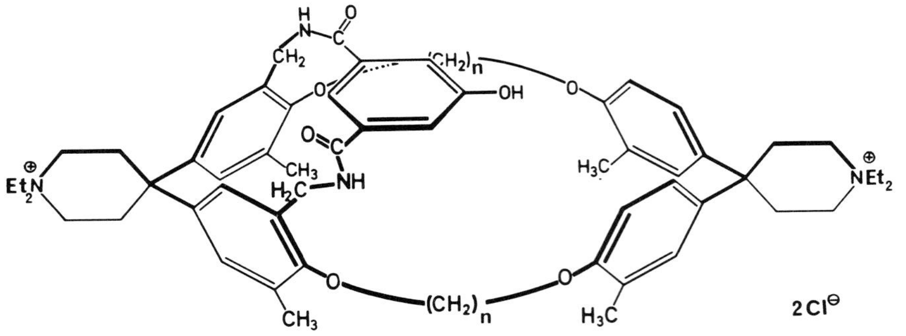

14 n = 4
15 n = 6

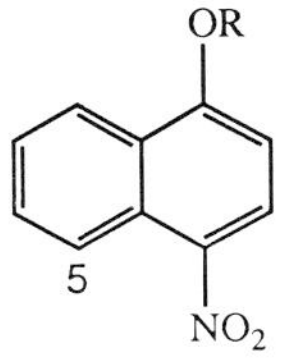

16 R = H
17 R = COCH$_3$

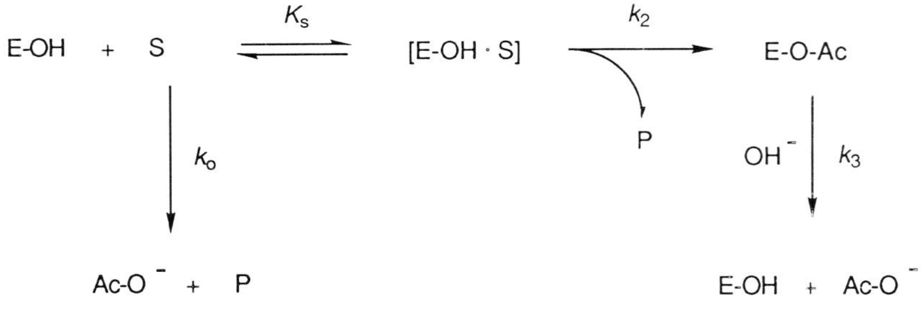

$$\text{E-OH} + \text{S} \xrightleftharpoons{K_s} [\text{E-OH} \cdot \text{S}] \xrightarrow{k_2} \text{E-O-Ac} \quad (8.6)$$

with k_o pathway giving Ac-O$^-$ + P, and k_3 (OH$^-$) hydrolysis of E-O-Ac giving E-OH + Ac-O$^-$.

hydrolysis of esters and amides shown in equation 8.6, and (c) operate like the enzyme by the addition–elimination mechanism in both transacylation (k_2) and deacylation steps (k_3) of equation 8.6.

The macrobicycles **14** and **15** were prepared by Diederich *et al.* as α-chymotrypsin mimics which possess phenolic nucleophiles that are active at near physiological pH.[58] Specific chemical shifts in the ^{1}H NMR spectra and characteristic signal enhancements observed by ^{1}H NOE difference spectroscopy showed that the nucleophiles are positioned atop the macrocyclic binding sites. The two cyclophanes form complexes of similar stability but different geometry with 4-nitro-1-naphthol (**16**). This naphthalene derivative represents the apolar binding moiety of the naphthyl acetate **17** which served

---

[58] F. Diederich, G. Schürmann, and I. Chao, *J. Org. Chem.*, 1988, **53**, 2744.

as the substrate in comparative studies of the esterase activity of the two macro-rings. Figure 8.4 shows the computer-assisted nonlinear curve-fitting of the experimental data of ^{1}H NMR binding titrations in D_2O/methanol-d_4 (70:30, v/v, $T = 298$ K) which gave $K_a = 1.5 \times 10^3$ L mol^{-1} ($-\Delta G = 4.35$ kcal mol^{-1}) as the association constant for the complex of **14** with substrate **16** and $K_a = 3.5 \times 10^3$ L mol^{-1} ($-\Delta G = 4.80$ kcal mol^{-1}) for the complex of **15** with the same guest.

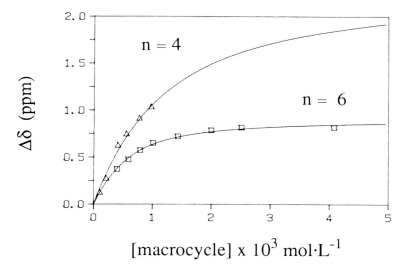

Figure 8.4 *Determination of the association constants of the **14**·4-nitro-1-naphthol complex ($\triangle$) and of the **15**·4-nitro-1-naphthol complex ($\square$) in D_2O/ methanol-d_4 (70:30, v/v, $T = 298$ K) by computer-assisted nonlinear curve-fitting of the experimental data from ^{1}H NMR titrations. The complexation-induced shifts $\Delta\delta$ observed for the guest proton 5-H are plotted against the host concentration*
(Reproduced with permission from reference 58. Copyright 1988 ACS.)

In contrast to their similar binding ability, the two macrocycles differ largely in their esterase activity. With 4-nitro-1-naphthylacetate (**17**) as a substrate, a large difference in transacylation rates was observed. In aqueous phosphate buffer/Me_2SO (99:1, v/v; pH 8.0, $T = 298$ K) with [cyclophane] = 5.0×10^{-4} mol L^{-1} and [**17**] = 2.0×10^{-5} mol L^{-1}, the acylation of the smaller macrocycle **14** is only 14 times faster ($k_{obs} = 2.43 \times 10^{-4}$ s^{-1}) than the hydrolysis in pure buffer. In contrast, the acylation of **15** by the complexed ester is 178 times faster ($k_{obs} = 3.21 \times 10^{-3}$ s^{-1}) than the hydrolysis in pure buffer. In addition, a modest catalytic turnover for the entire two-step hydrolysis (equation 8.6) was observed in the presence of **15**.

The reactions of the two very similar cyclophanes are entirely different in their kinetic profiles. Michaelis–Menten-type saturation kinetics is observed in the presence of **15** (Figure 8.5). The pseudo-first-order rate constant for

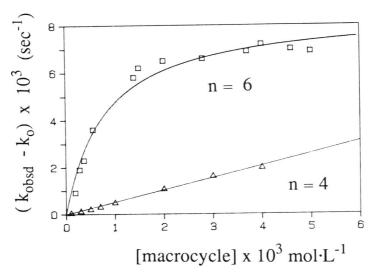

Figure 8.5 *Plots of the pseudo-first-order rate constants (k_{obsd} -k_o) for the acyl transfer step in the hydrolysis of 4-nitro-1-naphthylacetate (**17**) plotted as a function of host concentration. The reaction in the presence of host **15** (□) follows saturation kinetics (productive binding), and the reaction in the presence of host **14** (Δ) obeys second-order kinetics (nonproductive binding); T = 293 K, aqueous phosphate buffer, pH 8/Me_2SO (99:1, v/v) (Reproduced with permission from reference 58. Copyright 1988 ACS.)*

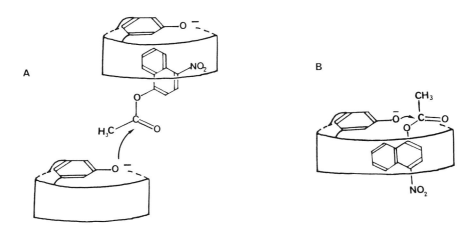

Figure 8.6 *(A) Nonproductive binding of **17** by host **14** leads to intermolecular acyl transfer; (B) Productive binding of ester **17** by host **15** leads to faster supramolecular acyl transfer*

acylation by fully bound substrate was calculated as $k_{complex} = 8.52 \times 10^{-3}$ s^{-1} and the Michaelis–Menten constant as $K_m = 7.6 \times 10^{-4}$ mol L^{-1}. Hence, the acylation by fully bound substrate is 472 times faster than the hydrolysis in pure buffer. In contrast, the cleavage of **17** in the presence of the smaller macrocycle **14** strictly follows second-order kinetics (Figure 8.5). The second-order rate constant for the bimolecular cleavage in the presence of **14** was calculated as $k_2 = 0.51$ s^{-1} L mol^{-1}. A comparison of second-order rate constants (k_2 *versus* $k_{complex} K_m^{-1}$) shows that the reaction in the supramolecular complex of **15** is 22 times faster than the transacylation in the presence of **14**.

The two macrobicycles **14** and **15** possess the same phenolic nucleophile and, according to the binding studies described above (Figure 8.4), form complexes of comparable stability with the aromatic unit of substrate **17**. The catalytic results clearly demonstrate that complexation alone is not sufficient to generate supramolecular catalysis. A proper orientation of nucleophile and ester residue is also required in the complex. Such a productive binding is only possible within the larger cavity of **15**. In the productive **15·17** complex, schematically shown in Figure 8.6B, the ester residue of **17** can extend out of the cavity on the phenol side and have a favorable orientation for intra-complex nucleophilic attack. In the supramolecular complex of the more rigid macrobicycle **14** with the smaller binding site, the phenol ring seems to completely block one side of the cavity, and only nonproductive complex conformations are possible. The ester residue of the complexed substrate **17** extends out of the cavity on the side opposite to the nucleophile and reacts in a slower intermolecular reaction with the phenoxide of another macrocycle (Figure 8.6A). The different findings with **14** and **15** provide a convincing example for the importance of a productive binding geometry; nonproductive binding is characterized by the complete absence of supramolecular reactivity.

8.4.2 Supramolecular Catalysis of the Benzoin Condensation

Thiamine pyrophosphate (TPP, **18**)[59] participates as the essential cofactor in numerous enzymatic reactions involving formation and breaking of carbon–carbon bonds, *e.g.* in the transketolase-catalyzed formation and cleavage of carbohydrates in the pentose phosphate pathway.[60] In pioneering work in the 1950s, Breslow established that the catalytic action of TPP is mainly due to the

18, TPP

---

[59] R. Kluger, *Chem. Rev.*, 1987, **87**, 863.
[60] L. Stryer, 'Biochemistry', Freeman, New York, 1988, 3rd ed., pp. 427.

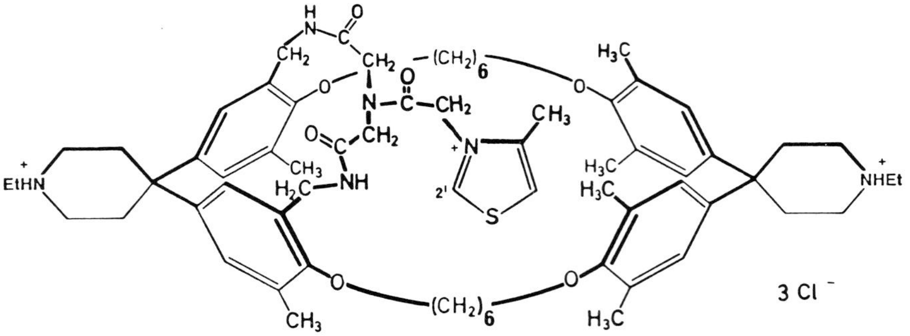

19

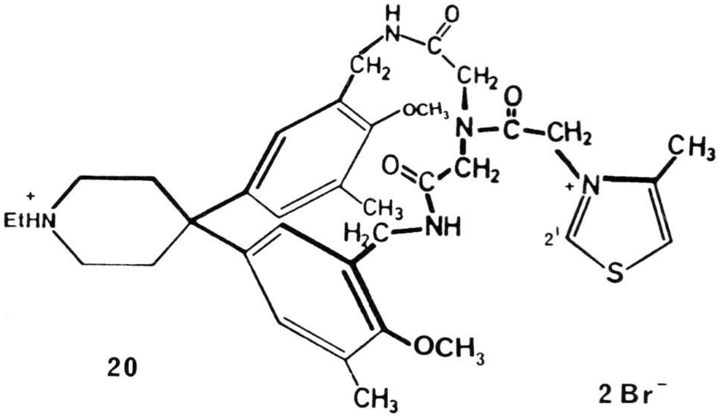

20

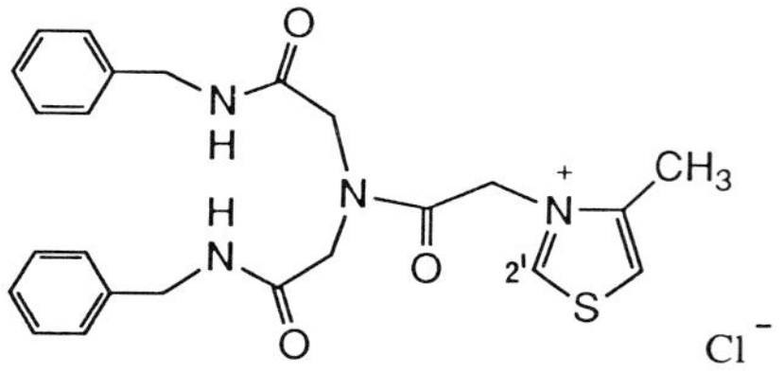

21

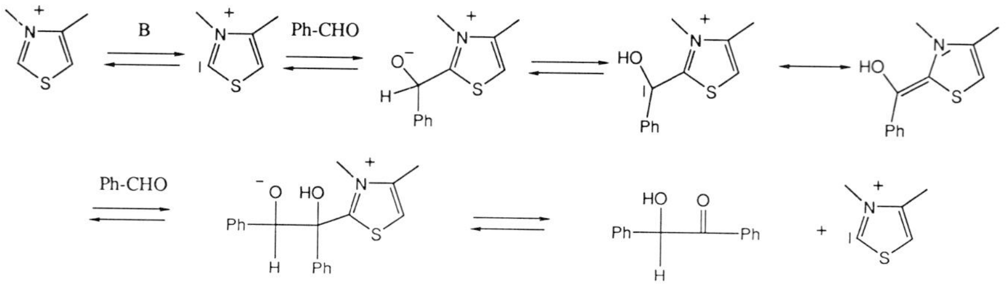

Scheme 8.4

thiazolium ring,[61,62] and that simple thiazolium ions catalyze many of the enzymatic transformations in the absence of the enzymes.

Since the formation and cleavage of carbon–carbon bonds are key conversions in organic chemistry, Lutter and Diederich became interested in exploring the supramolecular catalysis of the benzoin condensation (Scheme 8.4) by the thiazolium-cyclophane **19**.[63-66] With its long C_6-chains between the two diphenylmethane spacers, compound **19** provides a binding cavity large enough to accommodate the two benzaldehyde molecules that react to give benzoin. A thiazolium residue is connected by two side arms to the macrocyclic binding site. The double fixation atop the cavity was designed to ensure the location of the catalytic residue in favorable proximity to the aldehyde molecules in the supramolecular complex limiting its number of unproductive conformations. 2D NOESY ^{1}H NMR spectra provided experimental support for the preferred orientation of the thiazolium ring on the cavity side of **19**.

Cyclophane **19** is an efficient, very stable catalyst for the benzoin condensation and exhibits a large number of features typical for TPP-dependent ligases, C–C bond forming enzymes. Figure 8.7 shows the plots of product formation *versus* time for condensations catalyzed by **19** and the comparison compounds **20** and **21** in methanol-d_4 ([benzaldehyde] = 0.4 M, [NEt$_3$] = 60 mM, [catalyst] = 20 mM, T = 323 K, under N_2). The preparative work up after 12 h of the condensation catalyzed by **19** afforded pure benzoin in 93% yield. The reaction in the presence of **20** led to 74% isolated yield of benzoin, whereas the reaction catalyzed by **21** gave only a 27% yield of benzoin. The benzoin condensation does not proceed in the absence of thiazolium salt. A comparison of initial rates shows that the macrobicyclic catalyst **19** is 11 times more active than the nonmacrocyclic thiazolium salt **21**. Surprising at first, the macromonocyclic derivative **20** also is 5 times more active than **21**.

The catalytic advantage of macrocycles **19** and **20** over the thiazolium salt **21** results from collecting and orienting the reacting partners by complexation prior to the catalytic event. Characteristic complexation-induced shifts in the ^{1}H NMR spectra provided evidence for binding interactions between **19** and benzaldehyde in methanol-d_4 and in dimethylsulfoxide-d_6. In these solvents, sigmoidal saturation kinetics were observed which suggests the formation of a 1:2 Michaelis–Menten-type complex. High catalytic turnovers were obtained with **19**. As an example, for the reaction at T = 318 K in Me$_2$SO ([**19**] = 20 mM, [PhCHO] = 3.4 M, [NEt$_3$] = 0.16 M, and [NEt$_3$H$^+$ Cl$^-$] = 0.1 M), a catalytic turnover (mols of substrate converted by one mol of catalyst per time unit) of ≈ 1 min^{-1} was calculated.

[61] R. Breslow, *J. Am. Chem. Soc.*, 1958, **80**, 3719.
[62] R. Breslow and F. McNelis, *J. Am. Chem. Soc.*, 1959, **81**, 3080.
[63] H.-D. Lutter and F. Diederich, *Angew. Chem.*, 1986, **98**, 1125; *Angew. Chem., Int. Ed. Engl.*, 1986, **25**, 1125.
[64] F. Diederich and H.-D. Lutter, *J. Am. Chem. Soc.*, 1989, **111**, 8438.
[65] D. Hilvert and R. Breslow, *Bioorg. Chem.*, 1984, **12**, 206.
[66] R. Breslow and E. Kool, *Tetrahedron Lett.*, 1988, **29**, 1635.

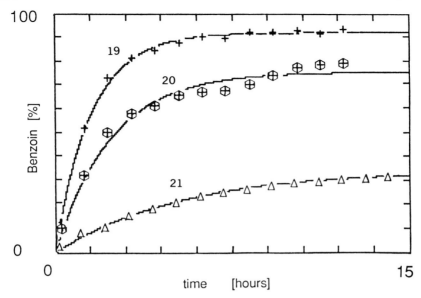

Figure 8.7 *Increase in benzoin versus time followed by ^1H NMR. [Catalyst] = 20 mM, [ArCHO] = 0.4 M, [NEt$_3$] = 60 mM, T = 323 K in methanol-d$_4$ (Reproduced with permission from reference 64. Copyright 1989 ACS.)*

Binding is also the origin of the very good catalytic performance of cyclophane **20** which lacks the macrocyclic binding and reactive site of **19**. The considerable activity of **20** can be explained by the fact that this macromonocycle, according to CPK model examinations, can provide a niche as a binding site for benzaldehyde. This assumption is supported by the observation in methanol of a curvature, characteristic of 1:1 Michaelis–Menten kinetics, in the plot of the initial rates of benzoin formation as a function of benzaldehyde concentration. Host–guest interactions with **20** are also supported by ^{1}H NMR binding studies.

In methanol, the thermodynamic equilibrium of the benzoin condensation apparently is far on the benzoin side as indicated by the 93% yield of benzoin isolated in the reaction catalyzed by **19**. In Me$_2$SO, however, the equilibrium is far less on the benzoin side, and by starting from pure benzoin (0.8 M), a 15% yield of benzaldehyde was obtained after 24 h in a reaction catalyzed by **19** (T = 318 K, [**19**] = 20 mM, [NEt$_3$] = 0.16 M, and [NEt$_3$H$^+$ Cl$^-$] = 0.1 M).

The apolar binding cavity of **19** produces a strong micropolarity effect on the kinetic acidity of the proton 2'-H of the thiazolium ring which was observed by ^{1}H NMR spectroscopy.[67-70] In deuterated buffers at various pDs

[67] P. Haake, L. P. Bausher, and W. B. Miller, *J. Am. Chem. Soc.*, 1969, **91**, 1113.
[68] J. A. Zoltevicz and S. Sridharan, *J. Org. Chem.*, 1978, **43**, 3785.
[69] A. A. Gallo and H. Z. Sable, *J. Biol. Chem.*, 1976, **251**, 2564.
[70] F. Vögtle and W. M. Müller, *Angew. Chem.*, 1984, **96**, 711; *Angew. Chem., Int. Ed. Engl.*, 1984, **23**, 712.

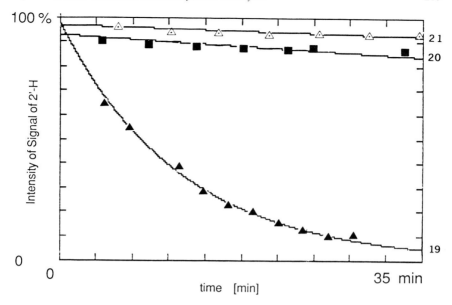

Figure 8.8 *H/D-exchange rates at C-2' of the thiazolium rings in **19–21** (c = 20 mM) in a KCl/DCl buffer, pD = 1.8, T = 303 K, monitored by 500-MHz ^{1}H NMR. The observed intensity of the NMR signal for 2'-H is plotted as a function of time*
(Reproduced with permission from reference 64. Copyright 1989 ACS.)

($\approx$ 1–5), the H/D exchange is considerably faster at C-2' of the thiazolium ring of **19** than at C-2' of nonmacrocyclic thiazolium derivatives (Figure 8.8). As an example, the observed first-order rate constant for H/D exchange at C-2' of **19** in DCl/KCl buffer, pD = 1.8, T = 303 K was 1.48×10^{-3} s^{-1}. The exchange rate constants for **20** ($k = 4.88 \times 10^{-5}$ s^{-1}) and **21** ($k = 2.25 \times 10^{-5}$ s^{-1}) are 30 and 65 times lower, respectively. The enhanced rate of H/D exchange was explained by a micropolarity effect of the cavity of **19** on the acidity of 2'-H of its thiazolium ring. The *ionic* thiazolium salt is stabilized by a polar environment, *e.g.* by the aqueous buffer. The *neutral* ylide however, formed upon deprotonation (Scheme 8.4) is stabilized by the less polar microenvironment provided by the macrocyclic binding cavity of **19**. This stabilization of the conjugate base leads to an enhanced acidity of 2'-H at the macrocyclic thiazolium ring which is also reflected in the product-resembling transition state of the deprotonation. A similar ylide stabilization is not present in **20** and **21**, and the kinetic acidity of their thiazolium protons at C-2' is reduced.

Micropolarity effects at active sites are believed to be the origin of the activation of TPP by holoenzymes.[71,72] The study with **19–21** shows that

[71] J. Crosby, R. Stone, and G. E. Lienhard, *J. Am. Chem. Soc.*, 1970, **92**, 2891.
[72] W. P. Jencks, *Adv. Enzymol. Relat. Areas Mol. Biol.*, 1975, **43**, 219.

enzymatic mechanisms such as pK_a-changes can be mimicked in supramolecular catalysis.

8.4.3 Supramolecular Catalysis of the Oxidation of Aromatic Aldehydes

In the continuation of their studies addressing the activation of coenzymes by the specific microenvironment of cyclophane binding sites, Diederich and coworkers became interested in mimicking the function of another TPP-dependent enzyme.[73] The flavin-dependent enzyme, pyruvate oxidase, transforms pyruvate into acetate (Scheme 8.5).[74,75] The active aldehyde, which is also the key intermediate in the benzoin condensation (Scheme 8.4), is oxidized. This is followed by rapid solvolysis resulting in the production of acetate and the regeneration of the thiazolium ylide. In a similar sequence, aldehydes are oxidized to carboxylic acids. A second coenzyme, flavin adenine dinucleotide (FAD) acts as the oxidizing agent in pyruvate oxidase,[74-76] whereas a variety of redox dyes or compounds like ferricyanide have been used in the oxidation of aldehydes by thiazolium model systems.[65,73]

The thiazolium cyclophane **22** was designed as a catalyst for the selective oxidation of aromatic aldehydes to carboxylic acids in the presence of potassium ferricyanide. In **22**, the thiazolium ring is attached to the periphery of a cavity complementary in size to one benzene or naphthalene molecule. The derivatives **24** and **25** were prepared to evaluate, by comparative studies, the catalytic advantages resulting from the formation of a supramolecular complex by **22**. Since aldehydic oxidation by ferricyanide does not occur in the absence of a thiazolium catalyst, no comparison was made between the reaction catalyzed by **22** and reactions without thiazolium ions. Only comparisons were made which show the catalytic advantage specifically provided by substrate complexation in favorable proximity and orientation to a thiazolium ring.

The ^{1}H NMR spectra in D_2O revealed that the thiazolium moiety of **22** is located within the cyclophane cavity. Considerable upfield shifts of all thiazolium protons were observed as compared to the same resonances in **25** (2-H: +0.70 ppm, 5-H: +0.39 ppm, 4-CH_3: +0.27 ppm). This raised concerns whether inclusion complexation by **22** would occur since the displacement of the thiazolium ring from the cavity by the entering substrate could possibly be an energetically very unfavorable process. Therefore, ^{1}H NMR binding titrations were performed to analyze how the thiazolium ion influences the binding of aromatic substrates by **22**. A comparison of the free energies for the complexation of various aromatic guests by cyclophanes **22** and **23** in D_2O/CD_3OD (60:40, v/v, $T = 293$ K) revealed that the binding to **22** is weaker by 0.4–0.7 kcal mol^{-1}. But the binding is still sufficiently large

---

[73] L. Jimenez and F. Diederich, *Tetrahedron Lett.*, 1989, **30**, 2759.
[74] L. Hager, *J. Biol. Chem.*, 1957, **229**, 251.
[75] G. Lienhard, *J. Am. Chem. Soc.*, 1966, **88**, 5642.
[76] N. Bergmann and F. P. Schmidtchen, *Tetrahedron Lett.*, 1988, **29**, 6235.

22 R = H₂C–[thiazolium-CH₃]

23 R = H

24 (with OMe groups, 2 Cl⁻)

25 (N-benzyl-4-methylthiazolium chloride)

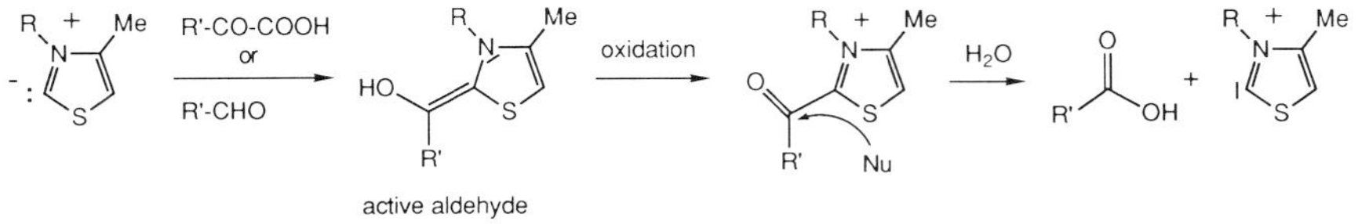

Scheme 8.5

Table 8.2 *Association constants, K_a, and free energies of complexation, $-\Delta G°$, determined by 500 MHz ^1H NMR titrations at 293 K for 1:1 complexes of cyclophanes **22** and **23** in $D_2O:CD_3OD$ (60:40, v/v). The $\Delta(\Delta G°)$ values indicate by how much the complexes of **23** are more stable than the corresponding complexes of **22***

Guest	K_a L mol^{-1}	$-\Delta G°$ kcal mol^{-1}	$\Delta(\Delta G°)$ kcal mol^{-1}
Complexes of **22**			
p-dicyanobenzene	3.49×10^2	3.41	
6-methoxy-2-naphthonitrile	4.81×10^3	4.94	
2-naphthaldehyde	1.97×10^3	4.42	
benzaldehyde	1.20×10^2	2.79	
Complexes of **23**			
p-dicyanobenzene	1.14×10^3	4.10	0.69
6-methoxy-2-naphthonitrile	1.50×10^4	5.60	0.66
2-naphthaldehyde	6.34×10^3	5.10	0.68
benzaldehyde	2.39×10^2	3.19	0.40

Table 8.3 *Kinetic data for the catalyzed oxidation of 2-naphthaldehyde to 2-naphthoic acid (T = 303K)*

Catalyst	Initial Velocities[a] M s^{-1}	Second-Order Rate Constants M^{-1} s^{-1}
22	4.18×10^{-6}	$k_{cat}/K_M = 2.8$
24	1.17×10^{-7}	$k_2 = 0.037$
25	3.90×10^{-8}	$k_2 = 0.0061$

[a] at [aldehyde] = 6.0 mM, [catalyst] = 0.5 mM. Rates are corrected for background rate of catalyst oxidation.

enough to assure high degrees of complexation in catalytic studies (Table 8.2).

The oxidation of 2-naphthaldehyde (2.0 mM–36 mM) by potassium ferricyanide (5 mM) in the presence of **22**, **24**, or **25** (0.5 mM) was studied at 303 K under Ar in Me$_2$SO/aqueous phosphate buffer, pH 7.5 (60:40, v/v).[73] The course of reaction was followed by monitoring the decrease in absorbance of the Fe(III) chromophore at $\lambda = 420$ nm. The reactions are zero order in

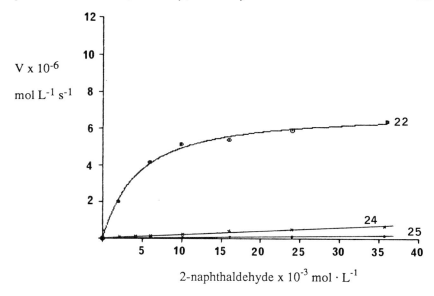

Figure 8.9 *Plot of initial rates vs. substrate concentration for the oxidation of 2-naphthaldehyde catalyzed by the thiazolium derivatives* **22**, **24**, *and* **25** *in Me$_2$SO/aqueous phosphate buffer, pH 7.5 (60:40, v/v), T= 303 K, [catalyst] = 0.5 mM*
(Reproduced with permission from reference 73. Copyright 1989 Pergamon Press.)

ferricyanide, and initial velocities were obtained using the initial linear portions of the absorbance *vs.* time plots. These rates were corrected by subtracting the background rates for slow thiazolium catalyst oxidation. A comparison of the corrected initial rates demonstrates that the oxidation catalyzed by **22** is much faster than those in the presence of **24** or **25** (Table 8.3).

Saturation kinetics were observed in the reactions catalyzed by **22** (Figure 8.9), and the kinetic data are shown in Table 8.3. For the oxidation of 2-naphthaldehyde, Lineweaver–Burke plots gave $K_M = 5.4$ mM, $V_{Max} = 7.5 \times 10^{-6}$ M s^{-1}, and a turnover number $k_{cat} = 0.90$ min^{-1}. In contrast, rates in the presence of **24** and **25** increased in a linear fashion with increasing substrate concentration. A comparison of the apparent bimolecular rate constant, k_{cat}/K_M, for the naphthaldehyde oxidation by **22** with the calculated second-order rate constants of the simple thiazolium salts **24** and **25** reveals that the rate acceleration is 75-fold between **22** and **24** and 460-fold between **22** and **25** (Table 8.3). The predominant formation of 2-naphthoic acid in the reactions catalyzed by **22** was confirmed by ^{1}H NMR spectroscopy.

The studies in aqueous solution demonstrated that **22** is an excellent selective catalyst for the oxidation of aromatic aldehydes to the corresponding carboxylic acids. The complexation of the aldehyde substrate in favorable proximity and orientation to the reactive thiazolium ring and the specific microenvironment of the cyclophane cavity provide a high catalytic advantage to **22** as compared to compounds **24** and **25** which lack binding sites.

8.4.4 A Porphyrin-bridged Cyclophane as a Model for Cytochrome P-450 Enzymes

In contrast to olefin epoxidations and alkane hydroxylations, the epoxidations and hydroxylations of aromatic hydrocarbons have not been the center of interest in studies with model compounds for cytochrome P-450 enzymes (Chapter 1.4.4).[77–80] Therefore, Diederich *et al.* designed compound **26** which has an efficient cyclophane binding site for arene complexation in close proximity to a porphyrin ring.[81] The synthesis of **26** is shown in Scheme 8.6. In addition to the free porphyrin cyclophane **26**, the zinc(II) and iron(III) porphyrinates **27** and **28** were prepared. The reader undoubtedly has noted that the synthesis of cyclophane receptors has almost entirely been left out from the previous Chapters. The author is well aware that, by leaving out synthesis in order to generate the space for a comprehensive treatment of molecular recognition studies, he has done injustice to the many scientists who invested months and years in the preparation of the molecular architecture shown. Synthesis is most often the major time-consuming component in a molecular recognition project that comprises the design, preparation, and study of cyclophanes or other designed receptors.

^{1}H NMR binding studies demonstrated the formation of inclusion complexes between **26** and aromatic hydrocarbons in $CD_3OD/D_2O/CD_3COOD$ (95:4.85:0.15, v/v). Table 8.4 gives the association constants and the free energies of formation for the 1:1 arene complexes. Figure 8.10 shows the remarkably large upfield shifts calculated for saturation binding for those guest proton resonances which could be evaluated in the NMR binding titrations. The upfield complexation shifts of the guest resonances in complexes of **26** are considerably larger than in the complexes of similar

---

[77] J. T. Groves and P. Viski, *J. Org. Chem.*, 1990, **55**, 3628.
[78] K. Korzekwa, W. Trager, M. Gouterman, D. Spangler, and G. H. Loew, *J. Am. Chem. Soc.*, 1985, **107**, 4273.
[79] C. K. Chang and F. Ebina, *J. Chem. Soc., Chem. Commun.*, 1981, 778.
[80] J. R. L. Smith and P. R. Sleath, *J. Chem. Soc., Perkin Trans. 2*, 1982, 1009.
[81] D. R. Benson, R. Valentekovich, and F. Diederich, *Angew. Chem.*, 1990, **102**, 213; *Angew. Chem., Int. Ed. Engl.*, 1990, **29**, 191.

Scheme 8.6 continued overleaf

Supramolecular Function, Reactivity, and Catalysis

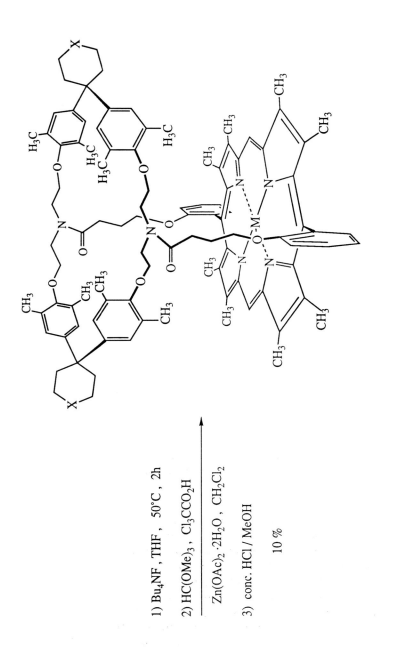

1) Bu$_4$NF, THF, 50°C, 2h
2) HC(OMe)$_3$, Cl$_3$CCO$_2$H
   Zn(OAc)$_2$·2H$_2$O, CH$_2$Cl$_2$
3) conc. HCl / MeOH

10 %

26 M = 2 H   X = EtN
27 M = Zn   X = EtN
28 M = Fe-Br   X = EtHN$^+$ Br$^-$

Scheme 8.6

Table 8.4 *Association constants, K_a, and free binding energies, $-\Delta G°$, for the arene-complexes of porphyrin-cyclophane **26** in methanol-d_4/ D_2O/acetic acid-d_4 (95:4.85:0.15, v/v), T= 293 K*

Guest	K_a [L mol^{-1}][a]	$-\Delta G°$ [kcal mol^{-1}]
Phenanthrene	1330	4.19
Acenaphthylene	1050	4.05
Anthracene[b]	460	3.57
Naphthalene	330	3.38
Pyrene	160	2.95

[a] Accuracy of the K_a-values: ± 10%; [b] In methanol-d_4/Me$_2$SO-d_6/D$_2$O/acetic acid-d_4 (90:5:4.85:0.15, v/v).

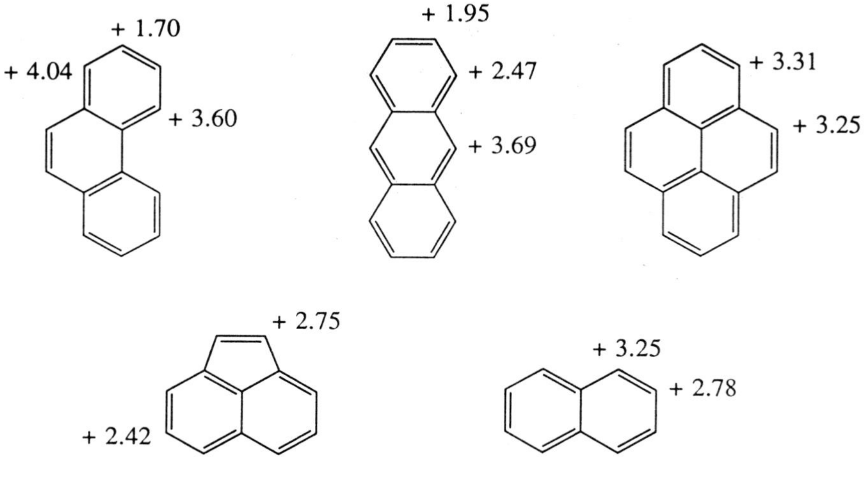

Figure 8.10 *Calculated upfield complexation shifts ($\Delta\delta$ values) of observable guest proton resonances in complexes of **26** at saturation binding.*

cyclophanes that are not bridged by a porphyrin (Chapter 2.5). In the complexes of **26**, the aromatic substrates are exposed not only to the shielding cyclophane cavity but also to the strongly shielding region of the porphyrin ring. According to computer modeling studies using the MM2 force field in the MACROMODEL program,[82] acenaphthylene and phenanthrene come within van der Waals distance of the porphyrin ring upon the energetically favorable deep incorporation into the cyclophane cavity of **26**.

As predicted by the linear free energy relationship in Chapter 3.2.2, stronger complexation occurs in 2,2,2-trifluoroethanol compared to methanol. For the

[82] Macromodel Version 2.6, Professor W. C. Still, Columbia University, New York.

high spin **28**-phenanthrene complex in 2,2,2-trifluoroethanol at 293 K, the evaluation of UV/Visible titrations in which the decrease in intensity of the Soret band was monitored as a function of increasing guest concentration gave an association constant of $K_a = 3.95 \times 10^4$ L mol^{-1}. In 2,2,2-trifluoroethanol in the presence of iodosylbenzene as the oxygen transfer agent,[83,84] the iron(III) porphyrinate **28** catalyzes the oxidation of acenaphthylene specifically at the C(1)-C(2) bond with formation of acenaphthen-1-one as the only isolable product. Thus, stirring a homogeneous solution of [**28**] = 0.54 mM, [acenaphthylene] = 11 mM, and [iodosylbenzene] = 15 mM for 4 h under rigorous exclusion of oxygen afforded acenaphthen-1-one with a turnover number (mols of substrate converted per mol of catalyst) of ≈ 6. Iodosylbenzene alone does not epoxidize acenaphthylene. The acenaphthylene oxidation catalyzed by **28** occurs *via* the epoxide (Scheme 8.7) which forms as the primary product but rapidly reacts under the reaction conditions to give acenaphthen-1-one.

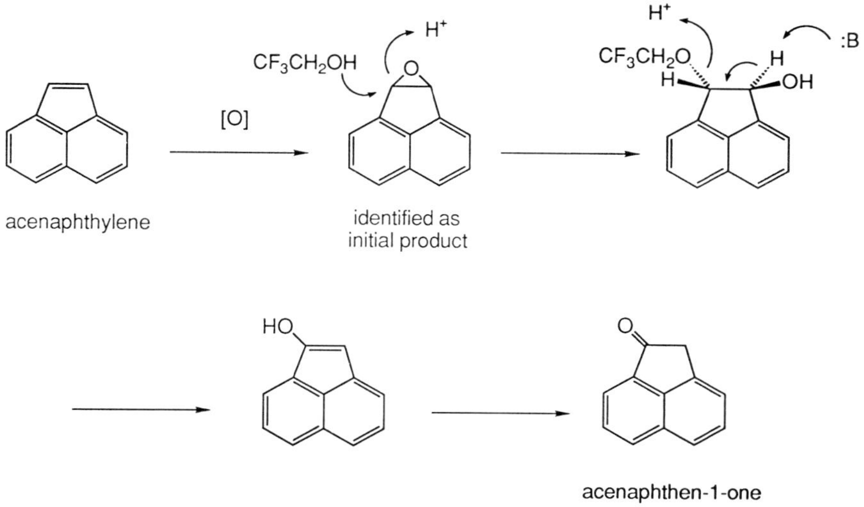

Scheme 8.7

It still needed to be shown that acenaphthylene oxidation proceeds *via* a supramolecular process in the cavity of **28** rather than at the open porphyrin face. The best evidence for supramolecular catalysis was obtained with the μ-oxo dimer **29** which was prepared by reacting **28** in dichloromethane with sodium hydroxide.[85] In methanol, the μ-oxo dimer is very stable, and acenaphthylene oxidation must occur in the cyclophane binding sites. In

---

[83] J. T. Groves, T. E. Nemo, and R. S. Myers, *J. Am. Chem. Soc.*, 1979, **101**, 1032.
[84] T. G. Traylor, J. C. Masters, Jr., T. Nakano, and B. E. Dunlap, *J. Am. Chem. Soc.*, 1985, **107**, 5537.
[85] D. R. Benson, Doctoral Dissertation, University of California, Los Angeles, 1990.

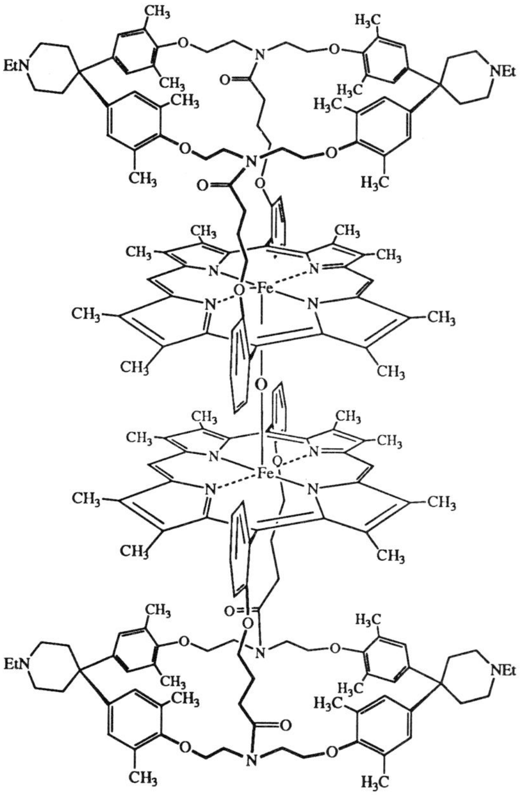

µ-oxo Dimer **29**

support of a supramolecular process, the dimer **29** shows a high catalytic activity in the oxidation of acenaphthylene. Stirring a homogeneous solution of [**29**] = 0.19 mM, [acenaphthylene] = 11.8 mM, and [iodosylbenzene] = 12 mM for 30 h afforded a 28% yield (9 turnovers) of acenaphthen-1-one, the remaining being mainly starting material. By showing strong complexation and selective epoxidation of acenaphthylene, the porphyrin-cyclophanes **28** and **29** represent true synthetic analogues of P-450 cytochromes.

8.5 Concluding Remarks

Cyclophanes represent the most universal type of synthetic molecular receptors. The aromatic rings that are bridged in cyclophane structures take a variety of important functions. They rigidify and organize binding sites, they participate in specific aromatic binding interactions, and serve as points of attachment for functional groups for recognition and catalysis. Novel, large cyclophane shapes are continuously being created[86-89] and their molecular recognition properties investigated. Undoubtedly, research with future generations of cyclophanes will continue to contribute to some of the most fascinating developments in the field of supramolecular chemistry.

---

[86] T. K. Vinod and H. Hart, *J. Am. Chem. Soc.*, 1988, **110**, 6574.
[87] P. R. Ashton, N. S. Isaacs, F. H. Kohnke, G. S. D'Alcontres, and J. F. Stoddart, *Angew. Chem.*, 1989, **101**, 1269; *Angew. Chem., Int. Ed. Engl.*, 1989, **28**, 1261.
[88] A. P. West, Jr., D. Van Engen, and R. A. Pascal, Jr., *J. Am. Chem. Soc.*, 1989, **111**, 6846.
[89] H. L. Anderson and J. K. M. Sanders, *J. Chem. Soc., Chem. Commun.*, 1989, 1714.

Indexes

Index A: Subject Index*

Acenaphthen-1-one, 295
Acenaphthylene, catalytic epoxidation of, 295
Acetylcholine binding, 150
Acetylcholinesterase, 150
Active aldehyde, 288
Acyl transfer, supramolecular catalysis of, 280
Adamantane, inclusion complexes, 101—102
Adams, R., 1
Adenine, cyclophane receptor for, 207
  recognition in molecular cleft, 183
Adenosine, inclusion complexation, 98—99
Aggregation of cyclophanes in water, 66—74
  Aldolase reaction, vitamin B_6-dependent, 31
Aliphatic substrates, apolar inclusion complexes of, 100—105
Allosteric effects, 158
AMBER force field, 89, 235, 238
Amine–arene exciplexes, 14
α-Amino acids, asymmetric synthesis of, 29—33
  α-deuterated, 29
Ammonium ion binding, 140
Androgenic steroids, inclusion complexation of, 103—104

1,8-ANS (8-Phenylamino-1-naphthalenesulfonate), 55—57, 143, 145
Anthraquinone-cyclophanes, 190, 192
Anion receptors, 153—157
Anion recognition, 153
  divalent anions, 154
  monovalent anions, 154
Ansa compounds, 1
Anson, F. C., 37
Apolar binding, 52
  dependence on solvent polarity, 106, 257
  dependence on van der Waals volume of guest, 124
  in binary solvent mixtures, 137—139
  predictability through linear free energy relationship, 108
  solvent effects, 246
  thermodynamics of, 251
Apolar inclusion complexation, 60
  in organic solvents, 106—139
Apolar surface interactions, 57
Arene epoxidation, supramolecular catalysis of, 290
Arene inclusion complexes, 75
Aromatic aldehydes, oxidation of, 286
Aromatic–aromatic interactions, 206, 247
Aromatic heterocycles, inclusion

*For guests in solution complexes for which stability constants are given, see Index B

complexation of, 98
Aromatic rings in cyclophane hosts,
 participation in ion–dipole
 interactions, 142
 role of, 62
Arylglycolic acids, by reduction of
 arylglyoxylic acids, 229
Association constants, definition of, 57
 from calorimetry, 81
 from ^{13}C NMR titrations, 100
 from competitive inhibition, 81
 from ESR titrations, 72—73
 from fluorescence titrations, 81,
  113—114
 from ^{1}H NMR titrations, 59, 80, 97,
  98, 102, 112—113, 193, 207, 244,
  279, 286
 from liquid–liquid extractions,
  74—75
 from optical titrations, 80, 113—114,
  165, 177
 from solid–liquid extractions, 74—75,
  196
 of TCNE-cyclophane complexes, 45
Asymmetric Diels–Alder reaction, 229
Asymmetric oxidation, by P-450
 models, 38
 of styrene, 40
Asymmetric reduction of carbonyl
 substrates, 26
Asymmetric synthesis, in solid state, 51
 of α-amino acids, 29—33
Atherosclerotic plaque, 105
Atrolactic acid, 228
Atropisomerism, 3
Axial base in porphyrins, 34, 40—41
Axial inclusion geometry, 60, 61, 82—
 83, 103, 272
Azaparacyclophanes, 176—181

Barbiturates, detoxification, 214
 recognition, 214—217
BATCHMIN Monte-Carlo program, 89
Bender, M. L., 265
Benesi–Hildebrand equation, 55, 81
Ben-Naim, A., 261
Benzene derivatives, complexation of,
 86—95
Benzidine, cyclophane spacer, 53
Benzoin condensation, 281

supramolecular catalysis of, 283—284
1-Benzyl-1,4-dihydronicotinamide
 (BNAH), 22—24
Bile acids, inclusion complexation,
 103—105
1,1′-Binaphthyl crown ethers, 222
Binary solvent mixtures, apolar
 complexation in, 98, 106, 137—139
 LFER between composition and
  apolar binding strength, 137
Binding titration, calorimetric, 81
 ^{13}C NMR, 100
 ESR, 72—73
 fluorescence, 81, 113—114
 ^{1}H NMR, 59, 80, 106, 113, 193, 207,
  244, 279, 286
 optical, 80, 113—114, 165, 177—178
Binding titration curves, boundary
 conditions, 80
 evaluation of, 80—81, 279
BIOGRAPH, computer modeling program,
 66
Biomimetic cyclophane research, 3, 18
Biomolecules, complexation of, 198
Biphenyl macrocycle, 237—238
Bis(diphenylmethane) hosts, 60—68, 72,
 83, 101, 240
Bis(paracyclophane) hosts, 158
Bis(phosphine oxide) cages, 190
Bis(thymine) substrate, 212
Bowl-type hosts, 125, 150—153
 binding of trimethylammonium
  guests, 150—153
Breslow, R., 106, 157, 265, 281
Bromochlorofluoromethane, analytical
 resolution by cryptophane,
 224—226
 maximum molar rotation of
  enantiomerically pure, 225
Brown, C. J., 1
Busch, D. H., 37, 100
1-Butylthymine, 209

Capped tetraaza[3.3.3.3]para-
 cyclophane, 177, 179
Carbohydrate complexation, 198—201
Carceplexes, 133
Carcerands, 131—135
Carothers, W. H., 1
Carnitine binding, 153

Carrier-mediated transport, 265—269
Catalysis, 21
Catalytic cyclophanes, 62, 265, 273
Catalytic mechanisms, microenvironment of binding site, 290
  proximity, 264, 274, 286
  strain, 21
  transition-state binding, 264
Catalytic turnover, 279, 283, 295
[2]Catenane, 170—176
  cyclic voltammetry of, 174—175
  design of, 172
  dynamics of, 174, 176
  interchromophore interactions in, 171
  synthesis of, 171, 173
  template effect in synthesis of, 171, 173
[n]Catenanes, organic materials, 176
Cation receptors, 150—153
Cavitands, 125
  small linear guest complexation, 125—127
Cavity-filling conformations, 53, 234
  ^{1}H NMR evidence for, 190, 236
  in biphenyl macrocycle, 238
  origin of, 231
Cavity formation in liquids, energetics of, 254
Cavity inclusion complexation, 57
  evidence for, 59, 60
Chapman, K. T., 217
Charged aromatic guests, inclusion complexation of, 142—145
Charge-transfer, absorptions, 15—17, 45, 116, 120
  configurations, 7
  interactions, 44, 165
Charge-transfer complexes, 113
  cyclophane models for, 15
  of dimethoxybenzenes, 119—121
  of tetracyanoethylene, 44—46
Charge-transfer salts, three dimensional, 124
Chiral recognition, in aqueous solutions, 228
  in organic solvents, 224
  of ammonium salts, 222
  of bromochlorofluoromethane, 224
  of chiral amides, 226
  of chiral onium salts, 229

of naproxen and derivatives, 231, 241—245
Chiral spacers, 238, 241
Chiral stationary phase (CSP), 115
  hydrogen-bonding on, 182
Cholesterol sequestration, 105
Chromatographic separations, 264
α-Chymotrypsin, catalytic mechanism, 278
  mimics, 276
Clathrates, cage-type, 47
  channel-type, 47
  formation of, 46
  of cyclophanes, 46—51, 53
  optical resolution in, 48
  reactions in, 48
Cleft-type receptors, 182—183
Co-complexation, 188
Coenzyme activation, 273, 285
Coenzyme F_{420}, 23
Collet, A., 66, 91, 100, 120, 225
Collman, J. P., 34, 37
Competitive inhibitor, 267
Complexation–decomplexation kinetics, 79, 120, 131, 200
Complexation-induced changes in ^{1}H NMR chemical shift, 61, 76, 84, 96, 122, 145, 147, 200, 212, 232, 294
Complexation modes, extracavity, 190—195
  intracavity, 190—195
Complexation strength as a function of, guest solvation, 217—221
  H-donor functionality, 190
  solvent polarity, 108—113, 257—259
  solvent size, 217—221
  van der Waals surface of solvent molecules, 219—221
Computer modeling, 3, 62, 87, 93, 184, 190, 226, 235—241
Concave functionality, 189—197
Conformational analysis, 3, 235—241
  boat-shaped 1,4-dihydronicotinamide, 24
  face-to-face conformation, 55
  helical macrocycles, 234
  macroring inversion, 5
  of anisol, 92
  of cavity-filling macrocycles, 235

Index A: Subject Index 301

of cyclophanes incorporating Tröger's base, 93—94
of dibenzodiazocine units, 91
of 2,6-dimethylanisole, 92
of dioxaalkane bridges in cyclophane hosts, 91
of diphenylmethane units, 93—94
of macrorings, 224
of Tröger's base derivatives, 91
rhomboid conformation of cyclophane host, 118
toroid conformation of cyclophane host, 118
Constant-current electrolysis, 124
Converging functional groups, 182, 186, 189, 261
Cooperativity in binding, 158
Corey–Pauling–Koltum (CPK) molecular model examinations, 55, 90—91, 120, 177, 180, 200, 220, 233, 269, 272, 284
Corticoids, inclusion complexation of, 103—105
Cosolvent, in binary aqueous mixtures, 74
Coulombic charge–charge interactions, 141—143, 150—157, 177, 247
Cram, D. J., 1—3, 6, 44, 61, 91, 125, 183, 185, 222, 241, 265
Cramer, F., 52, 265, 276
Critical aggregation concentration (CAC), 68—74
of cyclophane hosts, 71
Crown–cation complexation, 183
Crown ether, complexes of C–H acidic guests, 183—185
Cryptand nitrogen atoms, 107
Cryptophanes, 100
diastereomeric complex formation, 225
radical cation, 124
shape-selective inclusion of methane derivatives, 120—125
Cubic azaparacyclophane, 180—181
Cyanocobalamine (vitamin B_{12}), 198
Cyclic voltammetry, 174—175
Cyclodextrins, catalysis by, 52, 265
chiral recognition by, 222
dimers, 157
inclusion complexes in organic solvents, 106
inclusion complexes in water, 52, 256
iodine inclusion, 46
pyridoxamine substituted, 274
thermodynamics of complexation, 255
Cyclophanes, bridges in, 3, 65
electrophilic substitutions, 6
new shapes, 296
nomenclature, 1
optically active, 224, 238
racemization, 3
size of substituents, 5
Cyclophane cavity, depth of, 87
free space, 5, 91
micropolarity, 142, 284
solvation of, 110, 247—252, 260
Cyclophane receptor design, 60
Cyclotricatechylene, 46, 49
Cyclotriveratrylene, 46, 49, 66
Cyclochrome C oxidase, 33, 37—39
Cytochrome P-450, bridged porphyrins as enzyme models, 38—41, 290—296
carbon monoxide complex of, 41
enzymes, 33, 38
Soret band of, 41
ternary complexes as models of, 101

Dale, J., 93
5-Deazaflavinophanes, 22—24
Degree of enantioselection, 224, 228, 244
Deranleau, D. A., 80
Desolvation, driving force for complexation, 96, 253
of apolar cavity binding site, 79, 250
of apolar surfaces, 79
of functional groups, 107
Deoxyribonucleic acid (DNA), 19, 207, 213
base pairing, 182
minor groove binding, 255
Diastereo-differentiating hydride transfer, 23, 24
Diastereomeric, complex formation, 228
reaction transition states, 229
Diazapyrene-cyclophane receptors, intercalative binding of nitrobenzene, 154—156

Dibenzocrown ether, 165—168
2,6-Dibutyramidopyridine, 209
Dicarboxylic acid complexation, 200
Diederich, F., 98, 137, 233, 278, 283, 286, 290
Diels–Alder reaction, 98
Differential steric interactions in diastereomeric complexes, 244
Dihedral angle drive, in MM2, 89
Dihydroisoalloxazine, geometry of, 269
Dihydronicotinamide adenine dinucleotide (NADH), 22
  models of, 22, 24, 26
Dihydronicotinamide adenine dinucleotide phosphate (NADPH), 20
Dihydropyridine-cyclophanes, 26—29
Dimer formation, in organic solvents, 127—131
Dioxaalkane bridges, in cyclophane hosts, 65
Dioxa[n]paracyclophane, 2, 3
Dioxygen, binding, 34, 37
  binding site, 34
  carrier of, 34
  reduction, 37
Diphenylmethane units, 62, 65
Dipolar alignment, 48
Dipole–induced dipole interactions, 112
Diquat, 142, 165—170
Distribution constants, 267—268
Ditopic binding, energetics of, 159
  complex geometry, 160, 162
Ditopic cyclophane receptors, 158
Diyne-bridged cyclophanes, 189—196
Dome-shaped anion receptors, 156
Donor–acceptor cyclophanes, 15—19
Donor–acceptor stacking, 169
Dougherty, D. A., 65, 98, 117, 146, 148, 256
Drug, delivery, 183
  detoxification, 183
Dyes, in optical binding studies, 177—178
Dynamic nuclear magnetic resonance (DNMR), 3, 174

Edge-to-face interactions, 82, 90, 171, 190, 247
Eftink, M. R., 256

Electron donor–acceptor (EDA) complexes, 44—46
Electron donor–acceptor (EDA) interactions, in complexes of 2,6-disubstituted naphthalenes, 115—117
  in diquat and paraquat complexes, 167—170
  in hydrogen-bonded complexes, 206
  in inclusion complexes of aromatic guests, 113—120, 247
  in second-sphere complexes, 165
Electronic absorption spectra, of [$m.n$] paracyclophanes, 6—8
  of [3.3]paracyclophane-quinhydrones, 18
  of TCNE-cyclophane charge-transfer complexes, 45
Electron spin resonance, determination of association constants by, 73
  determination of CAC by, 72
  of cryptophane radical cation, 124
  of nitroxide spin probes, 72, 177
Electron transfer, between porphyrins and quinones, 41
Electron tunneling, 43
Electrostatic interactions, 44, 91, 116, 206
Enantioface differentiation, 31
Enantiomer separations, 3, 182
Enantioselectivity,
  in asymmetric reduction of carbonyls, 26—29
Enthalpic driving force for apolar binding, 89, 252
Entropy, driving force for complexation, 254—257
  of chelation, 261
  of desolvation, 256
Enzymes, in organic solvents, 107
Equatorial inclusion geometry, 82—83
Esterase activity, 273, 276
Ester hydrolysis, catalysis of, 57, 273
Ethyl phenylglyoxalate, reduction of, 27
Excimers, 7—14
  distance dependence of excimer interactions, 10
  inhibition by inclusion complexation, 113, 115
  of 1,3-bis(1-pyrenyl)propane, 113, 115

of naphthalene, 10
of pyrene, 10
Exciplexes, 14—15
  n—π interactions in, 14
Exciton resonance, 7
Extraction constants, 268

Face-to-face, conformation, 54—55
  dimer, 127
  interactions, 171, 247
Farthing, A. C., 1
Fast atom bombardment (FAB) mass spectrometry, 132
Fastrez, J., 57
Faust, G., 53
Fischer, E., 61
Flavin adenine dinucleotide (FAD), 20, 21, 272, 286
Flavinophanes, 20—24, 269
  redox-dependent complexation by, 269—273
Flavoenzyme active sites, 272
Fluorescence, binding titrations, 55, 81, 113—115, 142, 177, 180
  lifetime, 42
  quantum yield, 142
  quenching, 42, 204, 271
Fluorescence spectra, of excimers, 10—14, 113, 115
  of [$m.n$]naphthalenophanes, 12
  of naphthalenopyridinophane-dienes, 16
  of [$n.n$]pyrenophanes, 13
Förster, Th., 10
Free energies of activation, of cyclophane ring inversion, 5
  of desolvation, 79
  of guest inclusion into cryptophanes, 123
  of hemicarceplex decomplexation, 136
  of interconversion between degenerate [2]catenane structures, 174, 176
Freudenberg, K., 46
Functional group, complexation-induced changes in solvation of, 107, 117
  orientation in inclusion complexes, 82—83
  solvation, 82, 259

unfavorable desolvation in complexation process, 98, 105, 261—263

Gable conformation, 93
Gaeta, F., 222
Glutathione, 20
  reductase, 20
Guest selectivity in inclusion complexes, 142

Haemoglobin, 33
Haenel, M. W., 7, 10, 14
Hamilton, A. D., 202, 203, 210, 214
Hammett substituent constant, 46
Heat capacity changes, 252
  from calorimetry, 257
  from statistical analysis, 256
Heats of complexation, 252
Hemicarceplexes, 136
Hemicarcerands, 135—137
Hemin, 198
High-spin complex, 295
Hilgenfeld, R., 53
^{1}H NMR, binding titrations, 80, 102, 207, 244, 279, 286
  chemical shifts of cavity-filling conformers, 236, 286
  complexation-induced changes in chemical shifts, 59, 61, 76, 84, 122, 145, 147, 193, 294
  determination of CAC, 68, 70
  differential complexation shifts in diastereomeric complexes, 228, 231, 232, 244
  dimerization shifts, 127
  evidence for inclusion complexation, 60, 75
  fast exchange kinetics, 59, 79
  slow exchange kinetics, 120
  sensitivity range, 80
Högberg, A. G. S., 125
Hoogsteen base pairing, 207, 213
Host–guest chemistry, 2
Host–guest complexation, 1:2 stoichiometry of, 169—171
  2:1 stoichiometry of, 152—153
Huenig, S., 63
Human c-H-*ras* oncogene, guanine binding to, 204

Hydride transfer, 23, 24, 27, 30
Hydrogen bonding, complexation in
    water, 261—263
  complex geometries, 183, 199, 202,
    205, 206, 211, 214—216
  edge in receptors, 207
  from guest to solvent, 79
  in clathrates, 49—50
  in diquat and paraquat complexes,
    167—170
  receptors, 182
  substrate specificity, 182
  to phosphines, 190
Hydrolase models, 57, 276
Hydrophilic interactions, 261
Hydrophobic, bonding, 254—257
  interactions, 67, 254, 261
  pockets in enzymes and antibodies,
    255
  vitamin B_{12}, 273
Hydrophobicity–hydrophilicity balance,
  71
Hyperfine splitting constants (hfsc), 72
Hypochromic effects, 19—20
Hypochromism, 19, 271

Imidazole,
  recognition, 217, 220
  solvation, 220—221
Inazu, T., 54, 118
Inclusion complex geometries, 76—78,
  82—83, 97, 99, 103, 109, 121, 145,
  153
Induced fit, 91, 180, 202, 204
Intercalation complex, 98, 156
Interchromophoric distance, 8, 13,
  42—43
Interstices in crystal lattice, 46—51
Intraannular acidic groups, 185, 187
Iodosylbenzene, oxygen transfer agent,
  38, 295
Ion–dipole effect, 146
  separation from apolar binding term,
    149
Ion–dipole interactions, 146, 167, 247
Ionophores, 265
Ion pairing, 138, 145
  energetics of, 157
Iron(II)cyclophane, 37
Iron-oxo-intermediate, 38

Iron(III)porphyrinates, 38, 290
Isoalloxazines, 20—24, 269
Isoalloxazinophanes, 20
Isoalloxazinometacyclophane, 20
Isosbestic points, 113—114
H–D Isotope effect, 24, 284—285

Janzen, E. G., 71, 73
Jefford, C. W., 48
Jorgensen, W. L., 207

Kellogg, R. M., 26, 265
Kinetic acidity, 285
Kite-type molecules, dimerization,
  127—131
Koga, K., 52, 59, 60, 62, 82, 149, 158,
  228
Kuzuhara, H., 29, 31

Lehn, J.-M., 154, 265
Lennard-Jones potential, 247, 254
Lepropre, G., 57
Ligases, 283
Light scattering, 68
Linear free energy relationship (LFER),
    between solvent polarity parameter
    and apolar complexation strength,
    108—113, 137—139, 257
Lindsey, J. S., 42
Lineweaver–Burke plots, 289
Liquid–liquid extractions, 74—75, 267
Lithocholic acid, axial inclusion, 103
Lock-and-key principle, 61
London dispersion interactions, 44, 52,
  112, 167, 247, 254
Lüttringhaus, A., 1, 3
Lutter, H.-D., 283

Macrobicyclic, catalysts, 278
  hexa-ammonium ions for anion
    recognition, 154
  receptors, 62—63, 65—66, 107—109,
    217—221, 226
Macrocyclic, lactam hosts, 84, 87
  pK_a, 185—187
MACROMODEL, computer modeling
  program, 62, 89, 93, 294
Macroring inversion, 5
Major groove, of 1,1'-binaphthyl unit,
  162, 241

of DNA, 213
Mandelic acid, 228
Mansuy, D., 38
Matrix isolation at room temperature, 137
Mauzerall, D., 42
[2.2]Metacyclophane, 2
[12]Metacyclophane, 2
Metal insertion reaction, 40
Metal ion binding, 140
Metalloporphyrin-cyclophanes, 37
Methane derivatives, shape-selective inclusion by cryptophanes, 120—125
N-Methylacridinium iodide, 24
Methylviologen (paraquat), 150
Michaelis–Menten, constant, 281
   kinetics, 279, 283, 284
Microcalorimetry, 252
Micropolarity effects on C—H acidity, 284—285
Miller, S. P., 98
Minor groove, of 1,1'-binaphthyl unit, 162, 241
   of DNA, 213, 255
Misumi, S., 7, 18, 19
Mitchell, R. H., 5
MM2 force field, 77, 82, 87—89, 233, 235, 238, 240, 241, 294
   MNDO calculations, 206, 209
Molecular, architecture, 290
   anion complexation, 153—157
   cation complexation, 150—153
   dynamics siumlations, 226—227, 235—239
Molecular-dispersed water solubility, 62, 70
Monooxygenases, 38
Monosaccharide recognition, 198
Monte-Carlo, search of conformational space, 89
   statistical mechanics simulations, 210
Multilayered cyclophanes, 18
Multiple recognition sites, 157
Murakami, Y., 57, 176, 177, 180, 273, 276
Myoglobin, 33

Naphthalene derivatives, complexation of, 81—86

Naphthalenocrown ethers, receptors for paraquat, 169—171
[$m.n$]Naphthalenophane, 10
Naphthalenophane host, 58, 98—99, 189—191
Naphthalenopyridinophane, 14
Naphthalenopyridinophane-diene, 14
Naproxen,
   cyclooxygenase inhibitor, 231
   optical resolution, 233
Nesting complex, 57, 101
Nicotinamide adenine dincleotide (NAD^+), 21
4-Nitro-1-naphthylacetate, substrate in transacylations, 279
Nitroxide spin probes, inclusion of, 72
NOE difference spectroscopy, 127, 226, 234, 278
NOESY spectroscopy, 217, 234, 283
Non-covalent interactions, distance and orientation dependence of, 7, 20
Nonlinear least-squares curve-fitting, 81
Nonlinear optical properties, 48
Nonlinear regression methods, 81
Nonproductive binding, 276—281
Nucleic acid bases, stacking of, 19
Nucleotide base recognition, 202—217
   induced-fit mechanism for, 203

Octahydroxy[1.1.1.1]metacyclophane,
   dicarboxylic acid binding by, 200, 202
   hydrogen-bonding sites in, 198
   selectivity in monosaccharide binding, 200—201
   sugar extraction with, 200
Octamethoxycyclophane hosts, 71—73, 86—90
Octol host, 125, 198
Octopus cyclophane, 177, 179, 273
Onium ion–aromatic π-system interactions, 148
Onium ion complexation, 150
Onium substrates, chiral recognition, 229
   inclusion complexation, 146—148
OPLS/AMBER force field, 226
Optical resolution, by chromatography on chiral stationary phases, 115
   through clathrate formation, 48, 243

Optical titrations, 80, 113, 165, 177
Orange, G, 177
Orthocyclophanes, 46—49
Outer-sphere complexes, 165
Oxidation of aromatic aldehydes, catalysis of, 286—290
μ-Oxo dimer, of porphyrins, 34, 295—296
Ozonation, of pyrene, 234

Pallas, M., 53
[m.n]Paracyclophanes, 2, 3, 7, 44
  inclusion complexation of, 95—98
[n.n]Paracyclophane-quinhydrones, 15, 18
[2.2]Paracyclophane ([2.2]PCP), 1, 2, 6
[2.2.2.2]Paracyclophane, 54
[3.3.3.3]Paracyclophane, electron-deficient, 118—119
  electron-rich, 118—119
[4.3]Paracyclophanecarboxylic acid, 3
Paraquat (1,1'-dimethyl-4,4'-bipyridinium), 63, 142, 167—171
Paraquat-cyclophanes, charge-transfer inclusion complexation, 119, 121
  in [2]catenane, 170—176
Park, C. H., 153
Pedersen, C. J., 140, 182, 183, 222
Pentose phosphate pathway, 281
Peptide α-helix, 106
μ-Peroxo-dimer, 37
Person, W. B., 80
Perturbation theory, 210
Phenol complexation, 189
Phenol sticky hosts, 190
(ω-Phenylalkyl)ammonium binding by ditopic receptors, 158—162
2-Phenylpropionic acid, 228
Phosphines, hydrogen bonding to, 190
Phosphorescence spectra, of excimers, 10
  of [m.n]naphthalenophanes, 12
Photooxidation, flavin mediated, 204
Photosynthetic reaction centers, 34
Picket-fence porphyrins, 34
π—π complexes, 44—46
π—π-stacking interactions, between naphthalene and guanine, 207
  between naphthalene and thymine, 203
  between tyrosine and guanine, 203
  extra-cavity, 58
  in apolar inclusion complexes, 82, 85, 90
  in EDA inclusion complexes, 116
  in flavinophane–arene complexes, 272—273
  in flavinophane dimers, 271—273
  in host self-association, 59
  in intermolecular excimers, 10
  in isoalloxazine dimers, 271—273
  in paracyclophane inclusion complexes, 97—98
  in triple-hydrogen-bonded complexes, 204
  of flavins, 20
π—π-stacking intercalation, 240
Pirkle phases, 115
pK_a-values, complexation-induced changes of, 185—187
Polarizability, 254, 258
Polarization interactions, 44, 116
Pole–dipole interactions, 138
Polycyclic aromatic hydrocarbons (PAHs), complexation in organic solvents, 107—113
  complexation in water 74—80
  cytochrome P-450 oxidation of, 38
  solubility in water, 74
Polymolecular aggregation, 142
Polytopic receptors, 142, 157, 163
Porphyrin-cyclophanes, 34, 290
Porphyrinophanes, 33—43
Porphyrin-quinone cyclophanes, 41
Potassium ferricyanide, 286
Prelog, V., 1, 222
Preorganization of binding site, 61, 89, 94, 167, 168, 241
Principle of preorganization, 61
Productive binding, 241, 276—281
Productive complex, 27
Propellor conformation, 93
Protein, secondary structure stabilization by ion–dipole interactions, 147
Proton transfer in macrocycles, 186, 190
Proximity effects, 6
Pseudoaxial inclusion geometry, 82—83, 116
Pseudoequatorial inclusion geometry,

82—83
Pseudogeminal, 5, 6, 15
Pseudoortho, 5, 15
Purinophanes, 19—20
Pyrene, α-bands, 113
  p-bands, 113
  solvent dependent complexation strength, 108—113
[m.n]Pyrenophanes, 10—14
2,6-Pyridinium crown ethers, 186
Pyridiniophanes, 27, 30
Pyridinophanes, 24—33
Pyridoxal phosphate (PLP), 29, 31
Pyridoxamine (PMP), 31
Pyridoxamine-cyclophane, 31, 276
Pyridoxine (Vitamin B_6), 31
Pyruvate oxidase, 286

Quadrupolar interactions, 247

Rebek, J., Jr., 182
Redox-dependent complexation, 269—273
Reinhoudt, D. N., 185, 187, 188
Relaxation time T_1, paramagnetic effects on, 100
Riboflavin, 198
Ribonuclease T_1, guanosine binding site in, 203
Roos, E.-E., 53
Rotational entropy, 100, 256
Rotating-disk voltammetry, 37

Saenger, W., 53
Saigo, K., 158
Salt bridges, energetics of, 157
  in biological recognition, 157
Sanford, E. M., 98
Saturation kinetics, 283, 289
Scatchard, G., 81
Schneider, H. J., 137, 138, 148, 149, 157, 273
Schweitzer, D., 10
Scott, R. L., 81
Secondary electrostatic interactions, in hydrogen-bonded complexes, 210, 215
Second-harmonic generation (SHG), 48
Second-sphere coordination, 164—166
Self-aggregation, 67

Self-association, 59
  geometries, 171
  of flavinophanes, 269
Self-complexation, 234
Separation processes, 51, 77, 268—269
SHAKE algorithm, 235
Shape-selective inclusion, 120
Shell-closure reactions, 133
  solvation of transition states of, 133
Shinkai, S., 22, 24
Siegel, B., 106
Simmons, H. E., 153
Sinanoglu, O., 254
Size recognition in spherical cavity, 123
Smith, B. H., 1
Solid–liquid extractions, 74—75, 196
Solvation, characteristics of, 162, 259
  free energy of, 220
  shells, icelike, 250
Solvent, cohesive interactions, 250, 253, 258
  properties, expressed in E_T (30), 111, 259
Solvent-induced conformational changes, 107
Solvent polarity, correlation with apolar binding strength, 108—113
  effects on co-complexation by ditopic hosts, 164
Solvent polarity parameter E_T (30), correlation with Gibbs free binding energies, 110—113, 131, 137—139, 257
Solvophobicity parameter S_p, 138
Soret band, 41, 215
Space-filling molecular models, 3
Spacers in cyclophanes
  benzidine, 53
  1,1'-binaphthyl, 222
  9,10-bridged 9,10-dihydroanthracenes, 65
  cyclotriveratrylene, 66, 69, 120
  diacetylenes, 58
  4,4'-diamino-1,1'-binaphthyl, 53
  diphenylmethane, 62
  naphthylphenylmethane, 96
  paraquat (1,1'-dimethyl-4,4'-bipyridinium), 63—64
  4-phenyl-1,2,3,4-tetrahydroisoquinoline, 238

Spacers in cyclophanes (cont.)
  9,9′-spirobifluorene, 222
  2,2′,7,7′-tetrahydroxy-1,1′-binaphthyl, 162—163, 241
  2,2′,6,6′-tetrasubstituted biphenyl, 233
  triphenylmethane, 62—63
  Trögers base, 62
Spherands, 91
Spin probes, 71
Spin saturated transfer NMR, 27
9,9′-Spirobifluorene crown ethers, 222
Spiro piperidinium rings, 65
Staab, H. A., 7, 10, 15, 18, 20, 27, 41
Steinberg, H., 1
Stereochemical control in cyclophanes, 18
Stereoelectronic, complementarity, 61
  effects, 24
Steroids, hydroxylation by cytochrome P-450, 38
  inclusion complexation of, 102—105
Stetter, H., 53
Still, W. C., 217, 226
Stoddart, J. F., 63, 165, 170
Stoichiometry of complexes, determination of, 81
Strain, 3
Stuart space-filling molecular models, 3
Supramolecular catalysis, 264, 265, 273
  of arene epoxidation, 290, 295—297
  of benzoin condensations, 281
  of Diels–Alder reactions, 98, 274
  of H—D exchange, 284—285
  of nucleophilic substitution reactions, 274
  of oxidation of aromatic aldehydes, 286
  of photochemical carbon-skeleton rearrangements, 273
  of transacylation reactions, 279
  of transamination reactions, 274—276
  solvent dependence of, 107
Supramolecular structure, 140
Surface-area-based solvation treatment, 89
Surface tension measurements, for determination of CAC, 70
Synthesis, of porphyrin-cyclophane as P-450 model, 290—293

Tabushi, I., 54, 55, 250, 265, 276
Template effect, in cyclization transition state, 171, 173
Ternary complex, 27, 101, 162—163
Tetraaza[n.1.n.1]paracyclophanes, 59—63
  complexes with charged guests, 143, 145
  ion–dipole interactions, 148—149
  optically active, 228
  supramolecular catalysis by, 273
Tetraaza[3.3.3.3]paracyclophanes, clathrates of, 54
  inclusion complexes of, 54—57
  receptors derived of, 176
Tetracyanoethylene (TCNE), 44
7,7,8,8-Tetracyanoquinodimethane (TCNQ), 16
2,2′,7,7′-Tetrahydroxy-1,1′-binaphthyl, 162, 241
Tetra($meso$-phenyl)porphyrins, 34
N,N,N',N'-Tetramethyl-p-phenylenediamine (TMPD), 16
Tetraoxa[n.1.n.1]paracyclophanes, 2, 63—66
  clathrates of, 50
Tetraphenylene, 46
Tetrathia[3.3.3.3]paracyclophane, 55—56
Thiamine pyrophosphate (TPP), 281, 286
Thiazolium, cyclophane, 283, 286
  ions as catalysts in benzoin condensations, 283
Thermodynamic driving forces, for apolar complexation in water, 247—257
  for membrane and micelle formation, 255
Thiourea, solid state complex of, 183
Threonine, 31
Titration calorimetry, 81
2,6-TNS (6-(4-methylphenyl)amino-2-naphthalenesulfonate), 56, 57, 143
Transacylation, 279
Transaminase activity, 33, 274
Transamination, asymmetric of α-ketoacids, 29
Transannular, interactions, 6—15, 44
  substituent effects, 6

Index A: Subject Index

Transhydrogenases, 27
Transition metal complexes, 164—165
Transition states, of deprotonation, 285
  of hydride transfer, 26
  of racemization, 3
Transketolase, 281
Transport,
  active, 269
  carrier mediated through water, 265—269
  of arenes, 267
  of lipophilic substrates, 264
  passive, 269
  selectivity, 268
2,2,2-Trifluoroethanol, 106
Trihydroxybenzenes, complexation of, 196—197
Triple helix formation, 207
Triple hydrogen-bonding,
  in guanine complexes, 207, 211, 247
  in thymine complexes, 204
  with barbiturates, 214
  with 1-butylthymine, 204
  with cyclic imides, 203
  with 2,6-diamidopyridine, 203
  with isoalloxazines, 204, 206
Tripod hydrogen-bonding, 165
Tris(bipyridine) ligand, 193
Tri-o-thymotide (TOT), 46—49
  clathrates of, 48—49
Tröger's Base, 62, 91
TSP (sodium, 2,2,3,3-tetradeuterio-3-(trimethylsilyl)propionate), 70
Two-point hydrogen-bonding, 200
  energetics of, 202

Ultrasonic relaxation, 80
Unimolecular micelle, 177
Urea, co-complexation with electrophilic cations, 188
  complexation of, 187—189
  transport of, 187
U-type cell, 267

Van der Waals, distance, 203, 204
  interactions, 50, 52, 90, 123, 127, 235, 247, 255
  radius, 3, 91
  volume, 122, 124
Van't Hoff analysis, 252

Vase-type molecules, 127
  complexes of, 131
Verhoeven, J. W., 24
Vitamin B_6, 29
  model system of, 29
Vögtle, F., 5, 62, 101, 193, 268

Water, clusters, 250
  complexation of, 185—186
  role in apolar binding, 258
Watson-Crick base pairing, 207, 213
Werner, A., 165
Wilcox, C. S., 62, 91
Whitlock, H. W., 5, 58, 98, 189, 190
Woggon, W.-D., 40
Wudl, F., 222

X-ray crystal structure of
  acetone solvate of vase-type host, 132
  adenine receptor, 212
  barbiturate complex with hydrogen-bonding receptor, 217
  benzene clathrate, 53
  benzene-tetraoxa[6.1.6.1]paracyclophane inclusion complex, 65, 68
  bis(phosphine oxide) cages, 193, 196
  1-butylthymine complexed to synthetic thymine receptors, 205, 206, 208
  carbon disulfide–cavitand complex, 125, 127
  [2]catenane, 174
  cavitand complexes, 125
  chiral macrobicyclic hydrogen-bonding receptor, 226
  chloroform–cryptophane inclusion complex, 123
  complexes with tetraaza[3.3.3.3]paracyclophane, 55—56
  crown ethers and small acidic molecules, 184
  cytochrome P-450 complex with camphor, 38
  dichloromethane inclusion complexes, 56, 69
  dihydroisoalloxazines, 22
  dimer formed by kite-type molecules, 129

X-ray crystal structure (cont.)
dimethylformamide hermicarceplex, 136
p-dimethoxybenzene inclusion complex, 121
diquat-dibenzocrown complexes, 168
diquat receptor, 168
durene complex, 60, 61
glutathione reductase, 20
guanosine binding site of ribonuclease T_1, 203
imidazole receptor, 217
isoalloxazines, 22
isoalloxazinometacyclophane, 21
macrobicyclic hexa-ammonium ion salts, 154
naphthalene-tetraaza[n.1.n.1]-paracyclophane inclusion complexes, 83, 86
naphthalenophane-1,2-dichloroethane solvate, 194
naphthalenopyridinophanedienes, 14
nitrobenzene-bis(diazapyrene) receptor complex, 156
p-nitrophenol-naphthalenophane complexes, 194
nucleotide-binding enzymes, 203
octamethoxy-tetraoxa[5.1.5.1]-paracyclophane, 86—88
[m.n]paracyclophanes, 9
paraquat-cyclophane acetonitrile solvate, 63—64
paraquat-dibenzocrown complexes, 169—171
paraquat receptor, 170
porphyrin-quinone cyclophane, 42
[n.n](2,7)pyrenophanes, 13
pyridinophanes as NADH models, 24
2,6-pyrido-18-crown-6 complex with malononitrile, 184
purinophane, 20
second-sphere complexes, 165—166
TCNE-cyclophane complexes
tetraphenylene clathrates, 50
thymine receptor, 205
toluene-tetraoxa[6.1.6.1]-paracyclophane clathrate, 65, 67
TOT-chlorocyclohexane clathrate, 48
Tröger's base derivatives, 92
urea-phenolic crown ether complex, 187
urea-UO_2^{2+}-macrocycle ternary complex, 189
water complexed to 2-carboxyl-1,3-xylyl-24-crown-7, 185, 186
p-xylene-tetraoxa[6.1.6.1]-paracyclophane clathrate, 50

Ziegler, K., 1
Zinc(II) porphyrinate, 290

Index B: Guests in Solution Complexes for which Stability Constants are Given

Acenaphthylene, 294
Acetone, 122
N-Acetylalanine, benzylester, 227
N-Acetylbenzylamine, 227
Acetylcholine, 152
N-Acetyl-1-(1-naphthyl)ethylamine, 227
N-Acetyl-1-phenethylamine, 227
N-Acetylphenylglycine, methylester, 227
1-Adamantaneacetic acid, 101
1-Adamantanol, 101
Adenosine, 99
2-Amino-6-nitronaphthalene, 117
4-Aminopyridine, 219
1,8-ANS (8-phenylamino-1-naphthalenesulfonate), 55, 57, 145, 177
Anthracene, 294
Azide, 154
Azulene, 75

Barbital, 219
Benzaldehyde, 288
1,4-Benzenedicarboxylate, 154
Benzimidazole, 219
Benzoate, 148
Biphenyl, 75
1,5-Bis(dimethylamino)naphthalene, 85
2,7-Bis(dimethylammonium)naphthalene, 145
Bis(thymine) derivative, 212
1,3-Bis(trimethylammonium)propane, 152
Bornyl trimethylammonium, 231
Bromodichloromethane, 122
tert-Butane, 122
tert-Butanol, 153
tert-Butylchloride, 122
1-Butylthymine, 210
9-Butyladenine, 210

Camphor, 101
Carbon disulfide, 125
Carnitine, 152

Chenodeoxycholic acid, 102
Chloride, 154
Chloroform, 100, 122
Cholic acid, 102
Choline, 152
Copper(II) ions, 95
Cortisone, 102
p-Cresol, 90
p-Cyanophenol, 93, 193
trans-1,4-Cyclohexanedimethanol, 101
Cyclohexylacetate, 101, 148
Cyclohexylcarboxylate, 148

trans-Decaline, 101, 148
Deoxycholic acid, 102
p-Diaminobenzene, 90
2,6-Diaminonaphthalene, 117
Dibromomethane, 122
Dibromochloromethane, 122
Dichloromethane, 100, 122
2,2-Dichloropropane, 122
p-Dicyanobenzene, 90, 93, 288
2,6-Dicyanonaphthalene, 117
m-Dimethoxybenzene, 120
o-Dimethoxybenzene, 120
p-Dimethoxybenzene, 90, 93, 120
2,6-Dimethoxynaphthalene, 117
1-(Dimethylamino)naphthalene, 85
Dimethyl p-benzenedicarboxylate, 90
3,5-Dimethylcyclohexanol, 148
Dimethyl 2,6-naphthalenedicarboxylate, 117
1,5-Dimethylnaphthalene, 85
2,6-Dimethylnaphthalene, 85
3,5-Dimethylphenol, 148
p-Dinitrobenzene, 90, 93
2,6-Dinitronaphthalene, 117
4-(2',4'-Dinitrophenylazo)phenol, 193
Diquat, 167
Dithionate, 154
Durene, 75, 110

Ethyl anthranilate, 93

Fluoranthene, 75, 110
N-Formylbenzylamine, 227
N-Formyl-1-phenethylamine, 227

Glutaric acid, 200
Glutaric acid monomethyl ester, 202

n-Hexylisoalloxazine, 204
Hydrocortisone, 102
Hydroxynaphthoic acids, 57

Imidazole, 219, 220
Indole, 99
Iodomethane, 122
Isoquinoline, 99, 147

Lithocholic acid, 102

Mephobarbital, 219
6-Methoxy-2-naphthonitrile, 117, 164, 238, 242, 288
2-(6-Methoxy-2-naphthyl)-propionamide, 245
1-[2-(6-Methoxy-2-naphthyl)propionyl]-piperidine, 245
2-Methoxy-6-nitronaphthalene, 117
4-Methoxyphenol, 93
N-Methylacetamide, 227
trans-1,4-Methylcyclohexanol, 101, 148
2-Methylimidazole, 219
N-Methylindole, 99
1-Methylisoquinoline, 99
2-Methylisoquinolinium, 147
1-Methylnaphthalene, 85
4-Methylphenol, 93, 148
N-Methyl-phenylacetamide, 227
2-Methylquinoline, 99
4-Methylquinoline, 99
1-Methylquinolinium, 147
Myrtanyl trimethylammonium, 231

2-Naphthaldehyde, 288
Naphthalene, 75, 110, 145, 148, 294
2,6-Naphthalenedicarboxylate, 154
1,3-Naphthalenediol, 85
2,6-Naphthalenediol, 85, 117
2,7-Naphthalenediol, 59, 85, 145
1,5-Naphthalenedisulfonate, 145
2,6-Naphthalenedisulfonate, 145
1-Naphthalenesulfonate, 145

2-Naphthalenesulfonate, 145
1-Naphthol, 85
1-(1-Naphthyl)ethyltrimethyl-ammonium, 231
Naproxen, 245
Naproxen methyl ester, 245
Nitrate, 154
4-Nitro-1-naphthol, 279
p-Nitrophenol, 90, 95, 193
p-Nitrophenylacetate, 93
4-(4'-Nitrophenylazo)phenol, 193
p-Nitrotoluene, 90
Nitroxide spin probe, 72

Oxalate, 154

[4.2]Paracyclophane-2,3-dicarboxylic acid, 97
[4.2]Paracyclophane-2,3-diol, 97
[2.2]Paracyclophane-1-carboxylic acid, 97
[2.2]Paracyclophane-4-carboxylic acid, 97
[3.3]Paracyclophane-5-carboxylic acid, 97
[4.3]Paracyclophane-7-carboxylic acid, 97
[4.4]Paracyclophane-6-carboxylic acid, 97
Paraquat (methylviologen), 152, 167
Perylene, 75, 110
Phenanthrene, 294
Phenobarbital, 219
Phenylacetate, 148
(6-Phenylhexyl)ammonium picrate, 158
N-Phenyl-1-naphthylamine, 177
(5-Phenylpentyl)ammonium picrate, 158
(3-Phenylpropyl)ammonium picrate, 159
Phenyl ring of spin probe, 72
Phoroglucinol, 196
Pimelic acid, 202
Potassium cation, 164
N-Propionylbenzylamine, 227
$[Pt(bipy)(NH_3)_2]^{2+}$, 165
Pyrene, 75, 77, 110, 294
Pyridine, 99
4-Pyridone, 219
N-(4-Pyridyl)-2-(6-methoxy-2-naphthyl)-propionamide, 245

Index B: Guests in Solution Complexes for which Stability Constants are Given

Quinoline, 99, 147

Sulfate, 154

Testosterone, 102
Tetra(*n*-butyl)ammonium, 152
Tetrachloromethane, 122
Tetraethylammonium, 152
Tetralin, 148
Tetramethylammonium, 152
Tetra(*n*-propyl)ammonium, 152
2,6-TNS (6-(4-methylphenyl)amino-2-naphthalenesulfonate), 57, 145, 177
p-Tolunitrile, 90
Tribromomethane, 122
1,1,1-Trichloroethane, 122

2-Trimethylammonium-1-(3,4-dimethoxphenyl)ethanol, 231
1-(Trimethylammonium)adamantane, 147
2-(Trimethylammonium)ethyl benzoate, 152
1-(Trimethylammonium)naphthalene, 145
2,4,6-Trimethylphenol, 93
2′,3′,5′-Tri-*O*-pentanoylguanosine, 210

Urea, 188
Ursodeoxycholic acid, 102

p-Xylene, 90, 93